10 0156276 6
KU-625-919
WITHDRAWN
ROM THE LIBRAR

UNIVERSITY LIBRARY

Digital Signal Processing

Digital Signal Processing

Concepts and Applications

Bernard Mulgrew, Peter Grant and
John Thompson

Department of Electronics and Electrical Engineering
The University of Edinburgh

First published 1999 by
MACMILLAN PRESS LTD
Houndmills, Basingstoke, Hampshire RG21 6XS
and London
Companies and representatives throughout the world

ISBN 0-333-74531-0

A catalogue record for this book is available from the British Library.

This book is printed on paper suitable for recycling and made from fully managed and sustained forest sources.

10 9 8 7 6 5 4 3 2 1
08 07 06 05 04 03 02 01 00 99

Printed in Great Britain by
Antony Rowe Ltd
Chippenham, Wiltshire

Contents

Preface

Digital signal processing (DSP) provides a rapidly advancing portfolio of filtering and estimating techniques or algorithms which are used in signal analysis and processing. Significant current applications are in the development of mobile communications equipment, particularly for personal use and design of sophisticated radar systems. The aim of this book is to provide an introduction to the fundamental DSP operations of filtering, estimation and signal analysis as used in signal processing.

Most of the chapters include substantive numerical examples to illustrate the material developed, as well as self assessment questions which have been designed to help readers aid their comprehension of this material. All the chapters conclude with further problem questions for the student.

Chapters 1 to 3 cover basic analogue signal theory as a prerequisite to this DSP text. Chapter 4 extends these concepts to sampled-data systems and here discrete convolution is introduced. Chapters 5 and 6 explore digital filters, both infinite and finite impulse response, which implement the convolution operation and include both analytical design and software optimisation techniques. The two chapters conclude with a brief discussion on the problems of finite precision arithmetic. Chapters 7 and 8 introduce, at a more mathematical level, the concept of random signals, correlation and spectral density. Chapter 8 covers adaptive or self learning filters which alter their characteristics dependent on the signal scenario which is present. They find widespread application in communications systems as equalisers and echo cancellers. The final three chapters deal with spectral analysis techniques. Chapter 9 covers the discrete Fourier transform (DFT), its derivation and the design of DFT processors. This chapter then investigates the application of DFT processors in classical spectrum analysis equipment before introducing the modern analysers which are based on the adaptive filter technique of Chapter 8. Chapter 10 deals with the fast Fourier transform which is the most widely applied implementation of the DFT processing function. Chapter 11 introduces multirate techniques to extend the capabilities of these analysers to speech and image processing applications.

With this balance between signal theory, processor design and systems applications we hope that this text will be useful both in academia and in the rapidly growing commercial signal processing community. Advanced DSP is of fundamental importance in implementing many high performance systems such as personal communications and radar.

To aid the class instructor, the authors can provide a printed set of outline solutions to the end-of-chapter problems and these are available via the WWW on password access. MATLAB™ source code is also provided on an open access basis to assist the instructor with presentation of the material and the student in understanding of the material. In general the source code provides a computer animation of some of the figures in the book. For example the m-file "fig1_4.m" contains MATLAB code which produces an animation of the complex phasor of Figure 1.4. These are identified within the text by the □ symbol. These software and solutions to problems are available via the menu at the Edinburgh WWW server address: http://www.ee.ed.ac.uk/~pmg/SIGPRO/index.html. Subsequent corrections to this text will also be available at the same WWW address.

Edinburgh
May 1998

Bernard Mulgrew, Peter Grant and John Thompson

Acknowledgements

Parts of this book have been developed from BEng, MEng, MSc and industrial training courses provided by the Department of Electronics and Electrical Engineering at the University of Edinburgh. These courses were also taught by Professor Colin Cowan and Dr James Dripps, and we acknowledge their contribution to the initial shaping of these courses which is reflected in the book's content and structure. We are grateful to Professor Cowan for having provided a draft version of Chapter 5 and Dr Dripps for having provided a draft version of parts of Chapter 9 and for assistance with many of the problem solutions. We are also grateful to Dr Ian Glover at the University of Bradford and Prentice-Hall for permission to include the material on bandpass sampling within Chapter 11, and to the IEE for permission to reproduce Figure 9.22.

We would like to thank all those other colleagues at the University of Edinburgh who have provided detailed comments on sections of this text. Thanks must go to the many students who have read and commented on earlier versions of this material and helped to refine the end-of-chapter problems, particularly to Miss Oh who generated the initial version of many of the diagrams. We also gratefully acknowledge the generous assistance of Dr Jonathon Chambers of Imperial College in carefully reviewing and editing this text. In addition we acknowledge the assistance of Philip Yorke at Chartwell Bratt publishing and training in encouraging us to develop, in the 1980s, the preliminary version of this material.

Special thanks are due to Joan Burton, Liz Paterson and Diane Armstrong for their perseverance over several years in typing the many versions of the individual chapters, as they have evolved into their current form. We also acknowledge Bruce Hassall's generous assistance with the preparation of the final version of the text in the appropriate typefont and text format.

Finally we must thank our respective families: Fiona and Maeve; Marjory, Lindsay and Jenny; and Nadine – for the considerable time that we required to prepare this book and the associated WWW supporting material.

Bernard Mulgrew, Peter Grant and John Thompson

Abbreviations

AC	Alternating current (implying sinusoidal signal)
ACF	Autocorrelation function
A/D	Analogue to digital (converter)
ADPCM	Adaptive DPCM
AGC	Automatic gain control
AM	Amplitude modulation
AR	Autoregressive
BLMS	Block least mean squares
BP	Bandpass
BPF	Bandpass filter
BS	Bandstop
CD	Compact disc
CDMA	Code division multiple access
CELP	Codebook of excited linear prediction
CMOS	Complementary metal oxide silicon (transistor)
COFDM	Coded orthogonal frequency-division multiplex
D/A	Digital to analogue (converter)
DC	Direct current (implying a 0 Hz component)
DCT	Discrete cosine transform
DFS	Discrete Fourier series
DFT	Discrete Fourier transform
DIF	Decimation in frequency
DIT	Decimation in time
DM	Delta modulation
DPCM	Differential pulse code modulation
DPSK	Differential phase shift keying
DSB	Double sideband
DSP	Digital signal processing
DTFT	Discrete-time Fourier transform
ESD	Energy spectral density

EVR	Eigenvalue ratio
FFT	Fast Fourier transform
FIR	Finite impulse response
FM	Frequency modulation
FS	Fourier series; Federal Standard
FSK	Frequency shift keying
FT	Fourier transform
HP	High pass
I	Imaginary (quadrature signal) component
IDFT	Inverse DFT
IF	Intermediate frequency
IIR	Infinite impulse response
ISI	Inter-symbol interference
KCL	Kirchhoff's current laws
LMS	Least mean squares
LO	Local oscillator
LOS	Line of sight
LP	Low pass
LPC	Linear predictive coding
LS	Least squares
LTI	Linear time invariant
MA	Moving average
MAC	Multiply and accumulate
MATLAB	MATrix LABoratory commercial DSP software product
MFSK	Multiple frequency shift keying
MMSE	Minimum mean square error
MODEM	Modulator/demodulator
MOS	Mean opinion score (for speech quality assessment) Metal oxide silicon (transistor)
MPE	Multipulse excitation
MSE	Mean square error
NATO	North Atlantic Treaty Organisation
NPSD	Noise power spectral density
PAM	Pulse amplitude modulation
PCM	Pulse code modulation
pdf	Probability density function
PFE	Partial fraction expansion

PLL	Phase locked loop
PM	Phase modulation
PN	Pseudo-noise
PO	Percentage overshoot
PPM	Pulse position modulation
PR	Perfect reconstruction
PSD	Power spectral density
PSK	Phase shift keying
PWM	Pulse width modulation
Q	Quantiser
R	Real (in-phase signal) component
RAM	Random access memory
ROM	Read only memory
RLS	Recursive least squares
RMS	Root mean square
SAQ	Self assessment question
SBC	Sub-band coder
SG	Stochastic gradient
S/H	Sample and hold
SNR	Signal-to-noise ratio
TDM	Time division multiplex
THD	Total harmonic distortion
VLSI	Very large scale integrated (circuit)
WWW	World Wide Web
ZOH	Zero order hold

Principal symbols

$\mathbf{a}$	tap weight vector
a_i	digital filter weight coefficient value for tap i
A	A-law PCM compander constant
A_n	nth trigonometric Fourier component
$A(k)$	kth real Fourier coefficient
b_i	digital filter recursive weight coefficient value for tap i
B	signal bandwidth in Hz
B_n	nth trigonometric Fourier component
$B(k)$	kth imaginary (real valued) Fourier coefficient
c_n	FIR filter coefficient value
c'_n	windowed coefficient value
C	constant, capacitance (Farads)
$C_n(\omega)$	Chebyshev polynomial of order n
d	lag
$\mathbf{e}$	error vector
$e(n)$	scalar error signal at data sample n
E	energy of signal $x(t)$
f_0	centre frequency
f_1	passband cut-off frequency
f_2	stopband edge frequency
f_{3dB}	half power bandwidth
f_b	bit rate
f_c	centre frequency
f_H	highest frequency component
f_L	lowest frequency component
f_{LO}	local oscillator frequency
f_s	sample frequency (in Hz)
$F_m(z)$	z-plane reconstruction filter m response

F_N	speech formant N frequency component
$F(\omega)$	frequency response
$g(t)$	baseband signal
$\tilde{g}(kT_s)$	estimate of $g(t)$
$\hat{g}(kT_s)$	prediction of $g(t)$
G	amplifier gain
G_p	processing gain
$\mathbf{h}(n)$	filter impulse response vector
$h(n)$	filter impulse response for sample value n
$h(t)$	filter impulse response
$H_n/H(s)$	system transfer function
$H_A(\omega)$	real frequency response
$H_D(\omega)$	required or desired frequency response
$H(s)$	Laplace transfer function
$H(z)$	z-transfer function
$H(\omega)$	angular frequency response
$I_0[\ ]$	modified Bessel function of first kind and order zero
k	Boltzmann's constant
K	constant
L	upsampling ratio
$L(\omega)$	weighting function
M	number of feedback taps in an IIR filter; downsampling ratio
n	type of semiconductor material
N	order of DFT
N	number of feedforward taps in an IIR/FIR filter
N_0	noise power spectral density
p	type of semiconductor material
p_1	filter pole number 1
$p(v)$	probability density function of variable v
P	power
q	quantisation step size
Q	quality factor; f_H/B
$\mathbf{r}_{yx}$	cross-correlation vector
$\mathbf{R}_{yy}$	correlation matrix
R	resistance in ohms

R_D	dynamic range
RC	resistor–capacitor time constant τ_c
s	Laplace variable
$S_{xx}(\omega)$	power spectral density
t	time
Δt	sample period
$\mathbf{v}$	eigenvector of autocorrelation matrix
$v(t)$	information signal
$v_i(t)$	voltage waveform of symbol i
w_n	FIR filter/DFT window or weight coefficient n
W_N^k	kth value of the N roots of unity
$W_T(\omega)$	DFT of window function
$\mathbf{x}$	vector x
$x(n)$	sampled input signal
$\hat{x}(n)$	estimate of input signal
X_n	complex Fourier coefficient
$X(f/\omega)$	Fourier (voltage) spectrum of $x(n)$
$\lvert X(f/\omega)\rvert$	Fourier amplitude spectrum
$X(k)$	DFT output value for bin or sample number k
$X(s)$	Laplace transform of $x(t)$
$y(n)$	processed output signal
$y_D(n)$	decimated signal
$y_I(n)$	interpolated signal
z_n	filter zero number n
α	attenuation; forgetting factor or window taper
β	constant
δ	Dirac delta
$\delta(t)$	impulse function
$\delta_T(t)$	impulse train
ΔV	voltage difference
∇	gradient vector
ε	error voltage; Chebyshev design parameter

ε_q	quantisation error
$\overline{\varepsilon_q^2}$	mean-square quantisation error
ξ	filter damping factor
$\xi(n)$	MSE cost function
η	efficiency
$\eta(n)$	additive noise signal
λ_i	eigenvalue i
μ	step-size scaling constant
ρ	normalised correlation coefficient; sum of squared error cost function
$\rho(n)$	vector norm or deviation
σ	real part of Laplacian; standard deviation
σ_x^2	variance of signal x
τ	time constant
$\phi_{xx}(\omega)$	autocorrelation sequence $x(n)$
$\Phi_{xx}(\omega)$	autocorrelation matrix for $x(n)$
$\Phi_{xy}(\omega)$	cross-correlation vector
ω_0	filter centre frequency
ω_a	analogue filter cut-off frequency
ω_{cl}	lower bandpass/bandstop cut-off frequency
ω_{cu}	upper bandpass/bandstop cut-off frequency
ω_d	digital filter cut-off frequency
ω_s	(angular) sample frequency (in rad/s)
$\Delta\omega$	filter transition bandwidth, DFT bin spacing

Special functions

$E[\ ]$	statistical expectation operator
$[\ ,\ ]$	scalar product
$\langle\ \rangle$	time average
$*$	convolution operation
a^*	complex conjugate of a
$F[,]$	Fourier transform operator
$\Im$	operator corresponding to imaginary part of ...
$\int(.)$	integer part of ...
$L[,]$	Laplace transform operator
$\Re$	operator corresponding to real part of ...
sa (x)	sampling function
sgn (x)	signum (sign) function
sinc (x)	sinc function
$u(t)$	unit step function
Z^{-1}	inverse z transform
Z	z transform

Introduction

Real life signals are generally continuous in time and analogue, i.e. they exist at all time instances and can assume any value, within a predefined range, at these time instances. There are many kinds of analogue signals appearing in nature:

- Electrical signals: voltages, currents, electric and magnetic fields.
- Acoustic signals: mechanical vibrations, sound waves.
- Mechanical signals: displacements, angular motion, velocities, forces, moments, pressures.

Acoustic signals such as sound waves are converted to electrical voltages or currents by sensors or transducers, (i.e. a microphone) in order for them to be processed in an electronic system. Analogue processing involves linear operations such as amplification, filtering, integration and differentiation, as well as various forms of nonlinear processing such as squaring or rectification. This text does not cover the field of nonlinear signal processing, but, in practice, saturation in amplifiers or mixing of signals often introduces nonlinearities. Limitations of practical analogue processing operations are:

- Restricted accuracy.
- Sensitivity to noise.
- Restricted dynamic range.
- Poor repeatability due to component variations with time, temperature, etc.
- Inflexibility to alter or adjust the processing functions.
- Problem in implementing accurate nonlinear and time-varying operations.
- Limited speed of operation.
- High cost of storage for analogue waveforms.

However, in the 1970s, analogue signal processing in the form of surface acoustic wave and charge-coupled sampled-data devices delivered very sophisticated matched filter and correlator parts which were widely used in military equipment. At this time it was not possible to match the analogue speed/performance capability with digital devices so the accuracy of analogue processors was severely limited. The initial implementations of Dolby noise reduction systems also employed analogue filter techniques and the early V.21–V.29 data modems, for transmission of digital data over telephone lines, used exclusively analogue filters and modulators.

Digital signal processing (DSP) is achieved by sampling the analogue signal at regular intervals and representing each of these sample values with a binary number. These are subsequently passed to a specialised digital processor to perform numerical or computational operations on these signals. Operations in digital signal processing systems encompass additions, multiplications, data transfers, logical operations and can be extended to implementation of complex matrix manipulations. The essential operations of many DSP systems include:

- Converting analogue signals into a sequence of digital binary numbers, which requires both sampling and analogue-to-digital (A/D) conversion.
- Performing numerical manipulations, predominantly multiplications, on the digital information data stream.
- Converting the digital information back to analogue signal, by digital-to-analogue (D/A) conversion and filtering.

The basic components of the digital processor, which is equivalent to the analogue processing function, is considered later in Chapter 4. The DSP function is generally described as an algorithm or program which defines the in-built arithmetic operations. DSP is in fact an extension of the conventional microprocessor function except that a fast multiplier is added as a hardware accelerator element.

The main attraction of digital processors is that their accuracy, which is controlled by the quantisation step size or word length employed in the A/D converter, is extendable only at the cost of greater complexity for the ensuing processing operations, i.e. additions and multiplications. However, floating-point number representations permit the accuracy to be maintained over a wide dynamic range. Further, digital processors are generally repeatable and much less sensitive to noise than analogue processors as they are dealing with binary symbols and the processed outputs must always possess binary values. The microelectronics industry has been reducing continuously silicon VLSI circuit geometrics and hence improving the speed, complexity and storage capacity. The feature size in the 1970s was $5\mu m$ while in the year 2000 we will be designing VLSI circuits with $0.18\mu m$ feature sizes and single chip complexities of 40 M individual transistors, and by the year 2010 $0.7\mu m$ feature size circuits are predicted to have 3 GHz clock rates.

VLSI permits the design of application specific integrated circuits with exceptional cost/performance capabilities to execute sophisticated DSP processing functions. Current developments in microelectronics are delivering an order of magnitude increase in processor operating speed, coupled with a 30-fold reduction in power consumption, every eight years.

Some of the key benefits which derive from digital signal processing are:

- Efficient implementation of linear phase filters for communication and radar receiver design.
- Easy realisation of Fourier transform processors for signal analysis.
- Possibility of implementing complicated processing functions involving matrix manipulations.

These advances in DSP, complexity and functionality over analogue signal processing are not obtained without some penalties. High accuracy, high speed A/D and D/A conversion hardware is expensive and it can lead to noise and distortion problems. We always require bandlimiting filters before the sampling function and this introduces some loss of information. For some applications, e.g. RF signals, digital processors cannot achieve the necessary high speed sampling requirements. Even with these drawbacks the use of DSP is becoming ubiquitous, partly because of the 40% per year cumulative growth of the DSP since 1988, with predicted market sales of 6,000,000,000 dollars in 2000. The generic application areas are:

Speech and audio: noise reduction (Dolby), coding, compression (MPEG), recognition, speech synthesis.

Music: recording, playback and mixing, synthesis of digital music, CD players.

Telephony: speech, data and video transmission by wire, radio or optical fibre.

Radio: digital modulators and modems for cellular telephony.

Signal analysis: spectrum estimation, parameter estimation, signal modelling and classification.

Instrumentation: signal generation, filtering, signal parameter measurement.

Image processing: 2-D filtering, enhancement, coding, compression, pattern recognition.

Multimedia: generation, storage and transmission of sound, motion pictures, digital TV, HDTV, DVD, MPEG, video conferencing, satellite TV.

Radar: filtering, target detection, position and velocity estimation, tracking, imaging, direction finding, identification.

Sonar: as for radar but also for use in acoustic media such as the sea.

Control: servomechanisms, automatic pilots, chemical plant control.

Biomedical: analysis, diagnosis, patient monitoring, preventive health care, telemedicine.

Transport: vehicle control (braking, engine management) and vehicle speed measurement.

Navigation: Accurate position determination, global positioning, map display.

The key attraction of DSP devices is that their in-built programmability allows one standard part to be applied to most of these processing functions by changing the stored program or instruction set.

In terms of specific uses, the modern car has many tens of processors attached to sensors for fuel injection, engine management, passenger compartment climate control, braking system protection, detection of component failure, etc. In fact a major part of the modern automobile comprises electronics rather than mechanical engineering. Modern mobile and cellular telephone systems rely on advanced signal processing to detect and decode the received signals, to minimise received errors, to control the mobile transmissions, to enhance battery life, etc. The TV set and new set-top boxes contain a large number of DSP chips to detect the low power signals received and to decompress the video traffic efficiently.

In the late 1990s there are a tremendous number of applications for DSP and, in the future, we confidently expect these to increase as the continuous growth in processor power and novel algorithm development unlocks even more sophisticated processing algorithms which cannot be contemplated today. DSP is thus now firmly entrenched in the consumer marketplace.

CHAPTER 1

Signal representation and system response

1.1 Introduction

Perhaps the most familiar example of a signal is a musical sound. Plucking a string on a guitar causes that string to vibrate up and down as illustrated in Figure 1.1(a). Plotting the vertical position of one point on the string as a function, $x()$, of time, t, reveals the periodic motion of the string (Figure 1.1(b)). The motion of the string alters the pressure in the air around the string and the pressure of the air will oscillate in sympathy with the movement of the string. The pressure wave radiates from the guitar towards the microphone which converts the fluctuations in pressure to an electrical voltage oscillation (Figure 1.2(a)). All of these oscillations are examples of signals. All of them carry information. Listening to a recording of the electrical signal, most people would be able to identify the instrument being played, i.e. would be able to distinguish between a guitar and a piano. Some people might be able to identify the pitch and hence the note being played.

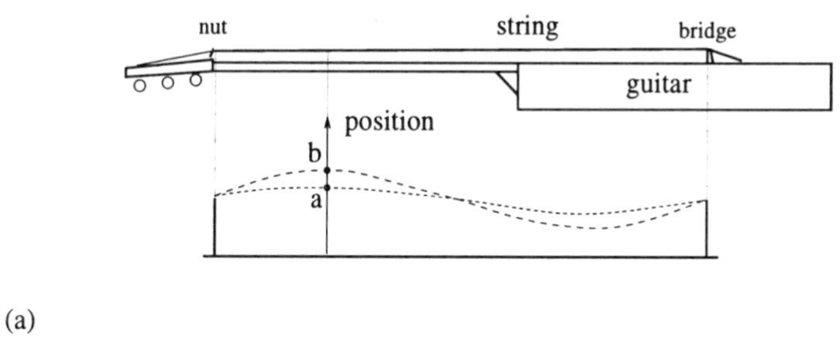

(a)

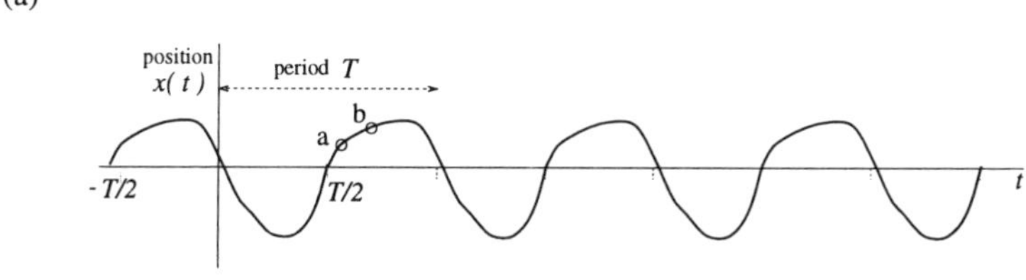

(b)

Figure 1.1 *Signal generation from a musical instrument: (a) standing wave on a string; (b) vertical position of one point on string as a function of time.*

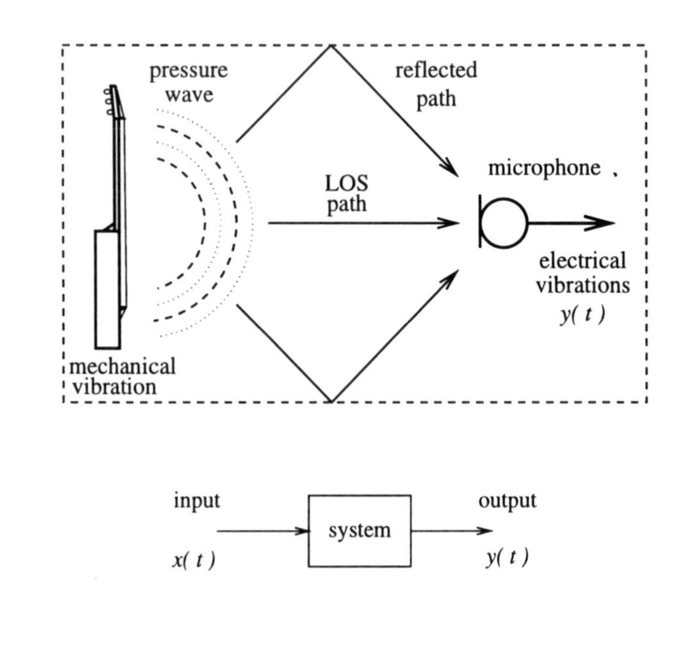

Figure 1.2 *Acoustic path from guitar string to microphone: (a) typical paths present in a room; (b) block diagram representation of signal paths.*

When the string is plucked, the pressure wave radiates out from the string in all directions. There are many paths that it can take from the string to the microphone. For example, there is a direct or line of sight (LOS) path from the string to the microphone; there is also a path from the front of the guitar, reflecting off a wall to the microphone. Since the speed of sound in air is approximately constant and the lengths of the paths are different, the signal from the guitar will reach the microphone at different times. These delayed versions of the signal will interfere constructively and destructively to produce the final electrical signal, $y(t)$, which comes from the microphone. The interference will alter the signal. Thus an audio engineer in a recording studio will move the microphone around to get a 'good sound' for the recording. The acoustical path between the string and the microphone is one example of a system and is often summarised in a block diagram as in Figure 1.2(b). The input is the position signal $x(t)$ and the output is the electrical signal $y(t)$.

The above example considers only one type of signal. In section 1.2 the issue of signal classification is introduced with consideration given to periodic, non-periodic, finite energy and finite power classes. Figure 1.1 provides one representation of a periodic signal as a function of time, known as a time-domain description. In section 1.3 the frequency domain representation is introduced through the trigonometric and complex Fourier series. This representation is extended in section 1.4 with the introduction of the Fourier transform to handle non-periodic signals. Some of the technical difficulties with the Fourier transform lead naturally to the Laplace transform of section 1.5. These alternative representations of signals are in fact variations on a theme in that they describe the original time-domain signal as a weighted sum (or integral) of complex phasors. In section 1.6 the concepts of transform representation, superposition and easy methods for calculating the response of linear equations to complex phasors are drawn together to highlight the intimate link between such transform representations and the analysis of the response of linear systems to a wide range of input signals. Section 1.7 reviews the widely used Laplace transform method for calculating the response of a linear system to a given input. Finally, in section 1.8 the key concepts are summarised.

1.2 Signal classification

There are many ways that signals can be classified. The most straightforward classification is that of a periodic signal. Such a signal repeats itself after a fixed length of time known as the period T as in Figure 1.1(b). More formally, a signal $x(t)$ is periodic if:

$$x(t) = x(t + T)$$

The smallest positive value of T which satisfies this condition is the period. If a signal does not repeat itself after a fixed length of time it is non-periodic or aperiodic. As far as signal representation is concerned there are two particularly important classes: energy signals and power signals. These are now examined.

1.2.1 Energy signals

Consider a time-varying voltage signal $v(t)$. This voltage is applied to a resistor of R Ω. The instantaneous power developed in the resistor at time t seconds is:

$$\frac{v^2(t)}{R} \quad \text{watts (W)}$$

Since power is rate of change of energy, the energy dissipated in the resistor between a time t_1 and a time t_2 is found by integrating the instantaneous power over the time interval, i.e.:

$$\int_{t_1}^{t_2} \frac{v^2(t)}{R}\, dt \quad \text{joules (J)}$$

If the signal is the current $i(t)$ through the resistor, the energy would be:

$$\int_{t_1}^{t_2} i^2(t) R\, dt \quad \text{J}$$

Thus the energy is proportional to the integral of the square of the signal. It is conventional and convenient to normalise the value of the resistor to 1 Ω and define the energy of a signal as the 'energy per ohm' or the 'energy dissipated in a 1 Ω resistor'. A signal $x(t)$ is said to be an energy signal if the total energy, E, dissipated by the signal between the beginning and the end of time is non-zero and finite. Thus:

$$0 < E < \infty \tag{1.1}$$

where

$$E = \int_{-\infty}^{\infty} x^2(t)\, dt \tag{1.2}$$

EXAMPLE 1.1

Find the the energy in the decaying exponential signal $x_1(t)$ where: $x_1(t) = 5\exp(-2t)$ if $t \geq 0$ and $x(t) = 0$ if $t < 0$.

Solution

$$E = \int_{0}^{\infty} x_1^2(t)\, dt$$

$$= 25 \int_{0}^{\infty} \exp(-4t)\, dt$$

$$= \frac{25}{4}$$

Energy signals usually exist for a finite interval of time or have most of their energy concentrated in a finite interval of time. They are commonly used in radar and sonar systems to measure the distance or range of an object of interest from the transmitter.

1.2.2 Power signals

Another important class is defined in terms of the signal power. The average power associated with a signal in an interval of time between t_1 and t_2 is simply found by dividing the energy dissipated by the length of the time interval, i.e.:

$$\frac{1}{t_2 - t_1} \int_{t_1}^{t_2} x^2(t)\, dt$$

If the average power delivered by the signal from the beginning to the end of the time is non-zero and finite, the signal is classified as a power signal. More formally

$$0 < P < \infty \tag{1.3}$$

where

$$P = \lim_{T \to \infty} \frac{1}{2T} \int_{-T}^{T} x^2(t)\, dt \tag{1.4}$$

An example of a power signal is the unit step $u(t)$ which will be examined in more detail in Chapter 3. A unit step is defined such that $u(t) = 1$ if $t \geq 0$ and $u(t) = 0$ otherwise. The power in this signal is 0.5. Another example of a power signal is a periodic signal of period T such as:

$$x(t) = \sin\left(\frac{2\pi}{T}\, t\right)$$

In the case of periodic signals there is no need to let the limits of equation (1.2) go to plus and minus infinity. An average evaluated over one period will give the same result. Thus for a periodic signal the power is given by:

$$P = \frac{1}{T} \int_0^T x^2(t)\, dt \qquad (1.5)$$

Energy signals and power signals represent two distinct classes. If a signal has finite energy its power will be zero and by virtue of the lower limit on (1.3) it cannot be a power signal as well. If a signal has finite power its energy will be infinite and it cannot satisfy the upper limit on (1.1).

Self assessment question 1.1: Evaluate the energy in the signal: $x(t) = 5t$ for $0 \leq t < 1$ and $x(t) = 0$ otherwise.

1.3 Fourier series

Any finite power periodic signal $x(t)$ with a period of T seconds can be represented as a summation of sine waves and cosine waves. This representation is known as the trigonometric Fourier series:

$$x(t) = \frac{A_0}{2} + \sum_{n=1}^{\infty} A_n \cos(n\omega_0 t) + B_n \sin(n\omega_0 t) \qquad (1.6)$$

The fundamental frequency of $x(t)$ is $\omega_0 = 2\pi/T$ radians per second or $1/T$ Hz. In general, the signal also contains a second harmonic at $2/T$ Hz, a third harmonic at $3/T$ Hz, etc. The trigonometric Fourier coefficients (A_n and B_n) can be calculated directly from the signal using the following two equations:

$$A_n = \frac{2}{T} \int_{-T/2}^{T/2} x(t) \cos(n\omega_0 t)\, dt \qquad n = 0, 1, 2, \cdots \qquad (1.7)$$

$$B_n = \frac{2}{T} \int_{-T/2}^{T/2} x(t) \sin(n\omega_0 t)\, dt \qquad n = 1, 2, 3, \cdots \tag{1.8}$$

The particular limits on the integrations of equations (1.7) and (1.8) are not the only ones that could be chosen. Integration over one whole period of the signal will give the same result no matter what starting point is chosen.

EXAMPLE 1.2

As an example consider the square wave $x(t)$, one period of which is illustrated in Figure 1.3(a). The period T of the square wave is 2 seconds. The Fourier coefficients A_n and B_n are calculated using equations (1.7) and (1.8). Thus:

$$A_n = \int_0^2 x(t) \cos(n\pi t)\, dt$$

$$= \int_0^1 \cos(n\pi t)\, dt - \int_1^2 \cos(n\pi t)\, dt$$

$$= 0$$

For this particular waveform all the A coefficients are zero and hence there will be no cosine wave terms in the summation of equation (1.6). The B coefficients are calculated in a similar manner:

$$B_n = \int_0^2 x(t) \sin(n\pi t)\, dt$$

$$= \int_0^1 \sin(n\pi t)\, dt - \int_1^2 \sin(n\pi t)\, dt$$

$$= \frac{2}{n\pi} \left(1 - \cos(n\pi) \right)$$

For even values of n the B coefficients are also zero. Thus the trigonometric Fourier series representation of this waveform is:

$$x(t) = \sum_{n=1}^{\infty} \frac{2}{n\pi} \left(1 - \cos(n\pi) \right) \sin(n\pi t)$$

It is instructive to write out some terms in this series:

$$x(t) = \frac{4}{\pi} \sin(\pi t) \qquad \text{(fundamental)}$$

$$+ 0 \qquad \text{(second harmonic)}$$

$$+ \frac{4}{3\pi} \sin(3\pi t) \qquad \text{(third harmonic)}$$

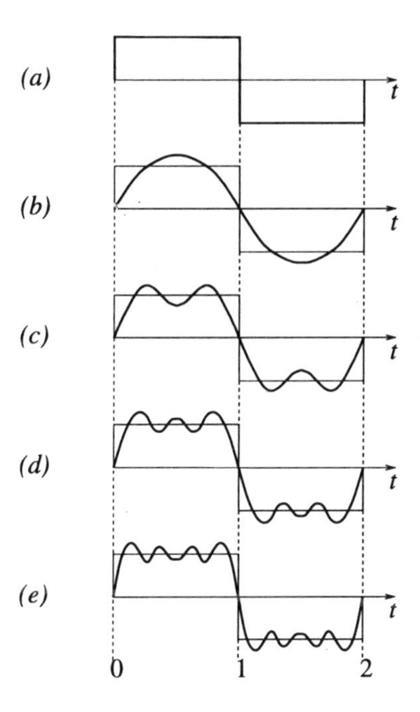

Figure 1.3 *One period of a square wave and its Fourier series approximation with increasing number of harmonic terms: (a) waveform $x(t)$; (b) fundamental; (c) summation of first three harmonics; (d) summation of first five harmonics; (e) summation of first seven harmonics.* □†

$$+\,0 \qquad \text{(fourth harmonic)}$$

$$+\,\frac{4}{5\pi}\sin(5\pi t) \qquad \text{(fifth harmonic)}$$

$$+\ \text{etc.}$$

Figure 1.3 illustrates the effect of adding the harmonic terms together to form the original waveform. The fundamental is shown in Figure 1.3(b). The fundamental (first harmonic), Figure 1.3(b), and the third harmonic are added together to give Figure 1.3(c). As more harmonics are added the summation gets closer and closer (converges) to $x(t)$.

The trigonometric Fourier series provides an alternative representation for the signal $x(t)$. This representation takes the form of sets of Fourier coefficients $\{A_n: n = 0, 1, 2 \cdots\}$ and $\{B_n: n = 1, 2, 3, \cdots\}$. Equations (1.6), (1.7) and (1.8) provide the mechanism for moving back and forward between these alternative representations. Given one period of the time-domain signal $x(t)$, the Fourier

† Note that the symbol □ is used throughout the text to indicate figures which are supported by computer animation.

coefficients can be calculated using equations (1.7) and (1.8). Given the Fourier coefficients, equation (1.6) can be used to evaluate the time-domain signal. This movement from one representation to another is known as a transformation. The trigonometric Fourier series is the simplest and most intuitively appealing of a family of Fourier and Laplace transform techniques for representing signals.

A key feature of many of these techniques is the use of complex phasors to describe sine waves and cosine waves. A complex phasor of amplitude A, frequency ω_0 rad/s can be split into real and imaginary parts:

$$A\exp(j\omega_0 t) = A\cos(\omega_0 t) + j\,A\sin(\omega_0 t) \tag{1.9}$$

As illustrated in Figure 1.4, the complex phasor $\exp(j\omega_0 t)$ can be interpreted as a vector of length A rotating anti-clockwise at ω_0 rad/s. The projection of this vector onto the imaginary axis when plotted as a function of time t produces a sine wave of amplitude A and frequency ω_0. Likewise the projection of the vector onto the real axis produces a cosine wave of the same amplitude and frequency. Thus we can write expressions for a cosine wave and a sine wave as the real ($\Re$) and imaginary ($\Im$) parts of the complex phasor:

$$A\cos(\omega_0 t) = \Re\{A\exp(j\omega_0 t)\}$$

and

$$A\sin(\omega_0 t) = \Im\{A\exp(j\omega_0 t)\}$$

Using the properties of complex numbers the cosine and sine terms in equations (1.6) to (1.8) can be written in terms of the sum or difference of two complex phasors rotating in opposite directions:

$$\cos(n\omega_0 t) = \frac{\exp(jn\omega_0 t) + \exp(-jn\omega_0 t)}{2} \tag{1.10}$$

and

$$\sin(n\omega_0 t) = \frac{\exp(jn\omega_0 t) - \exp(-jn\omega_0 t)}{2j} \tag{1.11}$$

Substitution of (1.10) and (1.11) into (1.6), (1.7) and (1.8) leads to an alternative form of the Fourier series.

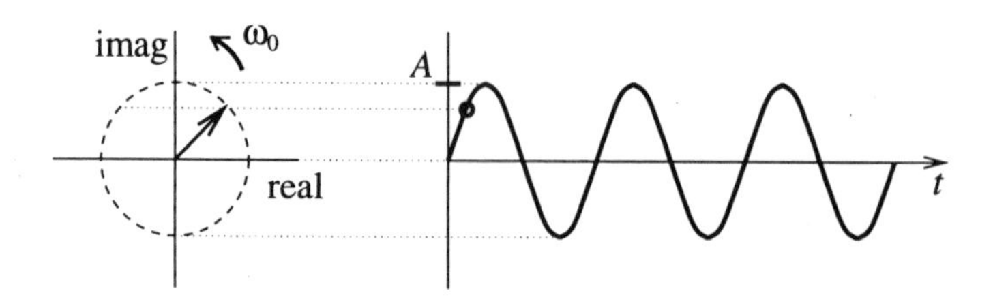

Figure 1.4 *A sine wave as the projection of a complex phasor onto the imaginary axis.* □

Thus any finite power periodic signal $x(t)$ with a period of T seconds can be represented as a summation of complex phasors:

$$x(t) = \sum_{n=-\infty}^{+\infty} X_n \exp(jn\omega_0 t) \tag{1.12}$$

The fundamental phasor frequency is $\omega_0 = 2\pi/T$ rad/s. This representation is known as the complex or exponential Fourier series. It is illustrated in Figure 1.5 where the signal $x(t)$ is viewed as being formed by a bank of phasor generators. Each phasor generator $(\cdots, e^{-jn\omega_0 t}, \cdots, e^{-j2\omega_0 t}, e^{-j\omega_0 t}, 1, e^{j\omega_0 t}, e^{j2\omega_0 t}, \cdots, e^{jn\omega_0 t}, \cdots)$ is amplified by a complex gain $(\cdots, X_{-n}, \cdots, X_{-2}, X_{-1}, X_0, X_1, X_2, \cdots X_n, \cdots)$. The product or outputs of the amplifier are summed to produce $x(t)$. The complex Fourier coefficients (X_n) can be calculated from the signal using the following equation:

$$X_n = \frac{1}{T} \int_{-T/2}^{T/2} x(t) \exp(-jn\omega_0 t)\, dt \tag{1.13}$$

Equations (1.12) and (1.13) together define the complex Fourier series. Notice in particular that the three equations, (1.6) to (1.8), required to define the trigonometric Fourier series have been reduced to two for the complex series. This simplification comes at a price, i.e. X_n is generally a complex number.

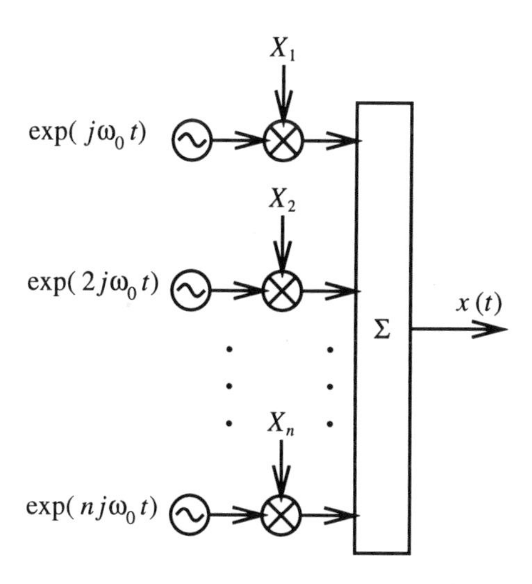

Figure 1.5 *Complex Fourier series representation of a periodic signal $x(t)$.*

EXAMPLE 1.3

As an example of the evaluation of the complex Fourier series, consider the periodic signal $x(t)$ illustrated in Figure 1.6(a).

$$X_n = \frac{1}{T} \int_{-T/2}^{T/2} x(t) \exp(-jn\omega_0 t)\, dt$$

$$= \frac{1}{T} \int_{-\tau/2}^{\tau/2} A \exp(-jn\omega_0 t)\, dt$$

$$= \frac{-A}{jn\omega_0 T}\left[\exp\left(\frac{-jn\omega_0\tau}{2}\right) - \exp\left(\frac{jn\omega_0\tau}{2}\right)\right]$$

$$= \frac{A\tau}{T}\, \frac{\sin(n\omega_0\tau/2)}{n\omega_0\tau/2}$$

Although in general these Fourier coefficients are complex, in this particular example they are real. Figure 1.6(b) illustrates the magnitude and phase of the complex Fourier coefficients as functions of the harmonic number n for the particular case where the pulse width is a fifth of the period, i.e. $T = 5\tau$.

There is a simple relationship between the complex and trigonometric Fourier coefficients, i.e.:

$$X_0 = \frac{A_0}{2}$$

$$X_n = \frac{A_n - jB_n}{2} \qquad n > 0$$

$$X_n = \frac{A_n + jB_n}{2} \qquad n < 0$$

From the above it is clear that the complex Fourier series of a real signal exhibits complex conjugate (Hermitian) symmetry:

$$X_{-n} = X_n^*$$

This is also evident from Figure 1.6 where the amplitude is symmetrical about the origin (even symmetry)

$$|X_{-n}| = |X_n|$$

and the phase is asymmetrical (odd symmetry)

$$\angle X_{-n} = -\angle X_n$$

If the symmetry is exploited, it is never necessary to calculate X_n for negative values of n. Using the above relationships it is also straightforward to move back and forward between a trigonometric Fourier series representation and a complex Fourier series. A

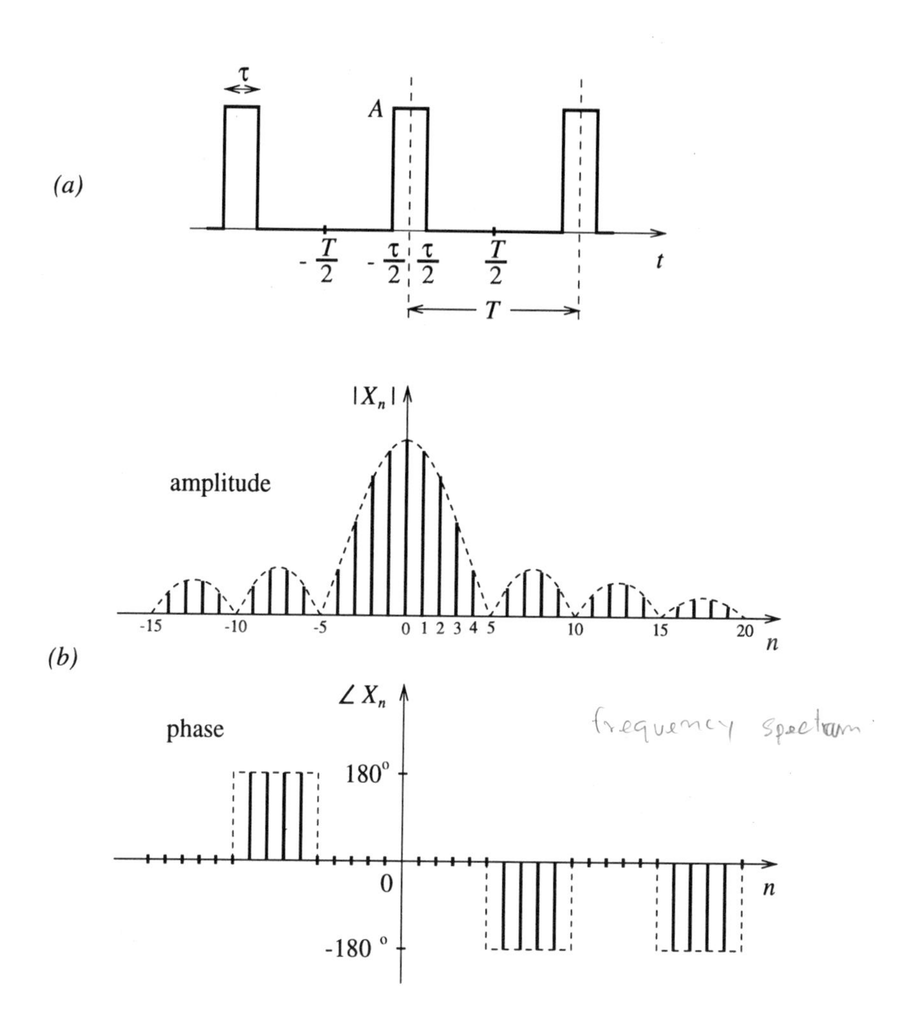

Figure 1.6 *Complex Fourier series representation of a square wave: (a) waveform $x(t)$; (b) amplitude and phase characteristics when $T = 5\tau$.*

variation of the Fourier series is the cosine series where:

$$x(t) = X_0 + \sum_{n=1}^{\infty} 2|X_n| \cos(n\omega_0 t + \angle X_n)$$

This form of the series can be obtained either by exploiting the Hermitian symmetry of the complex Fourier coefficients in equation (1.12) or by applying standard trigonometric identities to equation (1.6). It is seldom used but is included here because of the insight it gives into the meaning of the complex Fourier coefficients. It facilitates a physical interpretation of the complex Fourier coefficients.

By definition the series is used to represent a periodic signal as a sum of cosine waves. The frequencies of these cosine wave are: 0, ω_0, $2\omega_0$, $3\omega_0$, etc. The magnitude of the complex Fourier coefficient $|X_n|$ is half the amplitude of the nth harmonic. The angle of the complex Fourier coefficient $\angle X_n$ is the phase shift associated with the nth harmonic. Thus the nth harmonic might be written as:

$$2|X_n|\cos(n\omega_0 t + \angle X_n)$$

Figure 1.6(b) can be re-examined in this light. It represents the distribution of both amplitude and phase with harmonic number. It is commonly called a frequency spectrum. In this case it is a discrete spectrum because it only has values at the frequencies of the harmonics, i.e. $n\omega_0$.

Both the trigonometric and complex Fourier representations are examples of orthogonal expansions. In particular, two periodic signals $f_1(t)$ and $f_2(t)$ are said to be orthogonal if their product integrated over one period is zero, i.e.:

$$\frac{1}{T}\int_{-T/2}^{T/2} f_1(t)\, f_2^*(t)\, dt = 0$$

The superscript * denotes complex conjugate and is used to accommodate complex signals which can arise in radar and communications systems. In the complex Fourier series the basis functions $\exp(j\omega_0 t)$, $\exp(j2\omega_0 t)$, $\cdots$, $\exp(jn\omega_0 t)$, $\cdots$ are mutually orthogonal, i.e.:

$$\frac{1}{T}\int_{-T/2}^{T/2} \exp(jn\omega_0 t)\exp^*(jm\omega_0 t)\, dt = \begin{cases} 1 \text{ if } m = n \\ 0 \text{ if } m \neq n \end{cases}$$

Harmonically related sines, cosines and complex phasors are orthogonal functions. The orthogonality of the phasors in the complex Fourier series makes it possible to calculate a particular X_n without knowing any of the other X_n's, e.g. to calculate X_2 without knowing X_3 or X_4. If the phasors were not orthogonal then calculation of the Fourier coefficients would involve the solution of a large set of simultaneous equations.

EXAMPLE 1.4

As an illustration of the advantages of using orthogonal signal sets, consider calculating the power in a simple periodic signal $x(t)$ where:

$$x(t) = a_1 \sin(\omega_0 t) + a_2 \sin(2\omega_0 t)$$

The period is T, where $\omega_0 = 2\pi/T$. In fact this signal is so simple that its trigonometric Fourier coefficients can be obtained by inspection, i.e. $A_1 = a_1$, $A_2 = a_2$ – all other coefficients are zero. As in section 1.3 the power is calculated as follows:

$$P = \frac{1}{T}\int_0^T x^2(t)\, dt$$

$$= \frac{1}{T}\int_0^T a_1^2 \sin^2(\omega_0 t)\, dt + \frac{2}{T}\int_0^T a_1\, a_2 \sin(\omega_0 t)\sin(2\omega_0 t)\, dt$$

$$+\frac{1}{T}\int_0^T a_2^2 \sin^2(2\omega_0 t)\, dt$$

Because $\sin(\omega_0 t)$ and $\sin(2\omega_0 t)$ are orthogonal the second term on the right-hand side is zero. The total power in the signal is thus the sum of the power in the two sine waves, i.e.:

$$P = \frac{1}{T}\int_0^T a_1^2 \sin^2(\omega_0 t)\, dt + \frac{1}{T}\int_0^T a_2^2 \sin^2(2\omega_0 t)\, dt$$

$$= \frac{a_1^2}{2} + \frac{a_2^2}{2}$$

A direct consequence of orthogonality is Parseval's theorem which enables the power in a periodic signal to be calculated directly from the Fourier coefficients. There are two equivalent relationships based on the trigonometric and complex forms of the Fourier series respectively:

$$P = \frac{A_0^2}{4} + \frac{1}{2}\sum_{n=1}^{\infty} (A_n^2 + B_n^2) \quad (1.14)$$

$$P = \sum_{n=-\infty}^{\infty} |X_n|^2 \quad (1.15)$$

Calculating the power in a signal is also a good example of how two alternative descriptions of a signal can be used to do the same thing. The power can either be calculated directly from the time-domain description using equation (1.5) or from the Fourier coefficients using either (1.14) or (1.15).

Self assessment question 1.2: Evaluate the complex Fourier coefficients of the periodic signal $x(t)$ which has a period of 1. The signal is defined as: $x(t) = 4t;\ 0 \le t < 1$.

1.4 The Fourier transform

Although the Fourier series is a powerful and elegant concept, it suffers from one disadvantage, i.e. it is only applicable to periodic signals. Most signals of practical importance are not periodic. To see how a Fourier representation for a non-periodic signal might be developed, consider a periodic signal and examine what happens to the Fourier series as the period gets longer. Figure 1.7 illustrates the complex Fourier series of a pulse with increasing period. The waveform is similar to that illustrated in Figure 1.6. However this time, rather than plotting spectrum as a function of harmonic number n, it is plotted as a function of frequency ω. The frequency of the nth harmonic is now $n\omega_0$. As the period T is increased from 1/4 of a second to 1/2 of a second to 1 second, the harmonics (or Fourier components) get closer and closer together in frequency (radiancy) – the spectrum gets denser. In fact every time the period is doubled, the frequency spacing between the harmonics is halved. Note however that the general outline (shape or envelope) of the spectrum stays the same. It would not be unreasonable to conclude that the outline is a property of the pulse itself and not the period between pulses. Thus a Fourier representation of a single pulse might be obtained by letting the period T get very large. This process of increasing the period gives a good indication that there may be a way of representing non-periodic waveforms by adding complex phasors. It is worth noting however that the magnitudes of the Fourier components also halve as the period is doubled. The whole representation may disappear completely if the period gets too big.

To see how the problem may be circumvented consider the periodic pulse train of Figure 1.7(a). In section 1.3 the expression for its Fourier coefficients was reduced to:

$$X_n = \frac{1}{T} \int_{-\tau/2}^{\tau/2} A \exp(-jn\omega_0 t)\, dt$$

In the limit as $T \to \infty$, the Fourier coefficients will tend to zero and vanish. For a fixed pulse width τ the result of the integration is independent of the period T and hence will not change as T tends to infinity. However this fixed value is divided by T and hence in

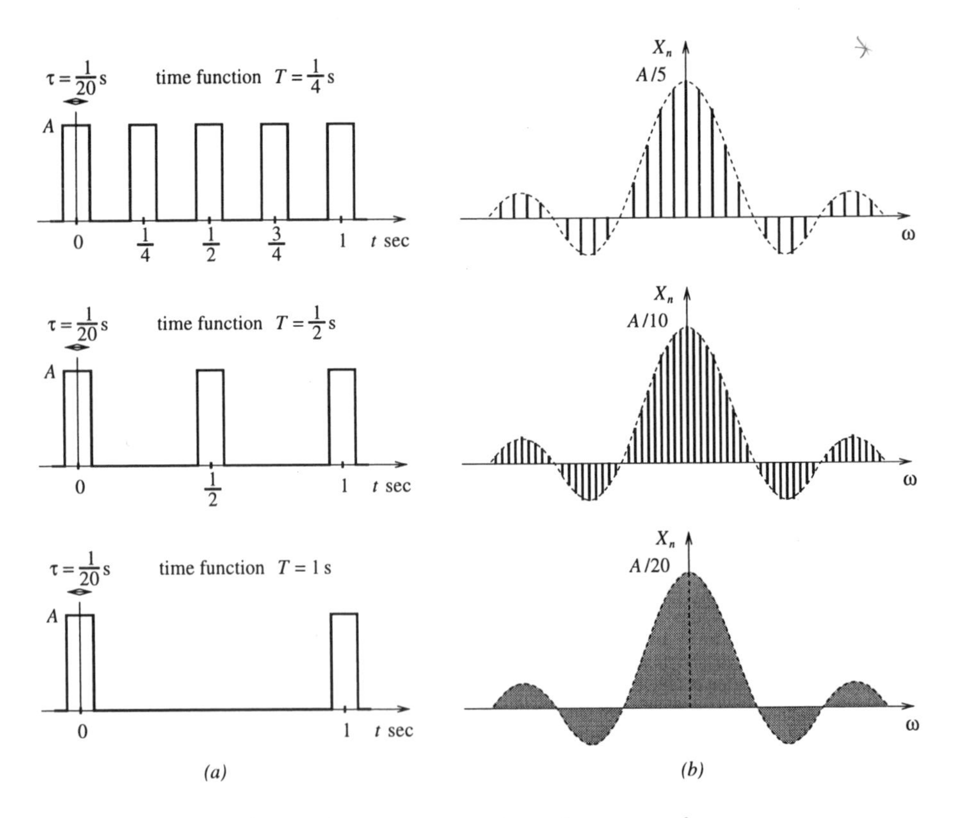

Figure 1.7 *From Fourier series to Fourier transform.*

the limit the whole thing will tend to zero. This problem could be avoided if the Fourier coefficients were redefined as:

$$X_n' = T\ X_n$$

$$= \int_{-T/2}^{T/2} x(t) \exp(-jn\omega_0 t)\ dt$$

The Fourier series would then become:

$$x(t) = \sum_{n=-\infty}^{\infty} \frac{X_n'}{T} \exp(jn\omega_0 t)$$

$$= \sum_{n=-\infty}^{\infty} X_n' \exp(jn\omega_0 t) \frac{\omega_0}{2\pi}$$

In the limit as $T \to \infty$, the spectral lines get closer together and the separation between ω_0 becomes the differential $d\omega$. The harmonic frequency $n\omega_0$ becomes the continuous frequency variable ω. The discrete spectrum X_n' becomes a continuous spectrum $X(\omega)$. The summation of all the discrete frequency components becomes an integration over all possible frequencies. Thus:

$$X(\omega) = \lim_{T \to \infty} X_n'$$

$$= \lim_{T \to \infty} \int_{-T/2}^{T/2} x(t) \exp(-jn\omega_0 t)\ dt$$

Therefore:

$$X(\omega) = \int_{-\infty}^{\infty} x(t) \exp(-j\omega t)\ dt \qquad (1.16)$$

and

$$x(t) = \lim_{T \to \infty} \sum_{n=-\infty}^{\infty} X_n' \exp(jn\omega_0 t) \frac{\omega_0}{2\pi}$$

Therefore:

$$x(t) = \frac{1}{2\pi} \int_{-\infty}^{\infty} X(\omega) \exp(j\omega t)\ d\omega \qquad (1.17)$$

Most finite energy signals can be represented in this manner. Equation (1.17) suggests that such a signal is a summation (integration) of complex phasors. The integration is present because the phasor frequency ω rad/s is continuous and can lie anywhere between plus and minus infinity. Equation (1.17) is known as the synthesis equation since $x(t)$ is synthesised from all possible complex phasors $e^{j\omega t}$ each with complex amplitude $X(\omega)d\omega/(2\pi)$. The complex Fourier transform or continuous *frequency spectrum* $X(\omega)$ can be calculated from the signal using equation (1.16). A short-hand form of these equations is often used:

- $X(\omega) = F[x(t)]$: $X(\omega)$ is the Fourier transform of $x(t)$.
- $x(t) = F^{-1}[X(\omega)]$: $x(t)$ is the inverse Fourier transform of $X(\omega)$.

EXAMPLE 1.5

As an example of the evaluation of the Fourier transform, consider the finite energy signal $x(t)$ illustrated in Figure 1.8.

$$\begin{aligned} X(\omega) &= \int_{-\infty}^{\infty} x(t)\exp(-j\omega t)\,dt \\ &= \int_{-\tau/2}^{\tau/2} x(t)\exp(-j\omega t)\,dt \\ &= \frac{-A}{j\omega}\left[\exp\left(\frac{-j\omega\tau}{2}\right) - \exp\left(\frac{j\omega\tau}{2}\right)\right] \\ &= A\tau\left(\frac{\sin(\omega\tau/2)}{\omega\tau/2}\right) \end{aligned}$$

The term inside the brackets occurs frequently and it is thus convenient to give it an abbreviation:

$$\mathrm{sa}(x) = \frac{\sin(x)}{x}$$

It is known as the *sampling function.* The Fourier series of example 1.3 could also be written in terms of the sampling function. An alternative to the sampling function is the sinc ('sink') function defined as:

$$\mathrm{sinc}(x) = \frac{\sin(\pi x)}{\pi x}$$

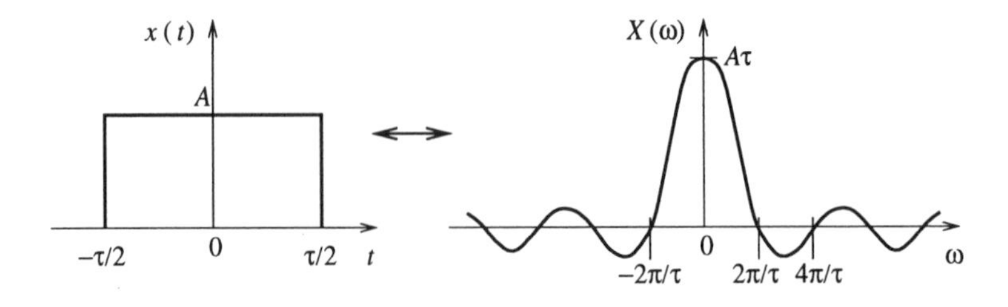

Figure 1.8 *Fourier transform of a rectangular pulse.*

The sinc function is more convenient when the Fourier transform is defined in terms of frequency f in Hz rather than angular frequency ω in rad/s.

Figure 1.8 shows the Fourier transform, $X(\omega)$, of the single pulse, $x(t)$. The pulse has a duration or width of τ seconds. Whereas the Fourier series of Figure 1.7 is a discrete function of frequency in that it is only specified at particular frequencies and each frequency has an amplitude, the Fourier transform is a continuous function of frequency and $|X(\omega)|$ represents the distribution of amplitude with frequency. For example, if the signal $x(t)$ was a voltage, then $|X(\omega)|$ would have units of volts per rad/s. The sa/sinc functions are very common in communications systems, control and signal processing. (It is left as an exercise for the reader to plot graphs of $\mathrm{sa}(x)$ as a function of x and $\mathrm{sa}(2x)$ as a function of x. Care should be exercised when evaluating $\mathrm{sa}(0)$ as this involves division by zero. Either use L'Hôpital's rule or try a small number instead of zero.) As with the Fourier series it is useful to have a physical interpretation of the transform:

> The complex Fourier transform is used to represent a signal as a 'sum' of cosine waves at all possible frequencies. All possible frequencies are needed because the signal is not periodic and hence cannot have harmonics. The magnitude of the complex Fourier transform $|X(\omega)|d\omega/(2\pi)$ is the amplitude of the sine wave with frequency ω rad/s. The angle of the complex Fourier transform $\angle X(\omega)$ is the phase shift associated with the cosine wave. The signal $x(t)$ is said to have a component in a small frequency band ω to $\omega + d\omega$ rad/s which might be written approximately as:

$$\frac{|X(\omega)|d\omega}{2\pi}\cos(\omega t + \angle X(\omega))$$

Further insight into the meaning of the Fourier transform can be derived from Parseval's theorem for finite energy signals. The energy can be calculated either from the time-domain signal itself or from the Fourier transform. The energy in a signal $x(t)$ is:

$$\begin{aligned} E &= \int_{-\infty}^{\infty} x^2(t)\,dt \\ &= \frac{1}{2\pi}\int_{-\infty}^{\infty} |X(\omega)|^2\,d\omega \end{aligned} \qquad (1.18)$$

In words, $|X(\omega)|^2/2\pi$ defines how the energy in the signal is distributed with frequency. Thus $|X(\omega)|^2/2\pi$ is known as the *energy spectral density* (ESD).

Self assessment question 1.3: Evaluate the Fourier transform of the signal $x(t)$, where: $x(t) = 4$ for $1 \le t < 4$ and $x(t) = 0$ elsewhere.

1.5 Laplace transform

It is not always possible to calculate the Fourier transform of a signal $x(t)$. For example, if the signal is of finite power rather than finite energy the Fourier transform does not exist. This can be problematic since several signals of practical importance are of finite power. One such signal is the unit step function $u(t)$, where $u(t) = 1$ if $t \geq 0$ and $u(t) = 0$ otherwise. This function is important because it can be used to characterise a sudden change in a signal. For example, $u(t)\cos(5t)$ is a cosine wave which has been switched on at $t = 0$. Because the step function is a finite power signal it cannot have finite energy and hence it will not have a Fourier transform. This can be illustrated by attempting to calculate the Fourier transform $U(\omega)$ of the unit step at $\omega = 0$:

$$\begin{aligned} U(0) &= \int_{-\infty}^{\infty} u(t) \exp(-j0t)\, dt \\ &= \int_{0}^{\infty} dt \\ &= \infty \end{aligned}$$

The solution is to multiply the signal $x(t)$ by a convergence factor $\exp(-\sigma t)$ to form a new signal $x_\sigma(t)$ whose Fourier transform does exist, i.e.:

$$x_\sigma(t) = \exp(-\sigma t)\; x(t)$$

where σ is a real number. The idea here is that with proper choice of σ, the Fourier transform of $x_\sigma(t)$ will exist even though the Fourier transform of $x(t)$ may not. The Fourier transform of $x_\sigma(t)$ is given by:

$$\begin{aligned} X_\sigma(\omega) &= \int_{-\infty}^{\infty} x_\sigma(t) \exp(-j\omega t)\, dt \\ &= \int_{-\infty}^{\infty} x(t) \exp(-(\sigma + j\omega)t)\, dt \end{aligned}$$

This can be written more compactly using the complex frequency variable $s = \sigma + j\omega$. The result is known as the *two-sided* or *bilateral* Laplace transform of $x(t)$:

$$X(s) = \int_{-\infty}^{\infty} x(t) \exp(-st)\, dt$$

The term *two-sided* or *bilateral* reflects the inclusion of both the positive and negative portions of the time axis. In a similar manner, consideration of the inverse Fourier transform of $X_\sigma(\omega)$ leads to the inverse Laplace transform:

$$\begin{aligned} x(t) &= \exp(\sigma t)\; x_\sigma(t) \\ &= \exp(\sigma t) \frac{1}{2\pi} \int_{-\infty}^{\infty} X_\sigma(\omega) \exp(j\omega t)\, d\omega \end{aligned}$$

$$= \frac{1}{2\pi} \int_{-\infty}^{\infty} X_\sigma(\omega) \exp((\sigma + j\omega)t)\, d\omega$$

Therefore:

$$x(t) = \frac{1}{2\pi j} \int_{\sigma - j\infty}^{\sigma + j\infty} X(s) \exp(st)\, ds \tag{1.19}$$

With the *two-sided* transform, the choice of σ and the convergence factor $\exp(-\sigma t)$ can cause difficulties in that different signals $x(t)$ can have the same transform. Thus there can be ambiguities when taking the inverse Laplace transform. A full consideration of these issues is beyond the scope of this book. The ambiguities/difficulties can be removed if the signal $x(t)$ is assumed to be *one-sided* or *causal* in nature, i.e. $x(t) = 0$ if $t < 0$. The one-sided Laplace transform can then be defined as:

$$X(s) = \int_{0^-}^{\infty} x(t) \exp(-st) dt \tag{1.20}$$

Since the signal is zero for negative values of the time axis, the integration is over the range $\{0 \le t < \infty\}$. The notation 0^- indicates that the origin itself is included in the integration. It is possible to compute this one-sided Laplace transform for most signals of practical interest. The one-sided nature is not particularly restrictive in that it simply asserts that the signal was switched on at some initial time $t = 0$. Two-sided transformed are required for dealing with random signals and will be discussed further in Chapter 7. Most positive values of σ will ensure that the one-sided transform exists and hence the choice of σ is transparent to the user of the transform.

As indicated by equation (1.19), the Laplace transform is essentially the same form of signal representation as the Fourier series and Fourier transform. However the Laplace transform is applicable to a wider range of signals than Fourier series or transforms. Most one-sided finite energy and finite power signals such as $x(t)$ can be represented by a 'summation' (integration) of growing phasors $\exp(st)$. The phasor $\exp(st)$ has both a growth rate controlled by σ and a frequency of oscillation ω. Thus the imaginary part of $A\exp(st)$ is a growing or decaying sine wave, depending on whether σ is positive or negative:

$$A\exp(st) = A\exp(\sigma t)\cos(\omega t) + j\, A\exp(\sigma t)\sin(\omega t)$$

The two cases are illustrated in Figure 1.9. Direct evaluation of such an integral such as (1.19) involves calculus of residues which is beyond the scope of this text. However partial fraction expansion techniques can provide straightforward solutions for many common problems, as will be seen later. Like the Fourier transform, the short-hand L is used to indicate the Laplace transform of a signal:

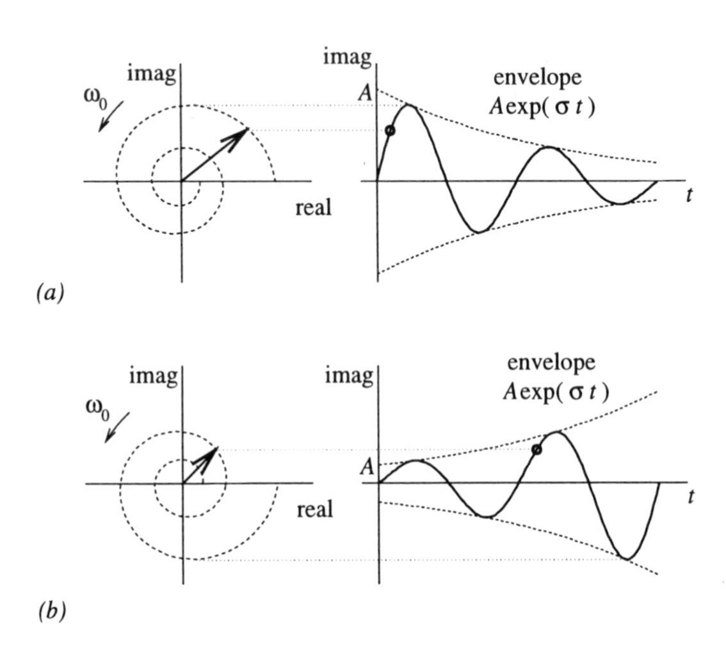

Figure 1.9 *Laplace basis functions: (a) decaying phasor, $\sigma < 0$; (b) growing phasor, $\sigma > 0$.* □

- $X(s) = L[x(t)]$: $X(s)$ is the Laplace transform of $x(t)$.
- $x(t) = L^{-1}[X(\omega)]$: $x(t)$ is the inverse Laplace transform of $X(\omega)$. Again a physical interpretation can provide insight:

> The Laplace transform is used to represent a signal as a sum of growing or decaying cosine waves over a range of frequencies. The magnitude of the Laplace transform $|X(s)|d\omega/(2\pi)$ is the amplitude of the growing or decaying cosine wave with a frequency ω rad/s. The angle of the Laplace transform $\angle X(s)$ is the phase shift associated with the cosine wave. The parameter σ, which is the real part of s, determines the rate of growth or decay. The signal $x(t)$ is said to have a component within a small frequency band ω to $\omega + d\omega$ and with a growth or decay rate of σ which might be written approximately as:
>
> $$\frac{|X(s)|d\omega}{2\pi}\exp(\sigma t)\cos(\omega t + \angle X(s))$$

EXAMPLE 1.6

As a simple example of the evaluation of a Laplace transform, consider a one-sided signal $x(t) = \exp(-\alpha t)$ where α is a positive real number. By definition:

$$X(s) = \int_0^\infty \exp(-\alpha t)\exp(-st)\,dt$$

$$= \int_0^\infty \exp(-(s+\alpha)t)\, dt$$

$$= \frac{-1}{s+\alpha}\Big[\exp(-(s+\alpha)t)\Big]_0^\infty$$

$$= \frac{1}{(s+\alpha)}$$

Self assessment question 1.4: Using the definition of equation (1.20), evaluate the Laplace transform of the signal $x(t)$, where: $x(t) = \exp(-t/5)$ for $t \geq 0$ and $x(t) = 0$ for $t < 0$.

1.6 Transform analysis of linear systems

Fourier series, Fourier transforms and Laplace transforms share a common theme in that they all represent signals as *weighted sums (integrals) of exponential orthogonal basis functions.* This form of representation is most evident in the complex Fourier series synthesis equation (1.12). The 'weights' are the complex Fourier coefficients $\{X_n\}$ and the complex phasors $\{\exp(jn\omega_0)\}$ are mutually orthogonal. In the Fourier transform of equation (1.17) the 'summation' has become an integration and the 'weights' have become densities, so that $|X(\omega)|$ is the energy spectral density. However the basic idea of adding up (or integrating) complex phasors $\{\exp(j\omega t)\}$, each with complex weight (or weighting function $X(\omega)$), remains the same. This particular form of signal representation is fundamental to the analysis of linear systems and in evaluating the response of such systems to a wide range of inputs. The reason why Fourier and Laplace transforms are so effective in the analysis and design of linear systems is because of the intimate relationship between the particular form of signal representation and the properties and nature of many linear systems. Two characteristics of linear systems and their relationship to Fourier and Laplace representation are considered below.

1.6.1 Superposition

An important property which is synonymous with linear systems is superposition. It can be defined in the following manner:

- the response of a linear system having a number of inputs can be computed by determining the response to each input considered separately and then summing the individual responses to obtain the total response.

EXAMPLE 1.7

A simple example of a linear system is a 5 Ω resistor. The current through the resistor is the input, and the output or response is the voltage across the resistor. If 2 A is applied to the resistor, the output is 10 V. If 3 A is applied to the resistor, the output is 15 V. There are two possible approaches to evaluating the output when the input is 5 A: (i) apply Ohm's law again; (ii) exploit the linearity of the system and the sum of the responses to 2 A and 3 A.

Using Fourier or Laplace techniques the inputs signal $x(t)$ can represented as a weighted sum of a number of separate inputs. If, for example, $x(t)$ was a periodic signal, the 'separate inputs' would be the complex phasors $\{\exp(jn\omega_0 t)\}$. Using superposition, the response to each phasor could be evaluated separately and the individual responses added to form the response to $x(t)$.

1.6.2 Linear ordinary differential equations

A second property of many linear systems is that they can be described or modelled by a linear ordinary differential equation. In general such a differential equation takes the form:

$$a_0\, y + a_1 \frac{dy}{dt} + \cdots + a_n \frac{d^n y}{dt^n} = b_0\, x + b_1 \frac{dx}{dt} + \cdots + b_m \frac{d^m x}{dt^m} \tag{1.21}$$

The input signal is $x(t)$, the output or response is $y(t)$ and the system is defined by the constant coefficients $a_0 \cdots a_n$ and $b_0 \cdots b_m$. As already indicated, if the system is linear and the input can be represented as a weighted sum of complex phasors, the key step is to evaluate the response to a typical phasor. This is straightforward for the simple reason that exponential functions are easy to differentiate.

EXAMPLE 1.8

Consider a typical basis function associated with a complex Fourier series, i.e. $\exp(jn\omega_0 t)$. Differentiating once gives:

$$\frac{d}{dt}(\exp(jn\omega_0 t)) = jn\omega_0 \exp(jn\omega_0 t)$$

Differentiating twice:

$$\frac{d^2}{dt^2}(\exp(jn\omega_0 t)) = (jn\omega_0)^2 \exp(jn\omega_0 t)$$

The response $y_n(t)$ to an individual basis function such as $x_n(t) = \exp(jn\omega_0 t)$ is found by solving the associated differential equation. For example, Figure 1.10 illustrates a simple RC circuit to which the nth harmonic phasor is applied. In common with the Fourier series representation it will be assumed that the complex phasor has existed since the beginning of time (i.e. $t = -\infty$). The system is described by the differential equation:

$$a_0 y_n(t) + a_1 \frac{dy_n}{dt} = b_0 \exp(jn\omega_0 t)$$

The constants a_0, a_1 and b_0 are 1, RC and 1 respectively, and can be calculated from the circuit using Kirchhoff's laws. Since the input has existed since the beginning of time, any transient response due to initial conditions will have decayed to zero long ago. The standard method of obtaining the steady-state solution of a linear differential

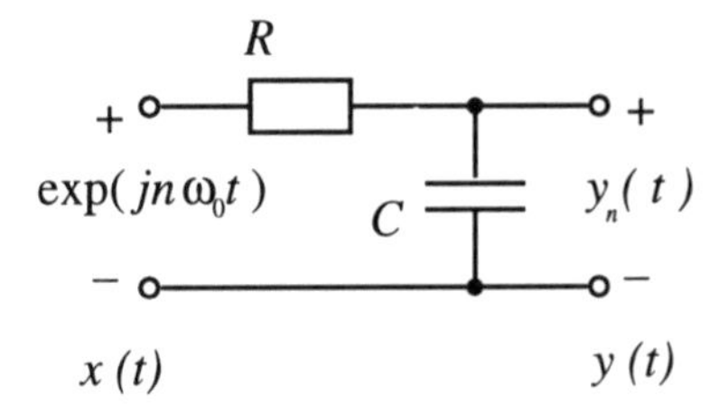

Figure 1.10 *RC circuit with applied phasor.*

equation with an exponential input is to adopt an assumed solution, i.e.:

$$y_n(t) = K \exp(jn\omega_0 t)$$

The only unknown is the complex constant K which can be evaluated by substituting the assumed solution back into the differential equation to get:

$$a_0 y_n(t) + a_1(jn\omega_0)y_n(t) = b_0 \exp(jn\omega_0 t)$$

Collecting terms gives the solution:

$$y_n(t) = \frac{b_0}{a_0 + (jn\omega_0)a_1} \exp(jn\omega_0 t)$$

i.e. the response to an individual basis function. Using exponential basis functions has made it easy to solve the differential equation.

The response of the system to this one basis function is characterised by the system transfer function H_n, i.e.:

$$y_n(t) = H_n \exp(jn\omega_0 t)$$

For the RC circuit of Figure 1.10:

$$H_n = \frac{b_0}{a_0 + (jn\omega_0)a_1} = \frac{1}{1 + (jn\omega_0)RC}$$

Self assessment question 1.5: For Figure 1.10, what would be the response to the phasor $\exp(jn\omega_0 t)$ if the positions of the resistor and capacitor were reversed?

1.6.3 Response of linear system to a periodic input

As an illustration of the relationship between Fourier and Laplace signal representation and the properties of linear systems, consider evaluating the response $y(t)$ of such a system to a periodic signal $x(t)$ as illustrated in Figure 1.11(a).

In Figure 1.11(b) the periodic input $x(t)$ is drawn in terms of its complex Fourier components. These can be viewed as a bank of complex (harmonic) generators,

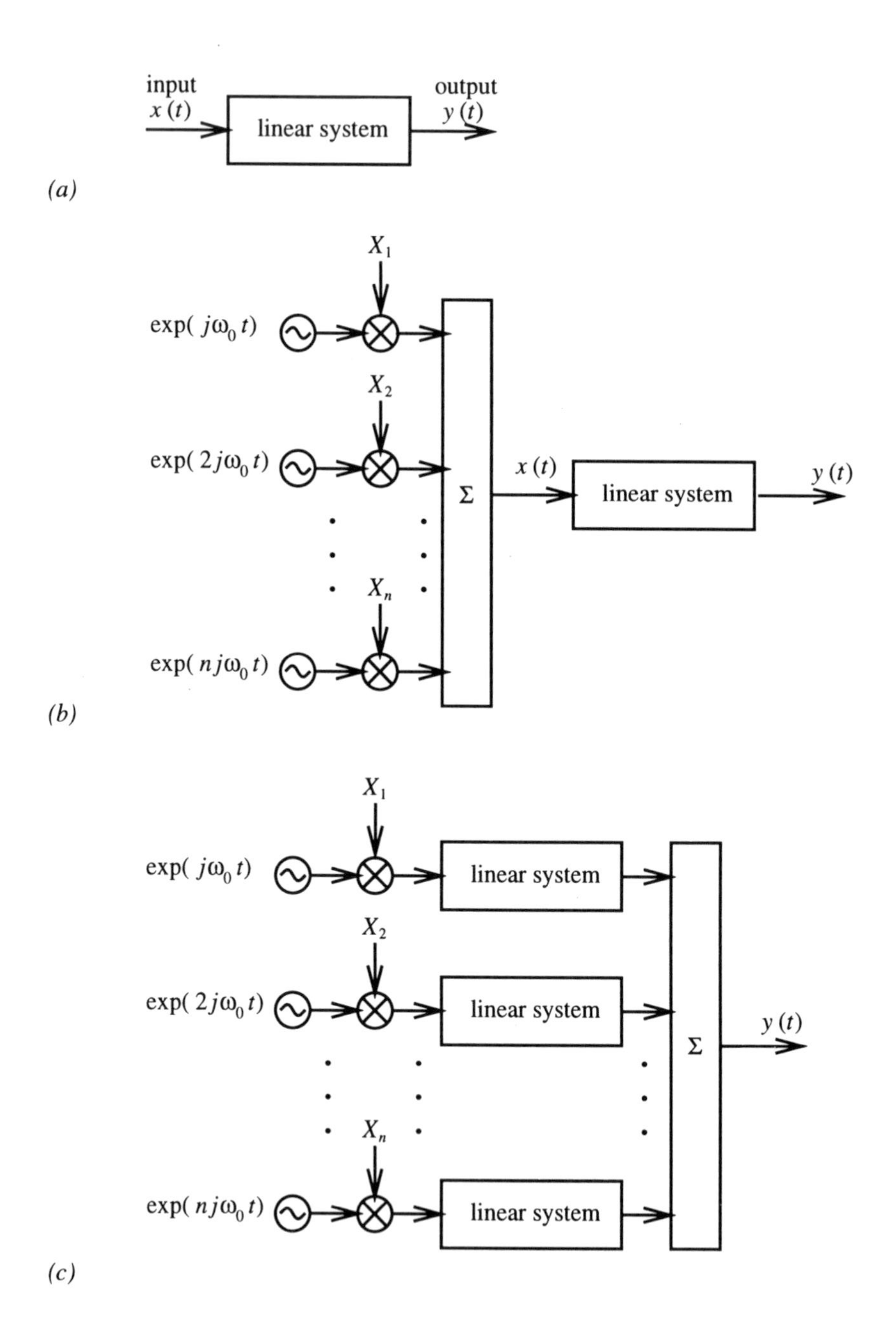
input
x (t)
linear system
output
y (t)
(a)
X1
exp(jω0 t)
X2
exp(2jω0 t)
Xn
exp(njω0 t)
Σ
x (t)
linear system
y (t)
(b)
X1
exp(jω0 t)
linear system
X2
exp(2jω0 t)
linear system
Xn
exp(njω0 t)
linear system
Σ
y (t)
(c)

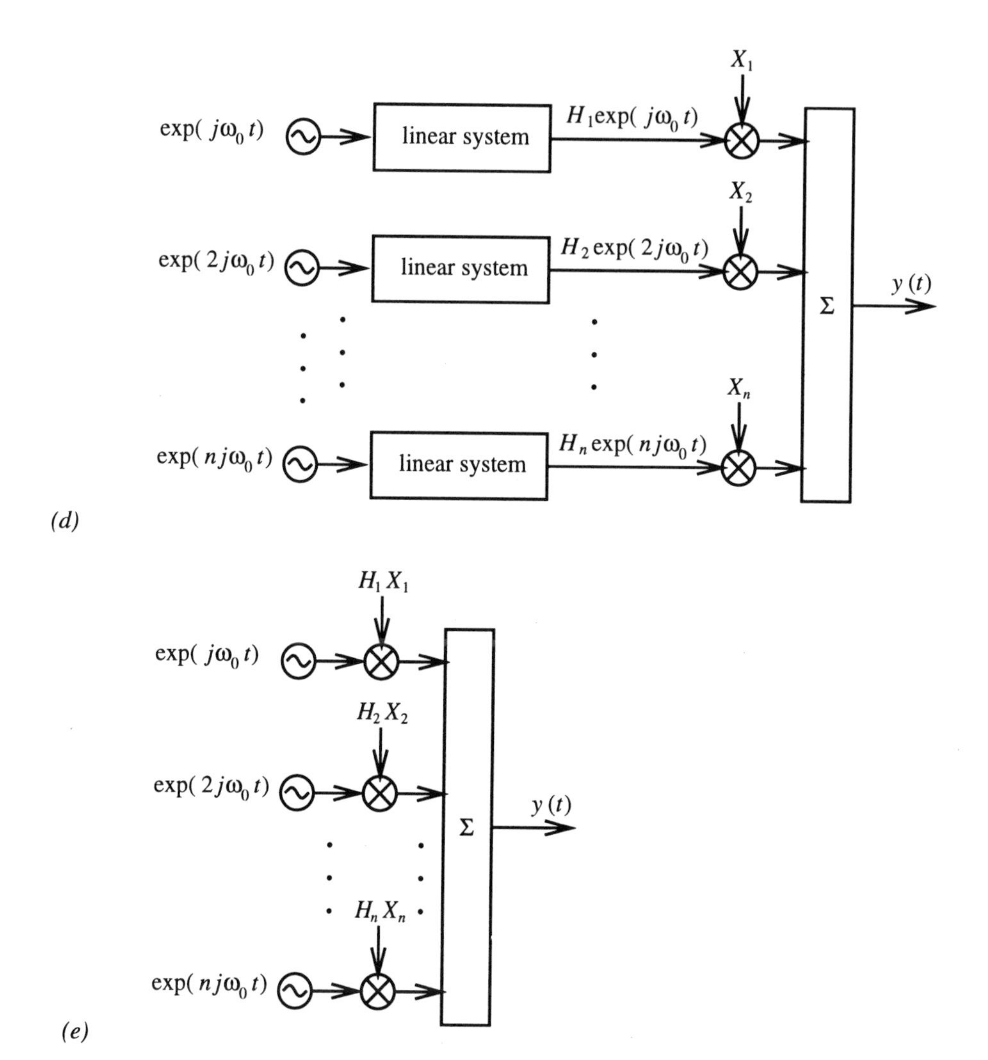

Figure 1.11 *Response of a linear system to a periodic signal: (a) linear system; (b) Fourier representation of input; (c) apply superposition; (d) apply superposition; (e) Fourier representation of output.*

$\exp(j\omega_0 t), \cdots, \exp(jn\omega_0 t)$. Each of these generators is amplified by a factor given by its associated complex Fourier coefficient, e.g. the 2nd harmonic generator is amplified by X_2. (Note: X_2 is complex; $|X_2|$ represents the gain and $\angle X_2$ represents the phase shift.) After each generator is amplified the harmonic components are added together to form the required input.

Application of the principle of superposition to Figure 1.11(b) enables it to be redrawn as Figure 1.11(c). Since the input $x(t)$ is formed by adding complex harmonic terms such as $X_2 \exp(j2\omega_0 t)$ and $X_3 \exp(j3\omega_0 t)$, these terms can be applied individually to the system and the individual responses added together to form the overall output $y(t)$. In Figure 1.11(c) a harmonic generator, such as $\exp(2j\omega_0 t)$, is amplified by a factor X_2 before it is applied to the system. Again, because the system is linear, it does not matter whether the signal is first amplified then applied to the system, or whether the signal is first applied to the system and the system output amplified. The result will be the same.

In Figure 1.11(d) the complex gain terms (e.g. X_2) have been moved from the inputs of the linear systems to the outputs. Because the linear system can be described by a differential equation, the response to one particular harmonic generator (e.g. $\exp(j2\omega_0 t)$), is easy to calculate and is simply $H_2 \exp(j2\omega_0 t)$ where H_2 is a complex number which depends on the linear system itself and the frequency of the harmonic applied to it (in this case the frequency is $2\omega_0$ and hence we have the subscript 2 on the H). The next step is to look more closely at the output $y(t)$. This is formed by taking harmonic terms such as $\exp(j2\omega_0 t)$, first multiplying (or amplifying) them by a complex term defined by the system (e.g. H_2), then multiplying by complex terms defined by the input (such as X_2) and finally adding all the harmonic components together. This is, of course, the definition of a Fourier series.

Figure 1.11(e) illustrates more explicitly how the output $y(t)$ could be generated. In this part of the figure the multiplications by, for example, H_2 and X_2 have been grouped together into one multiplication, i.e. multiplication by $H_2 X_2$. Comparison of Figure 1.11(e) with Figure 1.11(b) demonstrates that they are identical in structure, and hence Figure 1.11(e) is the Fourier series representation of the output $y(t)$ and the multiplication terms such as $X_2 H_2$ are the complex Fourier coefficients of the output. Thus there is a simple relationship between the Fourier coefficients of the input X_n and the Fourier coefficients of the output Y_n through the complex gain terms associated with the system H_n. That relationship is:

$$Y_n = H_n X_n. \tag{1.22}$$

Recall that each basis function is scaled by X_n to form the input $x(t)$ (see Figure 1.11(b)). Hence the response of the system to each Fourier component, $X_n \exp(jn\omega_0 t)$, is:

$$X_n H_n \exp(jn\omega_0 t)$$

Using superposition, the output $y(t)$ is obtained by summing the individual responses:

$$y(t) = \sum_{n=-\infty}^{\infty} H_n X_n \exp(jn\omega_0 t)$$

Then by definition $Y_n = H_n X_n$ is a complex Fourier coefficient of the output, see Figure 1.11(e).

1.6.4 General approach

A powerful general method for evaluating the response $y(t)$ of a linear system to a particular input signal $x(t)$ is to adopt the following procedure:

- represent the input signal as a *weighted sum of exponential basis functions* (e.g. $\exp(j\omega t)$ is the basis function in the Fourier transform);
- since these basis functions are *orthogonal*, the amplitude (and phase) of each one can be evaluated without reference to any of the others (e.g. for a periodic signal, the amplitude of the second harmonic $|X_2|$ can be calculated, without knowing anything about the amplitude of the third, fourth or fifth harmonics);
- the system can be described by a linear differential equation and hence it is easy to calculate its response to an exponential input (basis function);
- having calculated the response to one exponential input, the response to another exponential input can be deduced by simple analogy (e.g. if a second harmonic $\exp(j2\omega_0 t)$ produces a response $2\omega_0 \exp(j2\omega_0 t)$ then a fourth harmonic $\exp(j4\omega_0 t)$ will produce a response $4\omega_0 \exp(j4\omega_0 t)$);
- since the system is linear, superposition applies and the overall response is provided by adding up the individual responses to the exponentials which make up the input signal $x(t)$.

This method has been illustrated by considering the steady-state response of a linear system to a periodic input. More generally, if the complete response (both transient and steady-state) is required of a linear system to a non-periodic input, the Laplace transform can be used in a similar manner. The steps are summarised in Table 1.1. The system transfer function $H(s)$ is again defined by the response of the system to a typical basis function. In this case, if a Laplace basis function $\exp(st)$ is applied at the input to the system, the output or response will be $H(s)\exp(st)$.

Table 1.1 *Laplace transform system analysis.*

Definition	*Interpretation*
$X(s) = L[x(t)]$	Laplace transform the input signal
$H(s)$	Calculate system transfer function
$Y(s) = H(s)\ X(s)$	Calculate transform of the output by multiplication
$y(t) = L^{-1}[Y(s)]$	Inverse Laplace transform to obtain the output

1.7 Transfer function

The Laplace transfer function of a system is defined in a similar manner to the Fourier transfer function of equation (1.22):

$$H(s) = \frac{L[\text{ output }]}{L[\text{ input }]} = \frac{Y(s)}{X(s)} \tag{1.23}$$

The transfer function completely characterises the system. Armed with the transfer function, the response of the system to a wide variety of inputs and initial conditions† can be calculated. Thus the transfer function is of great value in systems analysis.

† Methods for dealing with initial conditions will not be dealt with in this text – the interested reader should consult [Balmer].

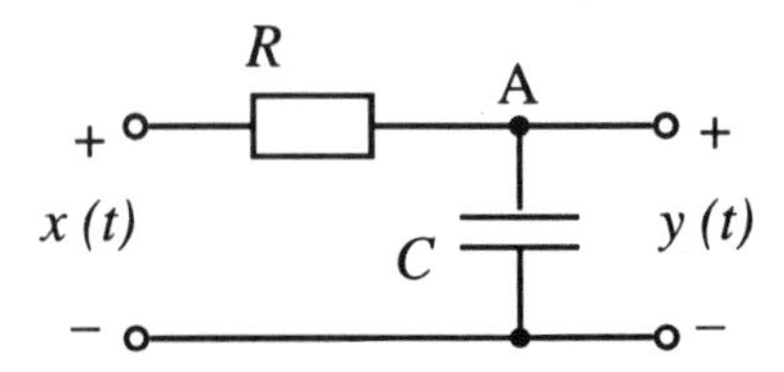

Figure 1.12 *RC circuit: time-domain analysis.*

Further, an understanding of the relationships between the transfer function and the characteristics of the system can be of assistance in solving the less direct questions associated with system design. The relationship between the transfer function and system characteristics will be explored in Chapter 3. As an example, the transfer function of the simple circuit of Figure 1.12 will be evaluated. The transfer function which relates the voltage across the capacitor $y(t)$ to the applied voltage $x(t)$ is required. Two methods of calculating the transfer function will be considered.

Method (i)

This method involves using Kirchhoff's laws to deduce the differential equation which describes the circuit and then solving the differential equation using Laplace transforms. The linear differential equation relates the input $x(t)$ to the output $y(t)$. Taking Laplace transforms of both sides of the differential equation and rearranging terms provides the desired ratio of $Y(s)$ to $X(s)$ – the transfer function.

EXAMPLE 1.9

Applying Kirchhoff's current law at point A in Figure 1.12 gives:

$$\frac{x(t) - y(t)}{R} = C\,\frac{dy}{dt}$$

Rearrange to put terms in $x(t)$ on one side and terms in $y(t)$ on the other:

$$\frac{x(t)}{R} = C\,\frac{dy}{dt} + \frac{y(t)}{R}$$

Take Laplace transforms of both sides:

$$\frac{X(s)}{R} = C\,\{sY(s) - y(0^-)\} + \frac{Y(s)}{R}$$

Assume that initial conditions such as $y(0^-)$ are zero. Thus:

$$\frac{X(s)}{R} = \left\{Cs + \frac{1}{R}\right\} Y(s)$$

A little rearrangement yields the desired transfer function:

$$H(s) = \frac{Y(s)}{X(s)} = \frac{1}{(1 + RCs)}$$

Method (ii)

The second method involves transforming the circuit elements before applying Kirchhoff's laws. The defining equation which relates the voltage across an inductor, $v(t)$, to the current through it, $i(t)$, is:

$$v(t) = L\,\frac{di}{dt}$$

If a complex phasor current $i(t) = \exp(st)$ is applied to the inductor, the resultant voltage phasor is:

$$\begin{aligned} v(t) &= L\; s\; \exp(st) \\ &= L\; s\; i(t). \end{aligned}$$

The ratio of the voltage phasor to the current phasor is the impedance of the inductor, Ls. In a similar manner the impedance of a capacitor is $1/(sC)$. Thus it is possible to transform the circuit itself to obtain Figure 1.13.

EXAMPLE 1.10

Apply Kirchhoff's current law at point A as before:

$$\frac{X(s) - Y(s)}{R} = sC\; Y(s)$$

$$\frac{X(s)}{R} = sC\; Y(s) + \frac{Y(s)}{R}$$

The desired result is as before:

$$H(s) = \frac{Y(s)}{X(s)} = \frac{1}{(1 + RCs)}$$

The concept of impedance is widely used in Electrical & Electronic Engineering because Method (ii) provides a fast and efficient method of obtaining the transfer function. However, in other applications of signal processing, the mechanism by which the signal is generated is more naturally described in terms of a differential equation, e.g. the population $x(t)$ of a city or cells in an organism might be described by:

$$\frac{dx}{dt} = -\,\lambda\; x(t)$$

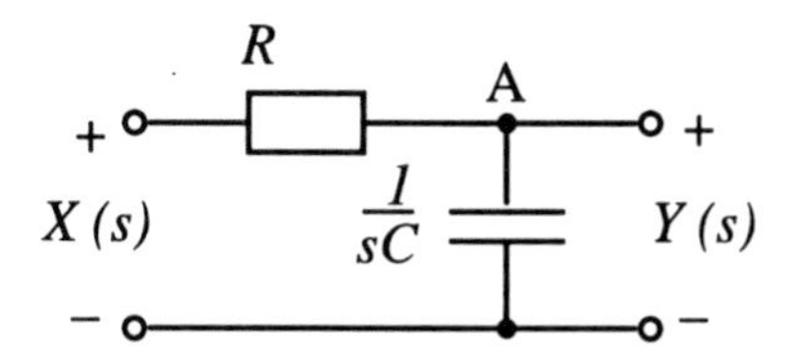

Figure 1.13 *RC circuit: transform-domain analysis.*

where λ is a positive constant. In such cases it is usual to work directly with the differential equation to obtain the transfer function.

Self assessment question 1.6: Consider the circuit of Figure 1.12 with the positions of the capacitor and resistor reversed. Evaluate the transfer function of the new circuit using both methods.

1.8 Summary

Usually the simplest representation of a signal is as a function of time. However other representations of the signal such as weighted sums or integrals of orthogonal basis functions, are possible. The weights or weighting functions are known as transforms. Representations of signals in terms of Fourier coefficients, Fourier transforms or Laplace transforms can be invaluable aids when analysing signals or considering the response of a system to a signal. Which representation is appropriate depends on the type or class of signals being considered. From a practical perspective the two most important classes are finite energy and finite power signals. Fourier series are used for periodic finite power signals, the Fourier transform is used for aperiodic finite energy signals and the Laplace transform for a wider range of finite energy signals. In subsequent chapters we shall see that Fourier and Laplace can also handle periodic signals.

The common theme which links the Fourier series, the Fourier transform and the Laplace transform is the use of *exponential* orthogonal basis functions. Since superposition is a fundamental property of linear systems, it is only necessary to calculate the response of a typical basis function to deduce the response of the system to the signal itself. Further, since many continuous-time linear systems can be described by sets of differential equations, the response to one exponential basis function is simply a scaled version of itself – the complex scale factor being the transfer function. By exploiting the above transforms, the calculation of the response of a system is reduced to multiplication. In particular, the transform of the output signal is the product of the transform of the input signal with the transfer function of the system.

1.9 Problems

1.1 A voltage signal $v(t)$ is connected across a 5 Ω resistor at $t = 0$. If the voltage is defined as:

$$v(t) = 2\exp(-3t) + 4\exp(-7t)$$

what is the total energy dissipated in the resistor? [0.68 J]

1.2 Calculate the power associated with the waveform shown below. [5 W]

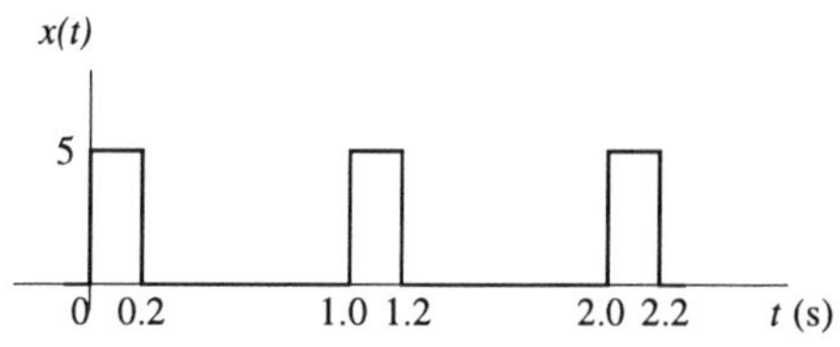

1.3 Develop expressions for the Fourier series representation of the periodic signal shown above (both trigonometric and complex forms). Plot graphs of the magnitude and phase of the complex form. [(Trigonometric) for $n = 0$, $A_0 = 2$, $B_0 = 0$; for $n > 0$, $A_n = 10\sin(2\pi n/5)/(2\pi n)$, $B_n = 10(1 - \cos(2\pi n/5))/(2\pi n)$; (complex) for $n = 0$, $X_0 = 1$; for $n > 0$, $X_n = -5j(1 - \cos(2\pi n/5) + j\sin(2\pi n/5))/(2\pi n)$]

1.4 Develop an expression for the Fourier transform and the Laplace transform of the signal illustrated below. Plot graphs of the magnitude and phase of the Fourier transform. [(Fourier) $X(\omega) = \exp(-j\omega/10)\text{sinc}(\omega/10)$; (Laplace) $X(s) = (1 - \exp(-s/5))5/s$]

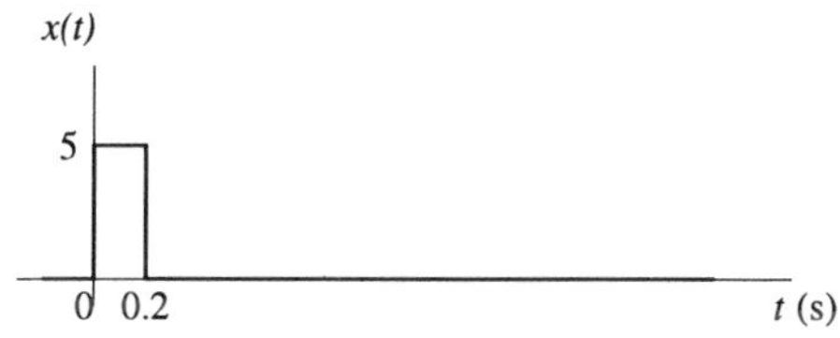

1.5 The voltage waveform illustrated below is a full-wave-rectified sine wave of amplitude 1 V and period 1 s.

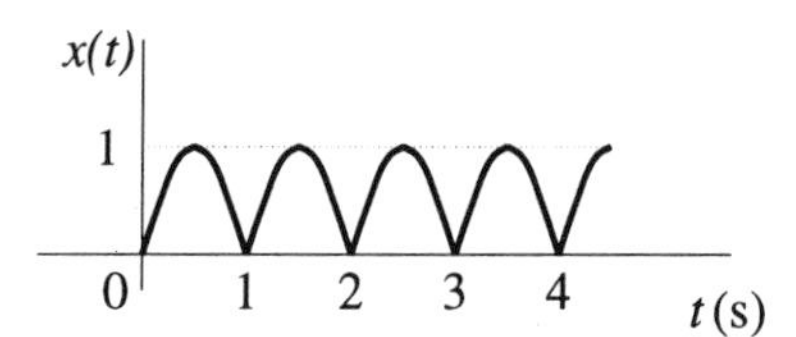

An inexpensive generator is required which will approximate this waveform and it is suggested that Walsh function generators might be used as these are easy to construct with digital logic circuits. One period of the first three Walsh functions, $w_0(t)$, $w_1(t)$ and $w_2(t)$, is illustrated below. Show that these three functions form an orthogonal set. Show how the outputs from these generators could be combined together using amplifiers and summers to form an approximation to the required waveform. Calculate the amplifier gains and sketch the waveform at the output of the generator. How might the quality of the approximation be improved? [(Amplifier gains) $A_0 = 0.637$, $A_1 = 0$, $A_2 = -0.264$]

Hint: Since Walsh functions form an orthogonal set, any periodic waveform can be expressed as a weighted sum of an infinite number of them:

$$x(t) = \sum_{n=0}^{\infty} A_n \, w_n(t)$$

$$A_n = \int_0^1 x(t) \, w_n(t) \, dt$$

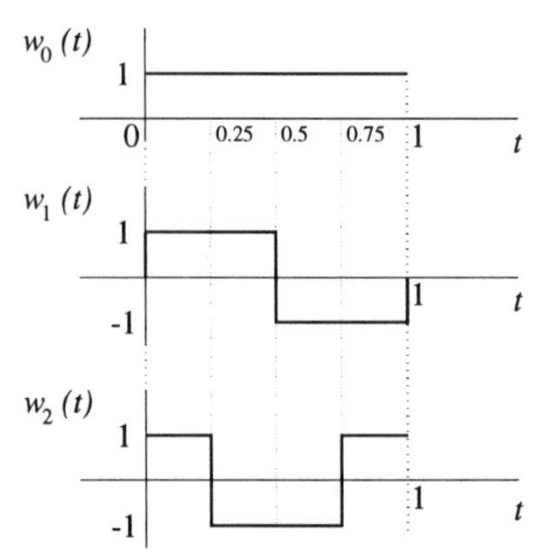

1.6 Develop an expression for the complex Fourier series of the following signals: (i) $x(t) = 3\cos(5t) + 4\sin(10t)$; (ii) $x(t) = \cos(\omega_0 t) + \sin^2(2\omega_0 t)$. Derive the complex form of the Fourier series from the trigonometric form. [(i) (Trigonometric) $A_1 = 3$, $B_2 = 4$, all the rest are zero; (complex) $X_1 = 3/2$, $X_{-1} = 3/2$, $X_2 = -2j$, $X_{-2} = 2j$, all the rest are zero; (ii) (trigonometric) (complex) $X_{-4} = -1/4$, $X_{-1} = 1/2$, $X_0 = 1/2$, $X_1 = 1/2$, $X_4 = -1/4$]

1.7 Determine the differential equation and transfer function of the following circuits. [(a) (Differential equation) $e/R_1 + Cde/dt = v(1/R_1 + 1/R_2) + Cdv/dt$; (transfer function) $H(s) = V(s)/E(s) = R_2(1 + R_1Cs)/(R_1 + R_2 + cR_1R_2s)$; (b) (differential equation) $e/R_1 + (CR_2/R_1)de/dt = v/R_1 + CR_2(1/R_1 + 1/R_2)dv/dt$; (transfer function) $H(s) = V(s)/E(s) = (1 + sR_2C)/(1 + sC(R_1 + R_2))$]

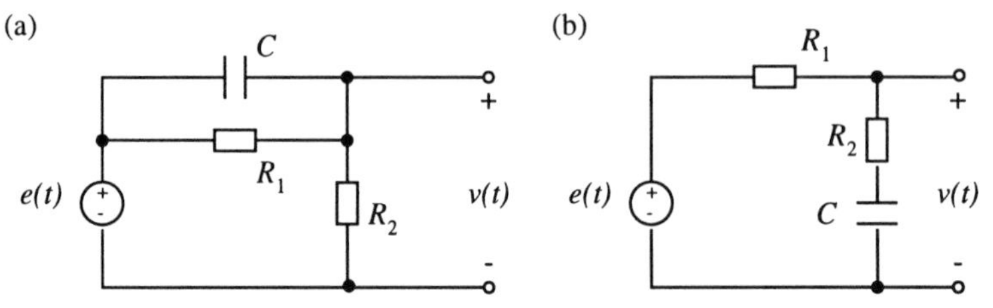

1.8 Determine the transfer function of the following operational amplifier circuit. [$H(s) = V(s)/E(s) = RCs/4$]

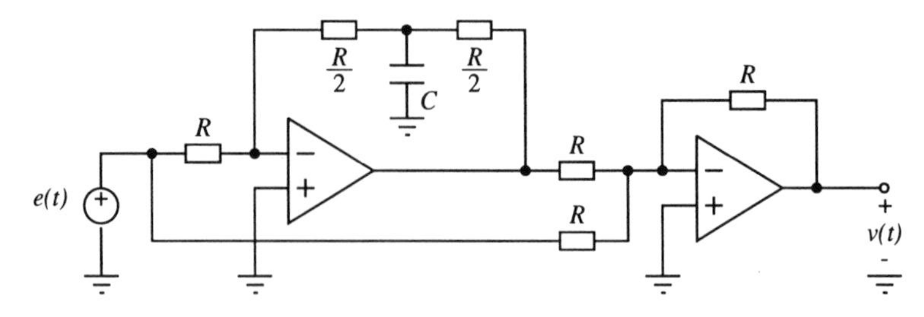

CHAPTER 2

Time-domain description and convolution

2.1 Introduction

In the previous chapter we saw how the Fourier and Laplace transforms can be used to represent signals and calculate the response of a linear system to a fairly general input signal. If we use Laplace or Fourier transforms to represent signals or systems we are said to be working in the 'frequency domain' or 'transform domain'. In contrast, the original signal $x(t)$ is said to be in the 'time-domain'. To calculate the response of a system to the signal $x(t)$ we first transform the signal to the frequency domain. We then multiply the transform of the input by the frequency response or transfer function of the system to get the output (in the frequency domain). Finally we inverse transform to get back to the time-domain.

In this chapter we explore the time-domain description of a linear system. This exploration begins in section 2.2 by defining the impulse function and by describing a signal by adding together suitably scaled impulse functions. The response of a system to one of these impulse functions, i.e. the impulse response, is defined. These two concepts are brought together in section 2.3 and lead to the description of the system output as the convolution of the input signal with the impulse response. A detailed example is provided to highlight the mechanics of performing a convolution integral. In section 2.4 some key properties of convolution are presented. An example of the analysis of a communication system demonstrates how convolution and its properties can be used in practice. Finally, in section 2.6 the key concepts are summarised.

2.2 The impulse response

From Chapter 1 it should be clear that there is both a time-domain and frequency/transform-domain description of the input signal, namely $x(t)$ and $X(s)$. There is also a time-domain and frequency-domain description of the output signal, i.e. $y(t)$ and $Y(s)$. The system itself can be described in the transform domain by $H(s)$. It is natural to ask: 'is there a time-domain description of the system?' The answer is 'yes'. A time-domain description of the system can be obtained by taking the inverse transform of the transfer function $H(s)$:

$$h(t) = L^{-1}[H(s)] \tag{2.1}$$

As such, $h(t)$ is a mathematical definition without physical interpretation. In fact $h(t)$ is an important time-domain description of the system and is called the *impulse response.* Before considering the impulse response in more detail it is appropriate to define the impulse function.

2.2.1 The impulse

An unusual but extremely useful function in the analysis of systems is the unit impulse. The definition of a unit impulse $\delta(t)$ is as follows. It occurs at $t = 0$, at which point it has infinite height. Everywhere else it is equal to zero.

$$\delta(t) = \begin{cases} \infty & \text{if } t = 0 \\ 0 & \text{if } t \neq 0 \end{cases} \tag{2.2}$$

The graphical notation for an impulse is shown in Figure 2.1. The arrow indicates that it is of infinite height at time zero. The other unusual thing about an impulse is that if we integrate it, we obtain one:

$$\int_{-\infty}^{\infty} \delta(t)\, dt = 1 \tag{2.3}$$

In other words, the area underneath it (if we can say that) is equal to 1. The impulse is said to have a strength of 1. If we amplify an impulse by a gain of 2 we obtain $2\delta(t)$. This will have exactly the same width and height as $\delta(t)$ but an area or strength of 2. All of the above is a bit much at first sight – fortunately there is a more practical way of looking at it. Consider the rectangular pulse, $d(t)$, shown in Figure 2.2. It starts at $t = 0$ and lasts for Δt seconds. The height of the pulse is $1/\Delta t$. Thus the area under the pulse $\Delta t(1/\Delta t)$ is one, i.e.:

$$\int_{-\infty}^{\infty} d(t)\, dt = 1$$

If we let the width of the pulse become smaller and smaller, the height will get bigger and bigger, yet the area will remain 1. Thus as the pulse width tends towards zero, $d(t)$ will tend towards an impulse:

$$\delta(t) = \lim_{\Delta t \to 0} d(t)$$

Self assessment question 2.1: On the same axis, make sketches of $d(t)$ for: $\Delta t = 1$; $\Delta t = 0.5$; $\Delta t = 0.25$.

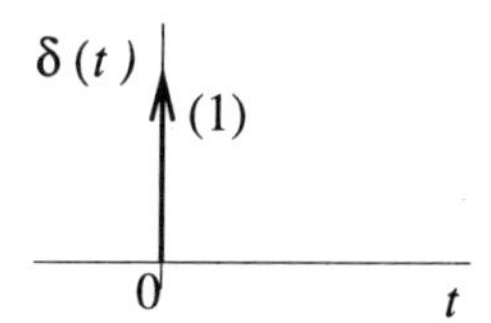

Figure 2.1 *The impulse function.*

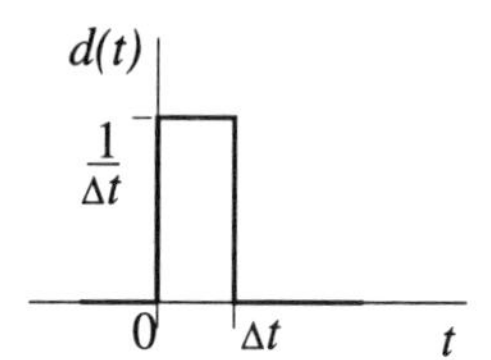

Figure 2.2 *From rectangular pulse to impulse: a limiting process.*

2.2.2 Signal representation

In Chapter 1 it was demonstrated that signals could be represented as the sum (or integrals) of exponential basis functions. In a similar manner a signal can represented as a summation (integral) of impulse functions.

Any signal can be approximated by packing pulses such as $d(t)$ closely together. Such an approximation is illustrated in Figure 2.3. The quality of the approximation will get better as the width of the pulses Δt becomes smaller. The pulse which occurs at $t = n\Delta t$ has a height of $x(n\Delta t)$ which is equal to the value of the signal at that point. This pulse can be written as:

$$x(n\Delta t)\ d(t - n\Delta t)\ \Delta t$$

This is the basic pulse of Figure 2.2 which has been delayed by $n\Delta t$ seconds and amplified by $x(n\Delta t)\Delta t$ to give it the height of $x(n\Delta t)$. Adding together all of the pulses gives the approximation $\hat{x}(t)$ to the original waveform:

$$\hat{x}(t) = \sum_{n=-\infty}^{\infty} x(n\Delta t)\ d(t - n\Delta t)\ \Delta t \tag{2.4}$$

In the limit as $\Delta t \to 0$, the pulse becomes an impulse at time $n\Delta t$, the product $n\Delta t$ becomes the continuous time variable τ, the time step Δt becomes the differential $d\tau$ and the summation becomes an integration. Thus:

$$x(t) = \int_{-\infty}^{\infty} x(\tau)\ \delta(t - \tau)\ d\tau \tag{2.5}$$

Therefore the signal $x(t)$ can be represented by a summation (integration) of impulses. Equations similar to (2.5) will be used again when convolution is discussed. The use of the second time variable τ can cause significant conceptual problems when first encountered. In this respect the relationship between (2.4) and (2.5) provides useful insights as the former is fundamentally easier to understand. The classic question is 'What does τ mean?' To which the answer is:

> Just as in equation (2.4), where the signal is approximated by a summation of pulses and $n\Delta t$ is the point in time at which one of these pulses occurs, the signal is exactly represented in equation (2.5) as an integration of impulses and τ is the point in time at which one of these impulses occurs.

Self assessment question 2.2: Illustrate using sketches how the signal $x(t) = 5\sin(2\pi t)$ can be approximated in the region $0 \le t < 1$ using pulses such as $d(t)$ where $\Delta t = 0.1$. Write down expressions for each of the first three pulses in term of $d(t)$ as defined in Figure 2.2.

2.2.3 System response to an impulse

What happens if we apply this impulse to a linear system? What will the output be? To answer these questions we apply transform techniques as before. This first thing we will need is the Laplace transform of the input $\delta(t)$. We do not know this, so we could use the definition of the Laplace transform to calculate it:

$$L[\delta(t)] = \int_{0^-}^{\infty} \delta(t)\exp(-st)\,dt$$

There is only one point in time, $t = 0$, where the impulse has any value. At that point the $\exp(-st)$ has a value of 1, i.e. $\exp(-0\,s) = \exp(0) = 1$. Thus the Laplace transform of an impulse is equal to the integral of an impulse which is 1:

$$L[\delta(t)] = \int_{0^-}^{\infty} \delta(t)\,1\,dt = 1 \tag{2.6}$$

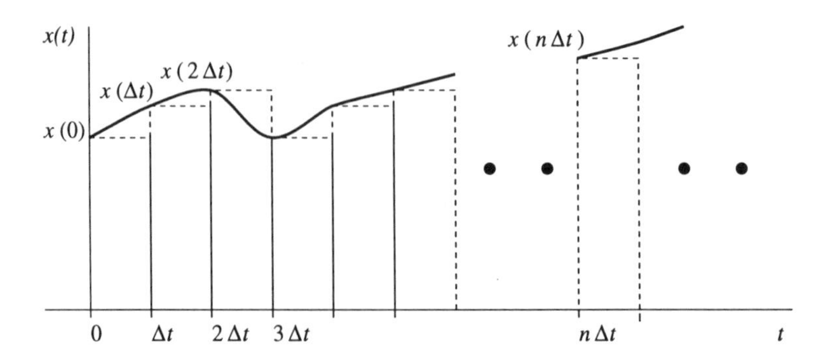

Figure 2.3 *Signal representation as a sum of pulses.*

To find the transform of the output we multiply the transform of the input by the transfer function of the system:

$$Y(s) = H(s)\,1$$
$$= H(s)$$

And to find the time-domain output we inverse transform $Y(s)$:

$$y(t) = L^{-1}[Y(s)]$$
$$= L^{-1}[H(s)]$$
$$= h(t)$$

Thus if we apply an impulse $\delta(t)$ to a system, the output will be the impulse response $h(t)$.

EXAMPLE 2.1

Calculate the impulse response of the *RC* circuit of Figure 1.12 using both a transform-domain and a time-domain approach.

Transform-domain solution

From section 1.7 the transfer function is:

$$H(s) = \frac{1}{(1 + RCs)}$$

and using equation (2.1) and the table of Laplace transforms (Appendix B):

$$h(t) = L^{-1}[H(s)]$$
$$= \frac{1}{RC}\exp\left(-\frac{t}{RC}\right); \quad t \geq 0$$

Time-domain solution

The input is an impulse. Thus:

$$x(t) = \delta(t)$$

The input only has a non-zero value at time zero. Thus at $t = 0$, the current through the resistor is:

$$i(t) = \frac{x(t) - y(t)}{R}$$

Assuming the voltage across the capacitor is initially zero, the expression for the current becomes:

$$i(t) = \frac{\delta(t)}{R}$$

This current only exists at $t = 0$. The relationship between current and voltage on the capacitor yields:

$$i(t) = C\,\frac{dy}{dt}$$

and hence

$$dy = \frac{1}{RC}\,\delta(t)\,dt$$

Since the impulse only exists at time zero, it is sufficient to integrate between $t = 0^-$ to $t = 0^+$ to evaluate the voltage left on the capacitor at $t = 0^+$ by the current impulse:

$$\begin{aligned} y(0^+) &= \int_{0^-}^{0^+} dy \\ &= \frac{1}{RC}\int_{0^-}^{0^+} \delta(t)\,dt \\ &= \frac{1}{RC} \end{aligned}$$

After $t = 0$ the input is zero, i.e. $x(t) = 0$. Applying Kirchhoff's current law again gives:

$$-\frac{y(t)}{R} = C\,\frac{dy}{dt}$$

This when rearranged gives a simple differential equation:

$$\frac{dy}{dt} + \frac{1}{RC}\,y(t) = 0$$

with initial condition $y(0^+) = 1/RC$. The solution is

$$y(t) = \frac{1}{RC}\exp\left(-\frac{t}{RC}\right)$$

which is the impulse response.

From the above example it is evident that for this type of system the transform-domain method is the most straightforward method of evaluating the impulse response. However the time-domain approach does provide additional insights. It emphasises that the impulse response is the output of a system when an impulse is applied at the input. The impulse exists very briefly at $t = 0$. Its effect is to set up initial conditions on the components within the system which store energy, e.g. the capacitor. After $t = 0$ the output is the natural response of the system with the input set to zero from the initial conditions induced by the impulse. Essentially the impulse is a 'kick' to the system which provides energy to observe the natural response.

Self assessment question 2.3: Derive an expression for the impulse response of the circuit in Figure 1.12 if the capacitor was replaced with an inductor of L H.

2.3 Convolution

In order to evaluate the response of a system to an input $x(t)$ using the impulse response $h(t)$, recall that: (i) the input $x(t)$ can be represented as a summation (integration) of impulses as in equations (2.4) and (2.5); (ii) the system is linear and superposition applies. Thus it is sufficient to evaluate the response of the system to one of the impulses in equation (2.5) and integrate the responses to all such impulses to obtain the output $y(t)$. However it is conceptually simpler to work with the summation of equation (2.4) rather than the integral of equation (2.5).

The response, $h'(t)$, to a single rectangular pulse $d(t)$ is illustrated in Figure 2.4(a). In the limit as Δt tends towards zero, $d(t)$ will tends towards the impulse $\delta(t)$ and $h'(t)$ will tend towards the impulse response $h(t)$. As indicated in section 2.2, the input $x(t)$ can be approximated by summing suitably scaled and delayed rectangular pulses. A typical one is:

$$x(n\Delta t)\; d(t - n\Delta t)\; \Delta t$$

which is illustrated in Figure 2.4(b). The response of the system to such a pulse is:

$$x(n\Delta t)\; h'(t - n\Delta t)\; \Delta t$$

which is illustrated in the same figure. Using superposition, the response to the input,

$$\hat{x}(t) = \sum_{n=-\infty}^{\infty} x(n\Delta t)\; d(t - n\Delta t)\; \Delta t$$

is simply

$$\hat{y}(t) = \sum_{n=-\infty}^{\infty} x(n\Delta t)\; h'(t - n\Delta t)\; \Delta t \tag{2.7}$$

In the limit as $\Delta t \to 0$, the pulse becomes an impulse at time $n\Delta t$, the product $n\Delta t$ becomes the continuous time variable τ, $h'(.)$ becomes the impulse response $h(.)$, the time step Δt becomes the differential $d\tau$ and the summation becomes an integration. Thus:

$$y(t) = \int_{-\infty}^{\infty} h(t - \tau)\; x(\tau)\; d\tau \tag{2.8}$$

This is known as the *convolution* equation. An alternative but equivalent form is:

$$y(t) = \int_{0}^{\infty} x(t - \tau)\; h(\tau)\; d\tau \tag{2.9}$$

In common with (2.5) these equations use two time variables t and τ. The first denotes the particular time at which we are calculating the output and the second is the variable used to do the integration. This convolution operation is often denoted with an asterisk:

$$y(t) = x(t) * h(t) \tag{2.10}$$

An important relationship between the transform-domain description discussed in Chapter 1 and this time-domain description of a linear system can be summarised as follows:

- *Convolution in the time-domain is equivalent to multiplication in the s domain (or frequency domain).*

It is possible to prove that the convolution integral is correct by taking the inverse Laplace transform of both sides of the Laplace expression $Y(s) = H(s)\,X(s)$. However a simpler approach is to verify that the convolution equation yields the same result when calculating the response to a complex phasor.

Consider a complex phasor input $x(t) = \exp(st)$ to a system with impulse response $h(t)$. The output can be evaluated using equation (2.9) thus:

$$\begin{aligned} y(t) &= \int_0^\infty x(t-\tau)\,h(\tau)\,d\tau \\ &= \int_0^\infty \exp(s(t-\tau))\,h(\tau)\,d\tau \\ &= \exp(st)\int_0^\infty h(\tau)\exp(-s\tau)\,d\tau \\ &= \exp(st)\,H(s) \end{aligned}$$

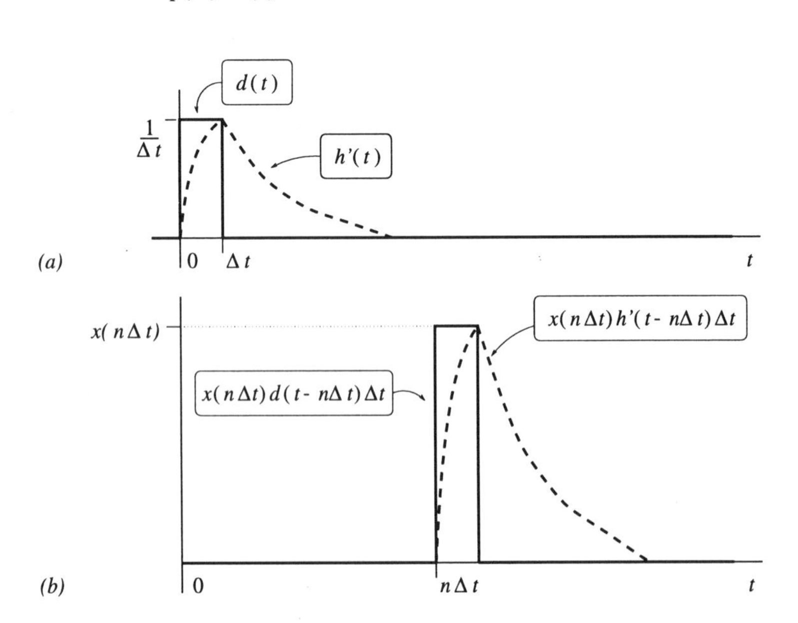

Figure 2.4 *System response to: (a) a single pulse $d(t)$ at $t = 0$; (b) a typical pulse $x(n\Delta t)d(t - n\Delta t)\Delta t$ used to construct the input $x(t)$.*

i.e. the response to a complex phasor is a complex phasor scaled by the transfer function $H(s)$.

The following examples illustrates many of the key features involved in performing a convolution operation analytically.

EXAMPLE 2.2

Recall the simple RC circuit of Figure 1.12. In section 2.2 its impulse response was calculated using both the time-domain and transform-domain approaches. This impulse response is illustrated in Figure 2.5(a). Using convolution, calculate the response of this circuit to the rectangular pulse $x(t)$ shown in Figure 2.5(b). The problem is to calculate the output $y(t)$ (using convolution) at every value of time t. The convolution integral is usually performed in pieces. The way to identify which bits are involved is to draw diagrams of the signals involved. This subdivides the problem into smaller problems which when solved describe the output at different regions on the time axis.

- *The first key step in convolution is to draw diagrams of the signals involved.*

For this particular example consider three particular regions of the time axis.

i) What is the output at time zero? Start by mechanically applying the convolution integral which states:

$$y(t) = \int_0^\infty x(\tau)\, h(t-\tau)\, d\tau$$

Replacing t with 0 gives an expression for the output at time zero:

$$y(0) = \int_0^\infty x(\tau)\, h(0-\tau)\, d\tau$$

$$= \int_0^\infty x(\tau)\, h(-\tau)\, d\tau$$

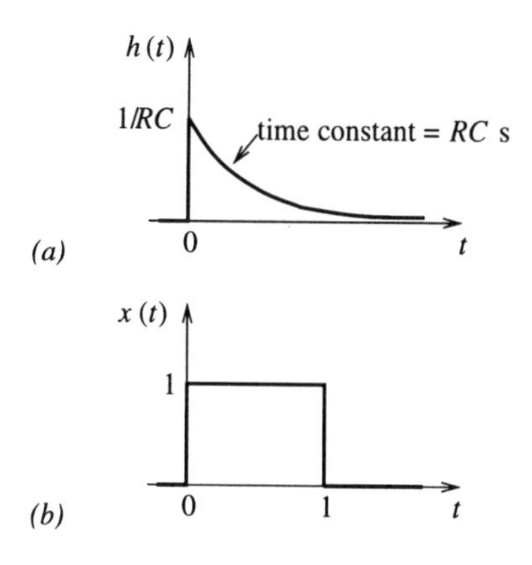

Figure 2.5 *Convolution example: (a) impulse response $h(t)$ and (b) input $x(t)$.* □

In words this means: multiply $x(\tau)$ by $h(-\tau)$ at every value of τ to get the product $x(\tau)\,h(-\tau)$, and then integrate the product from $\tau = 0$ to infinity. Figure 2.6 shows $x(\tau)$ drawn as a function of τ. Below it $h(-\tau)$ is drawn as a function of τ. $h(-\tau)$ is the same shape as $h(t)$ but it is the opposite way round, i.e. *time-reversed.* To calculate the product, multiply the two graphs together at every value of τ. For every value of τ less than zero, $x(\tau) = 0$. Thus the product will be zero when τ is less than zero. For every value of τ greater than zero, $h(-\tau) = 0$. Thus the product will be zero when τ is greater than zero. The product will be zero at every value of τ. The integral of the product is zero:

$$y(0) = 0$$

This result might have been expected since the input $x(t)$ does not start until $t = 0$.

ii) What is the output when t is between 0 and 1 second (i.e. $0 \leq t < 1$)? Again, start by mechanically applying the convolution integral:

$$y(t) = \int_0^\infty x(\tau)\,h(t-\tau)\,d\tau$$

Figure 2.7 shows $x(\tau)$ drawn as a function of τ. Below it $h(t-\tau)$ is drawn as a function of τ. Comparing Figure 2.7 with 2.6 it can be seen that $h(t-\tau)$ is obtained by *sliding* $h(-\tau)$ to the right by t seconds. Recall that t is constrained to lie between 0 and 1. Hence the vertical edge of $h(t-\tau)$ must lie somewhere between the rising

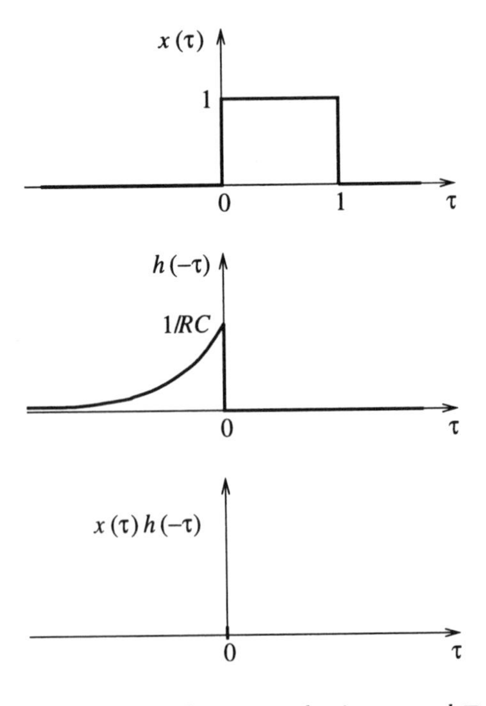

Figure 2.6 *Convolution example: time reversal.* □

edge and falling edge of $x(\tau)$. Again applying the convolution integral mechanically, multiply the two waveforms together at every value of τ to obtain the product waveform $x(\tau)h(t-\tau)$. Note that for values of τ less than zero, $x(\tau)=0$ and hence the product will be zero for these values of τ. Further, for values of τ greater than t, $h(t-\tau)$ is zero and hence the product will also be zero when τ is greater than t. Finally, integrate the product between zero and infinity. However it is clear that the product is zero for values of τ less than zero and greater than t. Thus the integration is simply the area under the product curve between 0 and t:

$$y(t) = \int_0^t x(\tau)\, h(t-\tau)\, d\tau$$

iii) Figure 2.8 illustrates the waveforms when t is greater than one. In this case the vertical edge of $h(t-\tau)$ has *slid* further to the right so that it is beyond the falling edge of $x(\tau)$. Multiply the waveforms together as before. Note that $x(\tau)$ is zero for all values of τ greater than one and hence the product will also be zero for these values of τ. Integrate the product; the integral is simply the area under the product curve between zero and one, i.e.:

$$y(t) = \int_0^1 x(\tau)\, h(t-\tau)\, d\tau$$

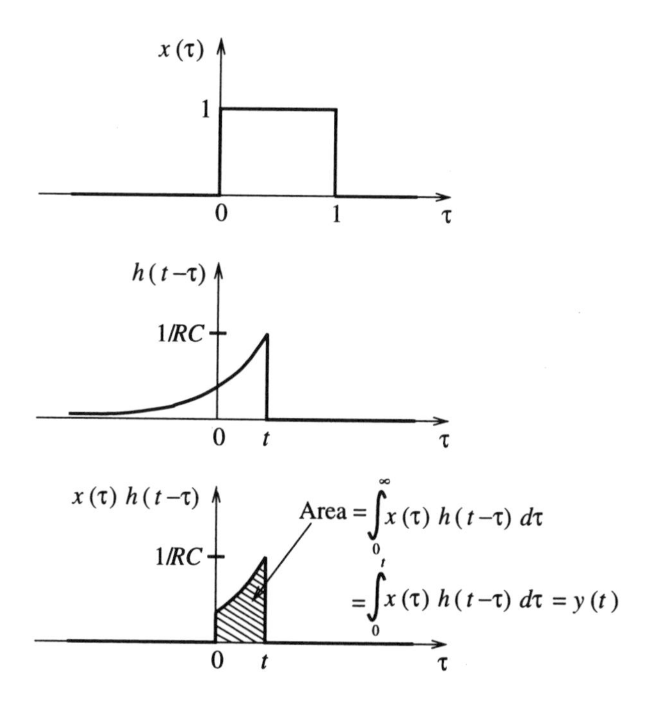

Figure 2.7 *Convolution example: slide into second region.* □

Thus it has been shown that the integration can be performed by separating the time axis (t) into three separate regions and simplifying the expression within each region. Thus it might be expected that three separate equations are needed to describe the output $y(t)$ – one for each region.

- *The second step in convolution is to divide the time axis into regions – the regions can be identified using the diagrams.*

We will now develop expressions for the output $y(t)$. Having identified the regions, it now remains to develop expressions for the output within each of these regions:

i) $t < 0$ – from the graphical arguments the answer is simply:

$$y(t) = 0$$

ii) $0 \le t < 1$ – start with the result obtained from the graphical arguments:

$$y(t) = \int_0^t x(\tau)\, h(t-\tau)\, d\tau$$

Within this region $x(\tau)$ is 1. An expression for the impulse response $h(t)$ was developed in section 2.2. An expression for $h(t-\tau)$ is obtained by simply substituting $t-\tau$ for t:

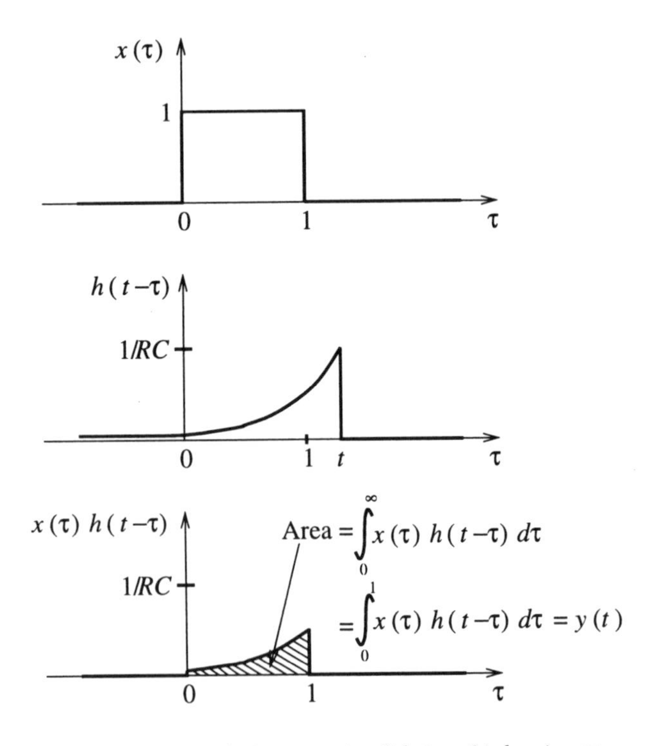

Figure 2.8 *Convolution example: slide into third region.* □

$$y(t) = \int_0^t 1 \; \frac{1}{RC} \exp\left(-\frac{t-\tau}{RC}\right) d\tau$$

It is important at this point to remember that integration is with respect to τ and not t. For the purposes of this integration, t behaves like a constant and can be taken outside the integral sign:

$$y(t) = \frac{1}{RC} \exp\left(-\frac{t}{RC}\right) \int_0^t \exp\left(\frac{\tau}{RC}\right) d\tau$$

The integration is now fairly straightforward:

$$y(t) = \frac{1}{RC} \exp\left(-\frac{t}{RC}\right) \left[RC \exp\left(\frac{\tau}{RC}\right) \right]_0^t$$

$$= 1 - \exp\left(-\frac{t}{RC}\right)$$

iii) $t \geq 1$ – from the graphical arguments:

$$y(t) = \int_0^1 x(\tau) \; h(t-\tau) \; d\tau$$

Within this region $x(\tau)$ is 1. Thus:

$$y(t) = \int_0^1 1 \; \frac{1}{RC} \exp\left(-\frac{t-\tau}{RC}\right) d\tau$$

$$= \exp\left(-\frac{t}{RC}\right) \left[\exp\left(\frac{1}{RC}\right) - 1 \right]$$

The three separate portions of the solution are illustrated in Figure 2.9.

- *The final step in convolution is to perform the integrations in each region.*

As a final check, it is worth verifying that the expression for each region should coincide at the boundaries. Thus for $t = 1$, both the result for region (ii) and the result for region (iii) give: $y(1) = 1 - \exp(-1/(RC))$.

Self assessment question 2.4: Rework example 2.2 for an input $x(t)$ where: $x(t) = 1$ when $1 \leq t < 2$ and $x(t) = 0$ elsewhere.

2.4 Properties of convolution

Linear systems exhibit many properties. Since the convolution operation describes a linear system in the time-domain, the properties of linear systems can be used to simplify what might seem like daunting convolution expressions.

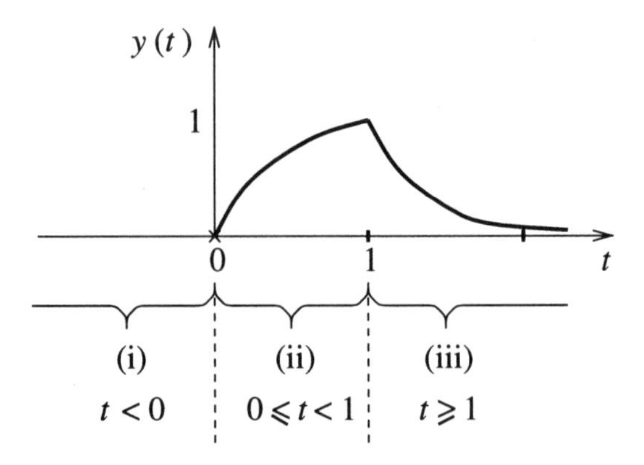

Figure 2.9 *Convolution example: complete solution.* □

- Convolution is cumulative:

$$x(t) * h(t) = h(t) * x(t) \tag{2.11}$$

If a signal $x(t)$ is applied to a system with impulse response $h(t)$ as in Figure 2.10, we get the same result as that from applying a signal $h(t)$ to a system with impulse response $x(t)$.

- Convolution is distributive over addition:

$$h(t) * [x_1(t) + x_2(t)] = [h(t) * x_1(t)] + [h(t) * x_2(t)] \tag{2.12}$$

This is a restatement of superposition as illustrated in Figure 2.11.

- Convolution is associative:

$$h_2(t) * [h_1(t) * x(t)] = h_1(t) * [h_2(t) * x(t)] \tag{2.13}$$

Applying a signal $x(t)$ to a system with impulse response $h_1(t)$ and then connecting the output to the input of a second system $h_2(t)$, the overall result will be the same as connecting to $h_2(t)$ first and then applying its output to $h_1(t)$ – Figure 2.12.

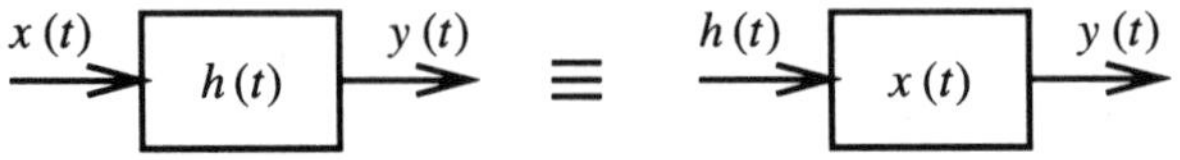

Figure 2.10 *Cumulative property of convolution.*

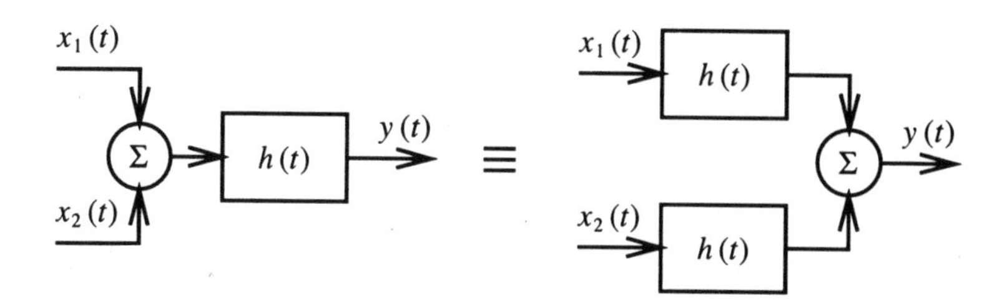

Figure 2.11 *Distributive property of convolution.*

Figure 2.12 *Associative property of convolution.*

An extremely important property, which is of particular application in communications systems as it forms a basis for the understanding of modulation, relates the product of two time-domain waveforms to the convolution of their transforms.

• *Multiplication in the time-domain is equivalent to convolution in the s domain (frequency domain):*

$$L[x_1(t)\, x_2(t)] = \frac{1}{2\pi j}\, X_1(s) * X_2(s) \tag{2.14}$$

$$F[x_1(t)\, x_2(t)] = \frac{1}{2\pi}\, X_1(\omega) * X_2(\omega) \tag{2.15}$$

In Chapter 4 this result will be used to study the effects of sampling on a signal.

2.4.1 Time delay

In section 1.6 we saw that many components of linear systems such as capacitors and inductors are defined by differential equations and hence are readily amenable to transform-domain description, e.g. impedance. Thus in section 2.4 it was easier to evaluate the transfer function of the *RC* circuit first and then perform an inverse Laplace transform to find its impulse response rather than attempt to work out the impulse response directly from the circuit. However, not all components of linear systems are easiest to approach in the transform domain. One important exception is a pure time delay. While it is easy to envisage what a time delay does to a signal in the time-domain, it is more difficult to determine what its Laplace transfer function is. An easier route is to ask what its impulse response is and then perform a Laplace transform

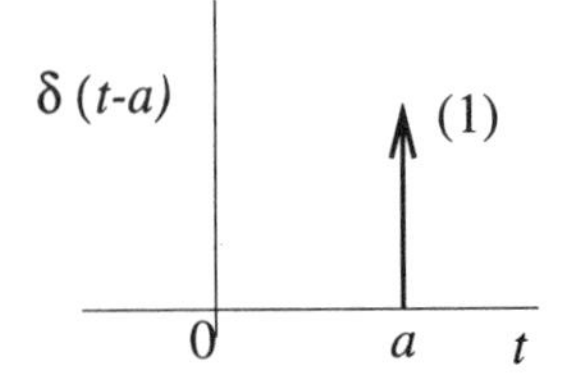

Figure 2.13 *The impulse response of a pure time delay of a seconds.*

on that to obtain the desired transfer function.

If we apply a unit impulse $\delta(t)$ to a pure time delay of a seconds, the output will be an impulse a seconds later, i.e. $\delta(t-a)$ as illustrated in Figure 2.13. Thus the impulse response is:

$$h(t) = \delta(t-a)$$

If we apply a signal $x(t)$ to such a system we can verify that convolution gives the same answer as intuition:

$$\begin{aligned} y(t) &= h(t) * x(t) \\ &= \delta(t-a) * x(t) \\ &= x(t-a) \end{aligned}$$

This is illustrated in Figure 2.14. Thus a system with impulse response $\delta(t-a)$ just delays the input by a seconds. If we have the impulse response we can obtain the transfer function by taking its Laplace transform:

$$\begin{aligned} H(s) &= L[\delta(t-a)] \\ &= \exp(-as)\ 1 \\ &= \exp(-as) \end{aligned}$$

Thus we have a complete description of a time delay in terms of its impulse response and transfer function. The following example illustrates how some of the concepts associated with convolution can be used to analyse a digital communications system.

EXAMPLE 2.3

A digital radio system consists of a transmitter on one building and a receiver on another building. The impulse response of the transmitter is illustrated in Figure 2.15. A logic 1 (HIGH) is transmitted by applying a unit impulse to the transmitter. A logic 0 (LOW) is transmitted by applying an impulse of strength -1 to the transmitter. The time taken for the transmitted pulse to reach the receiver is negligible. The receiver has an impulse response identical to that of the transmitter. Draw a block diagram of the system. Using convolution, make a detailed sketch of the waveform at the output of the receiver when a logic 1 is transmitted.

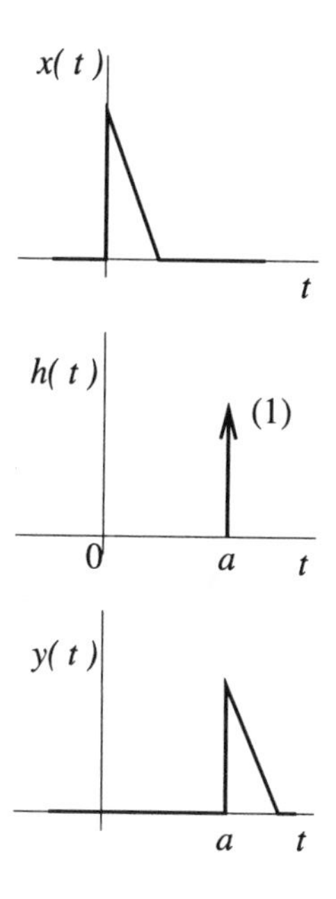

Figure 2.14 *Time delay as a convolution with an impulse.*

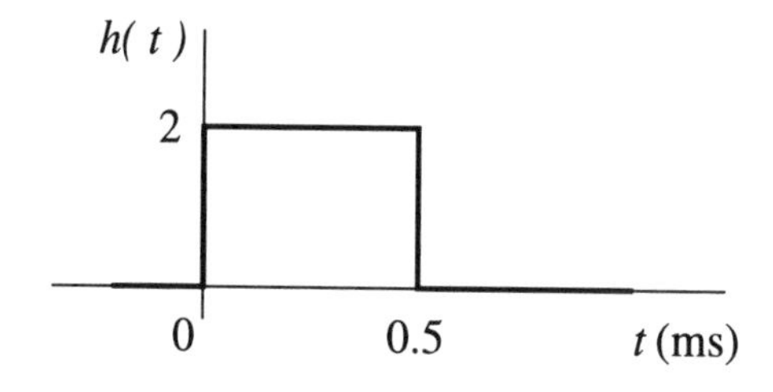

Figure 2.15 *Impulse response of transmitter.*

A logic 0 is transmitted 0.5 ms after a logic 1. In other words the signal

$$\delta(t) - \delta(t - 0.5 \times 10^{-3})$$

is applied to the transmitter. Make a detailed sketch of the waveform at the output of the receiver.

In bad weather there are two paths between the transmitter and the receiver. The first path has a negligible time delay and a gain of 1. The second path has a gain of −1 and involves a delay of 0.25 ms. The contributions from the two paths add together at the input to the receiver. Draw a diagram of the complete system. Make a detailed sketch of the waveform at the output of the receiver when a logic 1 is transmitted.

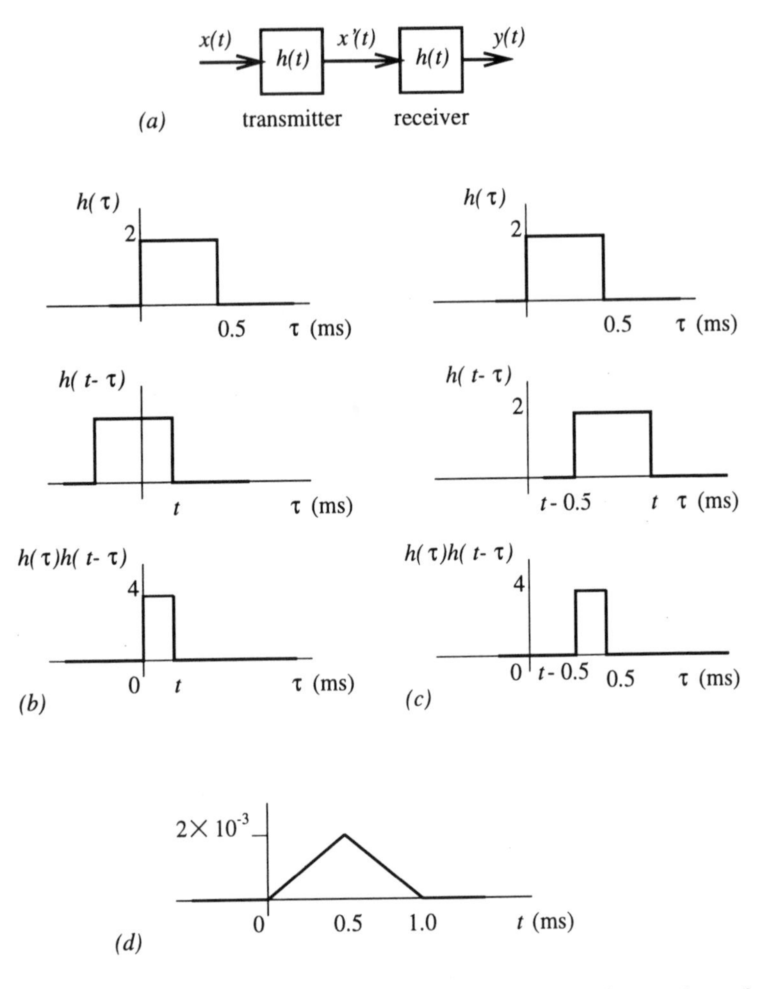

Figure 2.16 *Digital radio convolution example: (a) block diagram; (b) waveforms for region (ii); (c) waveforms for region (iii); (d) receiver output.*

Solution

A block diagram of the system is illustrated in Figure 2.16(a). When a logic 1 is transmitted

$$x(t) = \delta(t)$$

the output from the transmitter is

$$x'(t) = h(t) * \delta(t)$$
$$= h(t)$$

and the output from the receiver is

$$y(t) = h(t) * x'(t)$$
$$= h(t) * h(t)$$

Using the convolution integral:

$$y(t) = \int_{-\infty}^{\infty} h(\tau)\, h(t-\tau)\, d\tau$$

By drawing diagrams it is clear that there are four regions to the solution: (i) $t \le 0$; (ii) $0 < t \le 0.5 \times 10^{-3}$; (iii) $0.5 \times 10^{-3} < t \le 10^{-3}$; (iv) $10^{-3} < t$. The first and last are trivial since $y(t) = 0$ in these regions. The diagrams for region (ii) and (iii) are illustrated in Figures 2.16(b) and 2.16(c) respectively. For region (ii) the area under the product graph is

$$y(t) = 4\,t$$

For region (iii) the area under the product graph is

$$y(t) = 4(0.5 \times 10^{-3} - (t - 0.5 \times 10^{-3}))$$
$$= 4 \times 10^{-3} - 4t$$

The receiver output is illustrated in Figure 2.16(d). For the input

$$x(t) = \delta(t) - \delta(t - 0.5 \times 10^{-3})$$

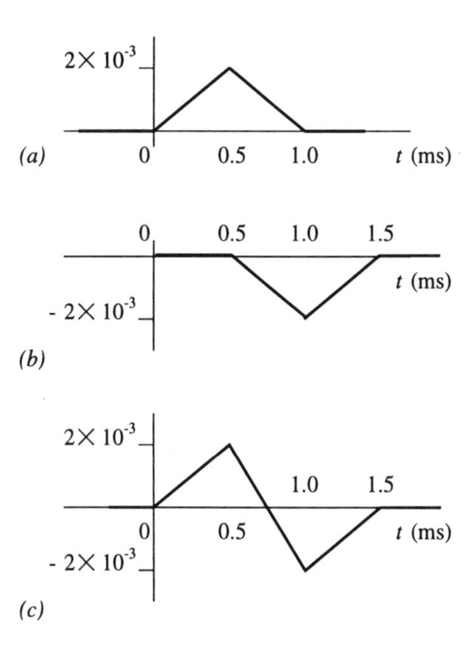

Figure 2.17 *Digital radio convolution example: (a) response of system to $\delta(t)$; (b) response of system to $-\delta(t - 0.5 \times 10^{-3})$; (c) total response.*

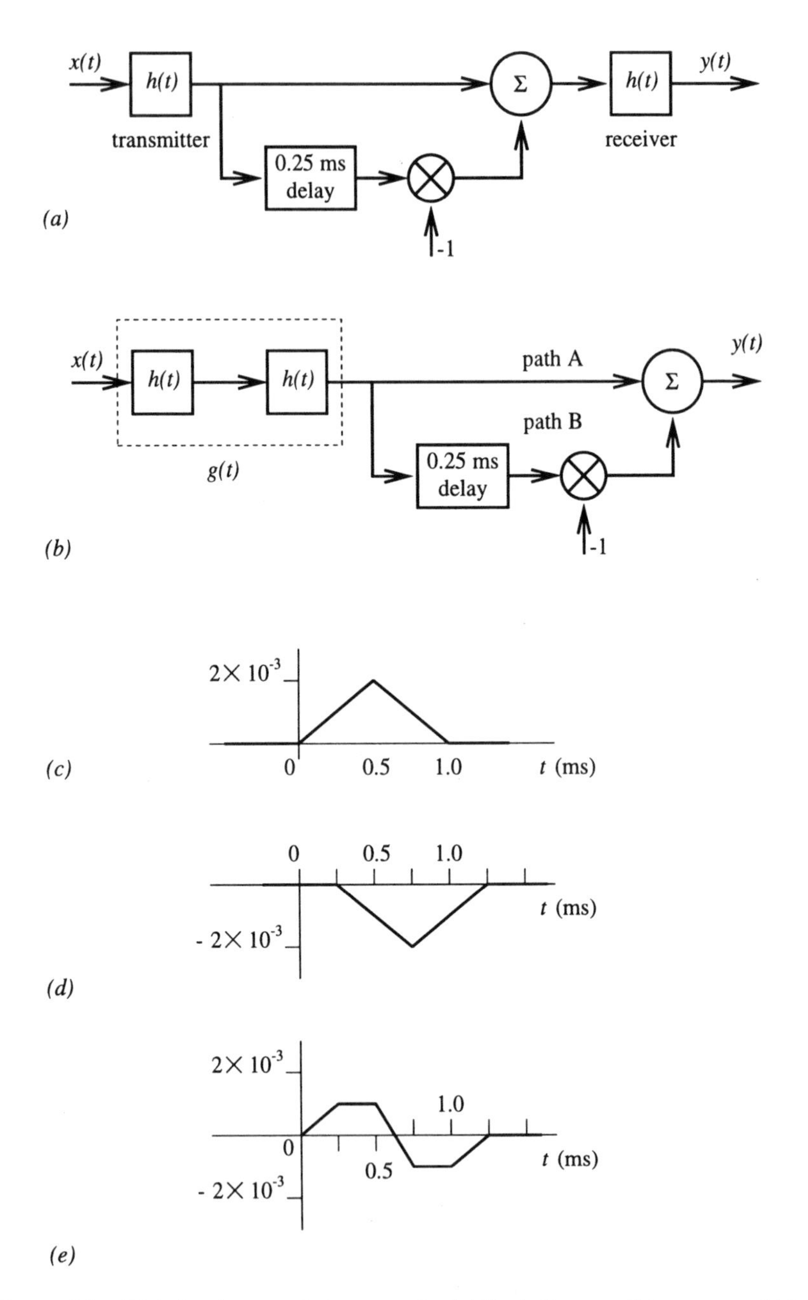

Figure 2.18 *Digital radio convolution example: (a) block diagram; (b) re-arranged block diagram; (c) path A; (d) path B; (e) received signal.*

the receiver output is

$$y(t) = (h(t) * h(t)) * x(t)$$

For convenience, define $g(t) = h(t)*h(t)$ which has already been evaluated, thus

$$y(t) = g(t) * (\delta(t) - \delta(t - 0.5 \times 10^{-3}))$$

and since convolution is distributive over addition:

$$y(t) = g(t) - g(t - 0.5 \times 10^{-3})$$

Figure 2.17 illustrates $g(t)$, $-g(t - 0.5 \times 10^{-3})$ and the receiver output $y(t)$.

The bad weather block diagram is illustrated in Figure 2.18(a). Using the cumulative property of convolution, this can be re-arranged to produce Figure 2.18(b). The combination of the transmitter and receiver has already been evaluated as

$$g(t) = h(t) * h(t)$$

The response at the output to transmitting a logic one is thus

$$y(t) = g(t) - g(t - 0.25 \times 10^{-3})$$

as illustrated in Figures 2.18(c) to (e).

2.5 Summary

Convolution provides an alternative, time-domain, description of a linear system. The output of a linear system can be evaluated by convolving the input signal with the impulse response of the system. This provides an alternative to the transform-domain multiplication discussed in the previous chapter. The impulse response of a system is the signal observed at the system output when a unit impulse is applied to the input. Since both the Laplace and Fourier transform of an impulse are one, the impulse response is the inverse transform of the system transfer function defined in Chapter 1. Many of the key properties of the convolution operation are a direct result of the fact that convolution is a description of a linear system. These properties allow for simplification of expressions involving convolution.

2.6 Problems

2.1 Convolve the following pairs of input signals and impulse responses. Sketch the outputs and label the horizontal and vertical axis with significant values.

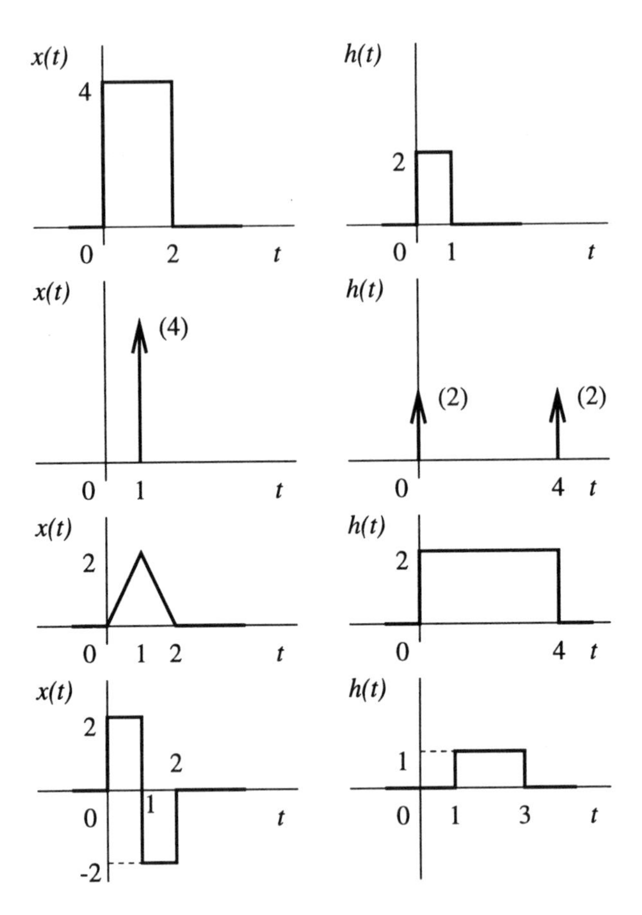

2.2 In seismic signal processing an explosion is used on the earth's surface to generate an input waveform similar to that shown in graph (a) below. Given that the impulse response of the earth itself is given by $h(t)$ where

$$h(t) = \begin{cases} 2t & \text{for } 0 \le t < 1 \\ 0 & \text{otherwise} \end{cases}$$

use convolution to develop an expression for the output $y(t)$. A problem can arise with the detonator such that the explosion hiccups, resulting in the waveform illustrated in graph (b) below. Derive an expression for the output in this case. [(a) $0 \le t < 1$, $y(t) = t^2$; $1 \le t < 2$, $y(t) = t(2-t)$; $t \ge 2$, $y(t) = 0$. (b) $0 \le t < ½$, $y(t) = t^2$; $½ \le t < 1$, $y(t) = t^2 - (t - ½)^2$; $1 \le t < 3/2$, $y(t) = t(2-t) - (t - ½)^2$; $3/2 \le t < 2$, $y(t) = t(2-t) - (t - ½)(2 - (t - ½))$; $2 \le t < 5/2$, $y(t) = -(t - ½)(2 - (t - ½))$; $5/2 < t$, $y(t) = 0$]

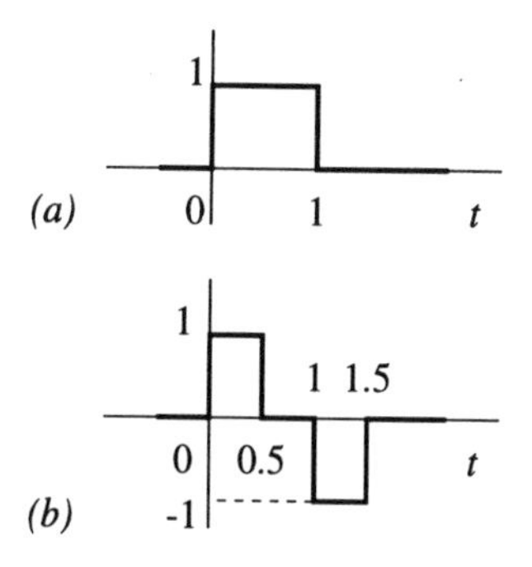

2.3 Sketch the impulse response of the operational amplifier circuit below:

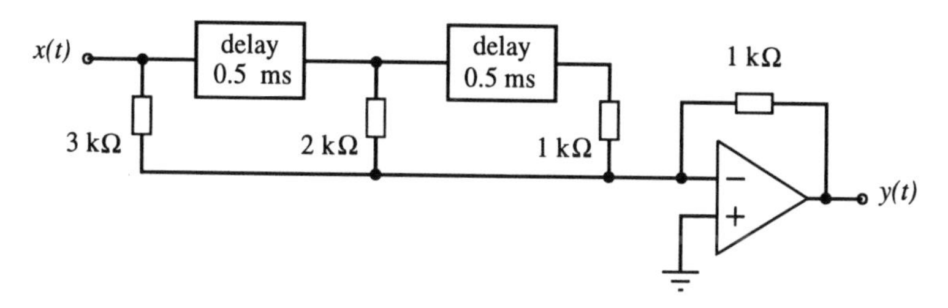

Write an expression for the impulse response and the Laplace transfer function. In order to measure the impulse response of such a system in the laboratory, the impulse is approximated by the pulse $d(t)$ shown in Figure 2.2. What is the largest value of the pulse width Δt that can be used and still ensure the success of the experiment? Why is the height of the pulse $1/\Delta t$? [Impulse response $h(t) = -\delta(t)/3 - \delta(t - 0.5 \times 10^{-3})/2$ $-\delta(t - 10^{-3})$; transfer function $H(s) = -1/3 - \exp(-s(0.5 \times 10^{-3}))/2 - \exp(-s(10^{-3}))$; $\Delta t < 0.5 \times 10^{-3}$]

CHAPTER 3

Transfer function and system characterisation

3.1 Introduction

This chapter is about connections and relationships between the alternative descriptions of continuous time linear systems. We are interested in questions such as

> What can we learn about the system (i.e. how does it respond or behave) by examining the transfer function?

The starting point for this discussion is the definition of the poles and zeros of the transfer function in section 3.2. The relationship between the position of the poles and the frequency response is examined in section 3.3. Several graphical methods are used to illustrate these points. An exploration of the relationship between the position of the poles and the impulse response is provided in section 3.4. This provides both links between the transfer function and the time-domain system description as well as insights into system stability. The speed of response of linear systems is considered in section 3.5 through the use of the step response. Speed of response is linked to bandwidth in section 3.6. Finally, in section 3.7, the key concepts are summarised.

3.2 Transfer function, poles and zeros

For the zero initial condition case, the differential equation (1.21) which describes a linear system can be transformed in the following manner:

$$a_0\, y + a_1 \frac{dy}{dt} + \cdots + a_n \frac{d^n y}{dt^n} = b_0\, x + b_1 \frac{dx}{dt} + \cdots + b_m \frac{d^m x}{dt^m}$$

$$a_0\, Y(s) + a_1\, sY(s) + \cdots + a_n\, s^n Y(s) = b_0\, X(s) + b_1\, sX(s) + \cdots + b_m\, s^m X(s)$$

$$A(s)\, Y(s) = B(s)\, X(s)$$

where $A(s)$ and $B(s)$ are polynomials in s, i.e. $A(s) = a_0 + a_1 s + \cdots + a_m s^n$ and $B(s) = b_0 + b_1 s + \cdots + b_m s^m$ respectively. Thus:

$$\begin{aligned} H(s) &= \frac{Y(s)}{X(s)} = \frac{B(s)}{A(s)} \\ &= \frac{b_0 + b_1 s + \cdots + b_m s^m}{a_0 + a_1 s + \cdots + a_n s^n} \end{aligned}$$

A pole is an infinite value of $H(s)$ obtained when s is set so that the denominator $A(s)$ has value zero, i.e. pole positions are obtained from the roots of $A(s)$. A zero is a zero value of $H(s)$ obtained when s is set so that the numerator $B(s)$ has zero value, i.e. zeros are obtained from the roots of $B(s)$. It is assumed that $A(s)$ and $B(s)$ share no common roots.

EXAMPLE 3.1

Start with a typical transfer function that might be obtained by applying Kirchhoff's laws to a circuit:

$$H(s) = \frac{s^2 + 2s + 2}{s^2 + 4s + 13}$$

Note that both the numerator and denominator are expressed in positive powers of s, i.e. s^0, s^1, s^2, etc. There are no terms in, for example, $1/s$ or $1/s^2$. You usually have to multiply the top and bottom lines by s until you get only positive powers. The zeros are calculated by setting the top line to zero and solving for s:

$$s^2 + 2s + 2 = 0$$

This is a quadratic which can be solved in the usual way:

$$s = \frac{-2 \pm \sqrt{2^2 - 4(2)}}{2}$$

$$= -1 \pm j$$

Note that there is a possibility of having complex roots. The poles are found by setting the bottom line to zero:

$$s^2 + 4s + 13 = 0$$

$$s = -2 \pm j3$$

An important way of visualising poles and zeros is to plot their position in the complex plane. Poles are usually denoted by × and zeros by o. Figure 3.1 is the pole/zero map for the example above. Zeros (or poles) exist in complex conjugate pairs for any real system.

For a simple example we can plot $|H(s)|$ against s. Note that s is complex and has a real part σ and an imaginary part ω. Hence we need a 3-D plot, Figure 3.2.

For every value of σ and ω, the magnitude of the transfer function will have a different value. In the particular example in Figure 3.2, $H(s)$ has a pole at $s = -1 + 2j$ and another at $s = -1 - 2j$. There are no zeros in this example. Recall that the poles are the values of s where the bottom line of the transfer function is equal to zero and hence, since we are dividing by zero, the size of the transfer function must be infinite at these points. This can be seen on the 3-D plot. There is a peak in $|H(s)|$ at the point where $\sigma = -1$ and $\omega = 2$. (What would happen if this point was a zero?)

Self assessment question 3.1: Make a pole zero plot of the transfer function: $H(s) = (s^2 - 4)/(10s^2 + 140s + 1000)$.

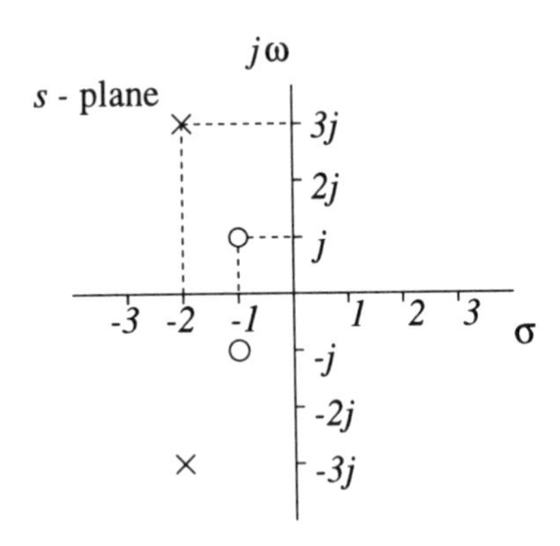

Figure 3.1 *Poles and zeros of transfer function.*

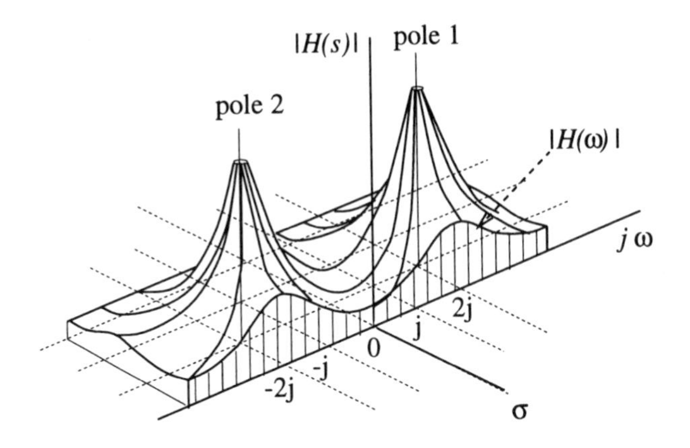

Figure 3.2 *3-D plot of transfer function magnitude for* $H(s) = 1/(s^2 + 2s + 5)$. □

3.3 Transfer function and frequency response

The frequency response of a system is defined as:

$$H(\omega) = \frac{F[\text{output}]}{F[\text{input}]} = \frac{Y(\omega)}{X(\omega)} \tag{3.1}$$

Again the transfer function can be viewed as the response of the circuit or system to one basis function, e.g. if the input is the basis function $\exp(j\omega t)$, then the output will be $H(\omega)\exp(j\omega t)$. In our simple RC circuit example, the frequency response (or Fourier transfer function) would be

$$H(\omega) = \frac{1}{1 + j\omega RC} \tag{3.2}$$

The Laplace and Fourier transforms are very similar. This similarity allows us to move easily from a Laplace to the Fourier representation of a signal or system.

- *Provided $H(s)$ has poles only in the left half of the s-plane, simply substitute $j\omega$ for s.*

The condition that the poles should lie in the left half plane is not excessively restrictive and will be explored later in this chapter when stability is discussed.

EXAMPLE 3.2

If the Laplace transfer function of a system is

$$H(s) = \frac{1}{s+7}$$

then the frequency response or Fourier transfer function is

$$H(\omega) = \frac{1}{j\omega + 7}$$

This relationship can also be seen by noting that the basis function used in the Laplace representation, i.e. $\exp(st)$, can be written out as $\exp(\sigma t + j\omega t)$. If $\sigma = 0$ the Laplace basis function becomes the Fourier basis function, $\exp(j\omega t)$. Looking again at the 3-D plot of $|H(s)|$, the magnitude frequency response $|H(\omega)|$ is obtained by letting $\sigma = 0$, in which case points on $|H(s)|$ directly above the $j\omega$ axis are considered alone. Thus the magnitude frequency response is a slice through $|H(s)|$ along the $j\omega$ axis as shown in Figure 3.2. This interpretation also provides an understanding of the relationship between the position of the poles in the s-plane and the shape of the frequency response. The peak in the frequency response occurs at $\omega = 2$ – directly opposite the pole. If the pole was moved closer to the $j\omega$ axis, the peak in the frequency response would get bigger. Likewise, if the pole was moved further from the $j\omega$ axis the peak in the frequency response would get smaller and eventually disappear. (What would happen if this pole was a zero?)

Self assessment question 3.2: Write down an expression for the frequency response of a system with transfer function: $H(s) = 1/(s^2 + 4s + 4)$.

3.3.1 Frequency response from pole/zero diagram

A pole/zero diagram can give us further information about the relationship between the positions of the poles and zeros and the nature of the frequency response. If a transfer function $H(s)$ has m zeros and n poles, it can be written as:

$$H(s) = \frac{A(s - z_1)(s - z_2) \cdots (s - z_m)}{(s - p_1)(s - p_2) \cdots (s - p_n)} \tag{3.3}$$

where A is a constant. The frequency response is obtained simply by replacing s with $j\omega$ thus:

$$H(\omega) = \frac{A(j\omega - z_1)(j\omega - z_2) \cdots (j\omega - z_m)}{(j\omega - p_1)(j\omega - p_2) \cdots (j\omega - p_n)}$$

To obtain the amplitude response of the system, take the magnitude of $H(\omega)$, which because of the properties of complex numbers, simplifies to:

$$|H(\omega)| = \frac{A|j\omega - z_1||j\omega - z_2| \cdots |j\omega - z_m|}{|j\omega - p_1||j\omega - p_2| \cdots |j\omega - p_n|} \tag{3.4}$$

In a similar way the phase response of the system is obtained by taking the argument of $H(j\omega)$. Again, because of the properties of complex numbers, this simplifies to:

$$\begin{aligned} \angle H(\omega) = {} & \angle(j\omega - z_1) + \angle(j\omega - z_2) + \cdots + \angle(j\omega - z_m) \\ & - \angle(j\omega - p_1) - \angle(j\omega - p_2) \cdots - \angle(j\omega - p_n) \end{aligned} \tag{3.5}$$

Calculating the gain and phase shift of a system at a particular frequency, ω_0, is a matter of repeatedly evaluating the modulus and argument of terms such as $(j\omega_0 - p_1)$ and then multiplying or adding, as appropriate, the resultant terms to get the desired gain or phase shift value. Since the operation has been reduced to repeatedly doing a particular type of calculation, it is useful to visualise the term $(j\omega_0 - p_1)$ to get some insight into how its magnitude and phase change as the value of ω_0 is varied. Consider a simple example of a single pole system:

$$H(s) = \frac{1}{(s - p_1)} \tag{†}$$

The frequency response is:

$$H(\omega) = \frac{1}{(j\omega - p_1)}$$

As a starting point, calculate the gain and phase shift at a frequency ω_0. Since complex numbers have both size and direction, they can also be viewed as vectors. Thus both the length and direction of $(j\omega_0 - p_1)$ are required. Figure 3.3(a) is the pole/zero map for this system. There is a pole at $s = p_1$. Viewed as a vector, the line from the origin to the point $s = p_1$ is the vector p_1. Hence the line from the point $s = p_1$ to the origin is the vector $-p_1$ (same size but opposite direction to the vector p_1). The line from the origin to the point on the imaginary axis where $\omega = \omega_0$ is the vector $j\omega_0$. If the vector $-p_1$ is simply added to the vector $j\omega_0$, the resultant vector represents the term $(j\omega_0 - p_1)$. In Figure 3.3(a), the vector $(j\omega_0 - p_1)$ is the line from the point $s = p_1$ to the point on the imaginary axis where $\omega = \omega_0$.

Choosing a second frequency ω_1 which is greater than ω_0, it is clear that the vector $(j\omega_1 - p_1)$ is longer than the vector $(j\omega_1 - p_1)$ and hence the gain $|H(\omega_1)|$ will be smaller than $|H(\omega_0)|$ – remember the system has one pole and hence the term is on the bottom line. The gains at ω_0 and ω_1 are plotted on the unusual graph of Figure 3.3(b) which also shows the results if this argument is repeated over the whole range of frequencies. Of particular interest is the frequency $\omega = \Im[p_1]$, at which point the vector is shortest and hence this is the frequency where the gain is a maximum.

† It is worth noting that for real systems, poles and zeros exist in complex conjugate pairs. A single complex pole at $s = p_1$ implies that this system is complex. Complex systems do exist, particularly in communications equipment. However, the choice of a complex pole here is purely to illustrate the more general case through a simple example.

The phase can be obtained in a similar way. The argument of the term $(j\omega_0 - p_1)$ is the angle of the vector $(j\omega_0 - p_1)$ measured anticlockwise from the horizontal. As illustrated in Figure 3.3(a), this angle is positive and hence the phase shift at ω_0 is negative. The angle of the vector $(j\omega_1 - p_1)$ is more positive than that of the vector $(j\omega_0 - p_1)$ and hence the phase shift at ω_1 is more negative than that at ω_0. The phase shift is illustrated in Figure 3.3(c). The frequency $\omega = \Im[p_1]$ is the point where the vector is horizontal and hence this is the frequency where the phase shift is zero.

In the general case, the amplitude response is obtained by taking the products of the lengths of the various vectors drawn from the zeros to the relevant frequency (ω_0) value on the imaginary axis and dividing this by the product of the lengths of the vectors from the poles – as indicated by equation (3.4). The phase is found by summing the phases of the individual vectors from the zeros minus the sum of the phases of the individual vectors from the poles – as indicated by equation (3.5). This is illustrated in Figure 3.4.

Thus it is obvious that a pole close to the imaginary axis gives rise to a maximum in the spectrum or frequency response as ω approaches a value equal to the pole's imaginary part (e.g. $\omega = \Im[p_1]$), while a zero gives rise to a minimum in the response. Examples illustrating this are shown in Figure 3.5. Note also that if a zero (pole) lies on the imaginary axis then $|H(\omega)|$ actually equals zero (infinity) at that frequency.

Self assessment question 3.3: Sketch the amplitude and phase response of a system with transfer function: $H(s) = (s^2 - 0.2s + 1.01)/(s^3 - 0.4s^2 + 9.04s)$.

3.3.2 Fourier transform of periodic signals

Chapter 1 started by examining the Fourier series representation of periodic signals and then extended the concept to non-periodic signals through the use of the Fourier transform. It is then reasonable to ask if there is a Fourier transform representation of periodic signals. The answer is affirmative but an impulse function is required. For example, a cosine wave, $x(t)$, with amplitude A and frequency ω_a rad/s, i.e.:

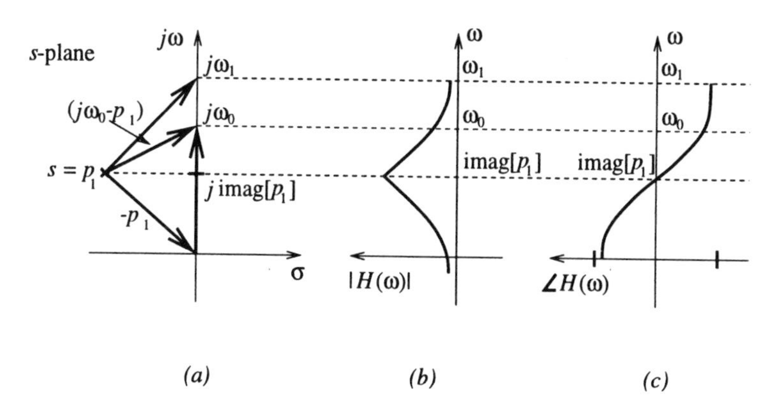

Figure 3.3 *Frequency response of a single pole system: (a) pole/zero map; (b) amplitude response; (c) phase response.* □

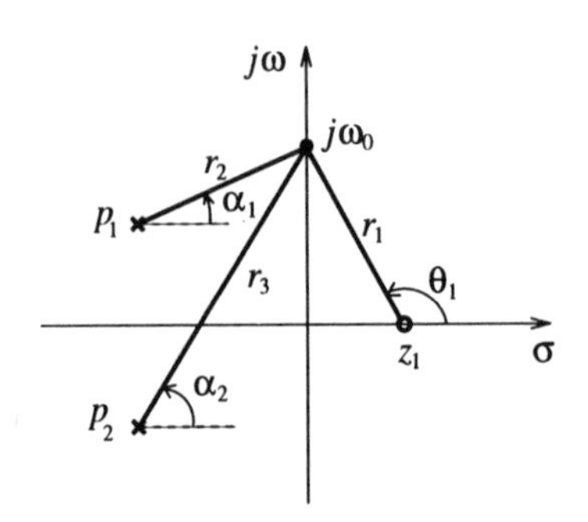

$$|H(\omega_0)| = \frac{r_1}{r_2 r_3} \qquad \angle H(\omega_0) = \theta_1 - \alpha_1 - \alpha_2$$

Figure 3.4 *General definition of $H(\omega)$ in the s-plane.*

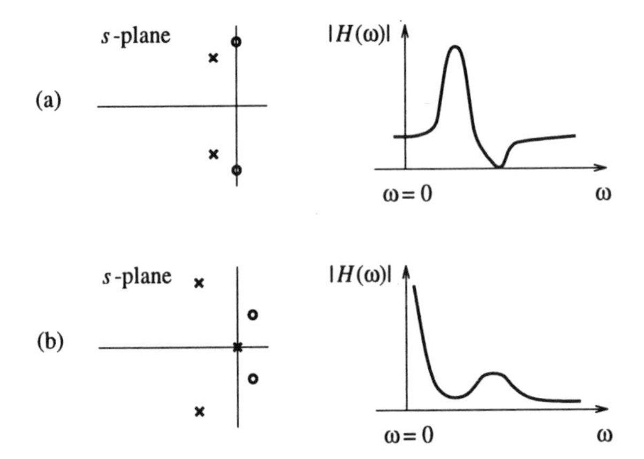

Figure 3.5 *Examples of pole/zero plots and corresponding amplitude frequency responses.* ❑

$$x(t) = A\cos(\omega_a t)$$

has a Fourier transform $X(\omega)$ where

$$X(\omega) = A\pi[\delta(\omega - \omega_a) + \delta(\omega + \omega_a)]$$

This is illustrated in Figure 3.6. Likewise, the Fourier transform of a sine wave of amplitude A and frequency ω_a rad/s is:

$$X(\omega) = -\,jA\pi[\delta(\omega - \omega_a) - (\omega + \omega_a)]$$

Both of these relationships can be justified by considering a signal with Fourier transform

$$X(\omega) = \delta(\omega - \omega_a)$$

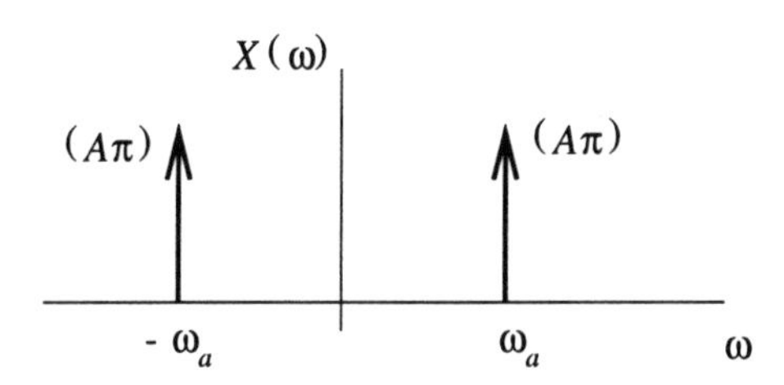

Figure 3.6 *The Fourier transform of a cosine wave of frequency ω_a rad/s and amplitude A.*

and using the inverse Fourier transform integral to obtain its time-domain representation. Thus:

$$
\begin{aligned}
x(t) &= F^{-1}[X(\omega)] \\
&= \frac{1}{2\pi} \int_{-\infty}^{\infty} X(\omega) \exp(j\omega t)\, d\omega \\
&= \frac{1}{2\pi} \int_{-\infty}^{\infty} \delta(\omega - \omega_a) \exp(j\omega t)\, d\omega \\
&= \frac{1}{2\pi} \int_{-\infty}^{\infty} \delta(\omega - \omega_a) \exp(j\omega_a t)\, d\omega \\
&= \frac{1}{2\pi} \exp(j\omega_a t) \int_{-\infty}^{\infty} \delta(\omega - \omega_a)\, d\omega \\
&= \frac{1}{2\pi} \exp(j\omega_a t)
\end{aligned}
$$

Having shown that the Fourier transform of a complex phasor is a single impulse in the frequency domain, it is straightforward to show the relationships for sines and cosines – an exercise for the reader. Recalling that the complex Fourier series expresses a periodic signal as a sum of complex phasors, the Fourier transform of a periodic signal is easily derived from the Fourier series, as the following example illustrates.

EXAMPLE 3.3

In the example of section 1.3, the complex Fourier series representation of the periodic waveform of Figure 1.6(a) was developed. Derive an expression for the Fourier transform of this signal.

In example 1.3 the complex Fourier series was shown to be:

$$X_n = \frac{A\tau}{T} \operatorname{sa}(n\omega_0\tau/2)$$

Combining this with equation (1.12), the signal is expressed as a weighted sum of complex phasors:

$$x(t) = \sum_{n=-\infty}^{\infty} \frac{A\tau}{T} \operatorname{sa}(n\omega_0\tau/2) \exp(jn\omega_0 t)$$

Taking the Fourier transform of both sides:

$$X(\omega) = F\left[\frac{A\tau}{T} \sum_{n=-\infty}^{\infty} \operatorname{sa}(n\omega_0\tau/2) \exp(jn\omega_0 t)\right]$$

$$= \frac{A\tau}{T} \sum_{n=-\infty}^{\infty} \operatorname{sa}(n\omega_0\tau/2)\, F[\exp(jn\omega_0 t)]$$

$$= \frac{2\pi A\tau}{T} \sum_{n=-\infty}^{\infty} \operatorname{sa}(n\omega_0\tau/2)\, \delta(\omega - n\omega_0)$$

3.3.3 Measurement of frequency response

It is useful in gaining an understanding of the complex frequency response $H(\omega)$ to consider how it could be measured from an actual system. The complex phasor $\exp(j\omega t)$ cannot be generated in the laboratory to measure the response $H(\omega)\exp(j\omega t)$ and hence deduce the transfer function $H(\omega)$. This is because the input is complex, not real. However, it is possible to generate a cosine wave or a sine wave of the same frequency. A cosine or sine wave can be thought of as the addition of two complex phasors:

$$\cos(\omega t) = \frac{\exp(j\omega t) + \exp(-j\omega t)}{2}$$

This can also be written using the symbol $\Re$ to denote 'real part of':

$$\cos(\omega t) = \Re[\exp(j\omega t)]$$

$$= \Re[\cos(\omega t) + j\sin(\omega t)]$$

For linear systems, superposition can be used to calculate the response to a cosine wave by adding the responses to the individual phasors. Thus the response to the cosine wave is:

$$\tfrac{1}{2} H(\omega) \exp(j\omega t) + \tfrac{1}{2} H(-\omega) \exp(-j\omega t) \qquad (\dagger)$$

$$= \tfrac{1}{2} H(\omega) \exp(j\omega t) + \tfrac{1}{2} H^*(\omega) \exp(-j\omega t)$$

$$= \tfrac{1}{2} H(\omega) \exp(j\omega t) + \tfrac{1}{2} (H(\omega) \exp(j\omega t))^*$$

which can also be written in short-hand using the $\Re$ notation:

$$\Re[H(\omega)\exp(j\omega t)]$$

At this point recall that $H(\omega)$ is a complex number and can be expressed in term of its

† The second term is obtained by symmetry.

magnitude $|H(\omega)|$ and its angle $\angle H(\omega)$ as $|H(\omega)|\exp(\,j\angle H(\omega)\,)$. Thus the response to the cosine wave can be rewritten as:

$$\Re[|H(\omega)|\exp(j\angle H(\omega))\exp(j\omega t)]$$

Since $|H(\omega)|$ is already real it can be moved to the left-hand side of the $\Re$ symbol:

$$|H(\omega)|\;\Re[\exp(j\angle H(\omega))\exp(j\omega t)]$$

The next step is to combine the two exponential terms:

$$|H(\omega)|\;\Re[\exp(j\omega t + j\angle H(\omega))]$$

And finally extracting the real part of the complex exponential:

$$|H(\omega)|\cos(\omega t + \angle H(\omega))$$

In summary, a cosine wave with frequency ω rad/s, an amplitude of 1 and a phase angle of 0 is applied to the system. The steady-state response is a cosine of the same frequency. The amplitude has been amplified by a factor of $|H(\omega)|$ and the phase has been increased by an angle of $\angle H(\omega)$. Once the magnitude and phase response at every frequency have been measured, the complex frequency response can be calculated using the rules of complex numbers:

$$H(\omega) = |H(\omega)|\cos(\angle H(\omega)) + j|H(\omega)|\sin(\angle H(\omega))$$

This sort of response is characteristic of a linear system – if a sine wave or a cosine wave of a particular frequency is applied to a system, then when the transients have decayed away, the output will be a sine wave or cosine wave of exactly the same frequency – the only thing that will be different is the amplitude and the phase. One way of measuring the linearity of a hi-fi amplifier is to apply a 1 kHz sine wave to its input. If it is perfectly linear (which it never is) the output will also be a 1 kHz sine wave (hopefully bigger!). If it is not linear, some 2 kHz sine wave (2nd harmonic), 3 kHz (3rd harmonic) etc. components will be present. One way of measuring the linearity of an amplifier is to calculate the total power in the harmonics and express that as a fraction of the power in the fundamental – this is the *total harmonic distortion* (THD).

3.3.4 Bode plots

While the graphical approach discussed previously can give a useful insight into the relationships between the positions of poles and zeros and the frequency response of simple systems, it is not usually adequate when we come to more complicated systems. A more powerful graphical technique is the Bode plot. With the availability of user friendly system software with flexible graphical output, the Bode plot techniques are no longer as necessary a drawing tool as they once were. However they are still an invaluable design tool because they give insights into the relationship between pole/zero position and frequency response. In any transfer function $H(s)$ there are basically four different kinds of factor which can occur:

(i) a constant gain, K;

(ii) poles (or zeros) at the origin, s;
(iii) poles or zeros on the real axes – terms of the form

$$(s + a) = a\left(\frac{s}{a} + 1\right) = \frac{1}{\tau}(\tau s + 1)$$

where a and τ are constants;
(iv) complex conjugate poles (or zeros), $(s^2 + As + B)$ where A and B are constants and the quadratic has complex roots.

The first three are fairly straightforward but the last term may require more detailed discussions. Usually the quadratic is expressed in normalised form in terms of the damping factor, ζ (zeta), and the undamped natural frequency ω_0. Thus it may be written as:

$$(s^2 + As + B) = (s^2 + 2\zeta\omega_0 s + \omega_0^2)$$

In the context of Bode plots a further scaling is often useful:

$$(s^2 + As + B) = \omega_0^2(\,(s/\omega_0)^2 + 2\zeta(s/\omega_0) + 1)$$

The roots of the quadratic are guaranteed to be complex if $\zeta < 1$.

As an example of how we might deduce the frequency response from the transfer function, consider the following system:

$$H(s) = \frac{K}{s\,(\tau s + 1)\,(\,(s/\omega_0)^2 + 2\zeta(s/\omega_0) + 1)}$$

This system has a constant gain term K, a pole at the origin, a pole on the real axis at $s = -1/\tau$ and a pair of complex conjugate poles. To obtain the frequency response, simply replace s with $j\omega$:

$$H(\omega) = \frac{K}{j\omega(j\omega\tau + 1)\,(\,(j\omega/\omega_0)^2 + 2\zeta(j\omega/\omega_0) + 1)} \qquad (3.6)$$

The logarithmic magnitude of $H(\omega)$ can be obtained by adding the gain in decibels (dB) of each individual factor:

$$\begin{aligned} 20\log_{10}|H(\omega)| &= 20\log_{10}(K) - 20\log_{10}(|j\omega|) - 20\log_{10}(|j\omega\tau + 1|) \\ &\quad - 20\log_{10}(|1 + (2\zeta/\omega_0)j\omega + (j\omega/\omega_0)^2|) \end{aligned}$$

The phase of $H(\omega)$ is the summation of the phase angles due to each individual factor:

$$\angle H(\omega) = -90^\circ - \tan^{-1}(\omega\tau) - \tan^{-1}\left(\frac{2\zeta\omega_0\omega}{\omega_0^2 - \omega^2}\right)$$

Thus if we knew the gain and phase response associated with each of the four types of terms, it would be simply a matter of adding them together to get the overall frequency response. Let consider each type of term in turn:

(i) The constant gain term is trivial, providing a gain of $20\log_{10}(K)$ at all frequencies and no phase shift.

(ii) The pole at the origin gives rise to a $j\omega$ term in the denominator; this provides -90° of phase shift at all frequencies because of the presence of j. The gain Bode plot is illustrated in Figure 3.7; on this dB/log frequency scale, the gain term $-20\log_{10}(|j\omega|)$ is a straight line with a slope of -20 dB/decade. The line cuts the point (1 rad/s, 0 dB). A zero at the origin would provide a similar result but the slope would be 20 dB/decade.

(iii) The pole at $s = -1/\tau$ gives rise to the term $j\omega\tau + 1$ on the bottom line; the gain and phase Bode plots are illustrated in Figure 3.8. The actual Bode plot is shown as a dotted line, and the asymptotic gain curve and the linear approximation to the phase curve are shown as continuous lines. At frequencies below the cut frequency of $\omega = 1/\tau$, the gain is nominally 0 dB. At the cut-off frequency, the gain is -3 dB. At frequencies above the cut-off, the slope of the gain curve has an asymptotic slope of -20 dB/decade. The phase Bode plot is -45° at the cut-off frequency; approximately 0° one decade below the cut-off frequency and approximately -90° one decade above the cut-off frequency.

(iv) The complex conjugate pair of poles give rise to the term $((j\omega/\omega_0)^2 + 2\zeta(j\omega/\omega_0) + 1)$ in the denominator of equation (3.6). The gain Bode plot for this term is illustrated in Figure 3.9 for a range of ζ values; the continuous line is the asymptotic curve. At frequencies below the undamped natural frequency ω_0, the gain is nominally 0 dB; at frequencies above the cut-off, the slope of the gain curve has an asymptotical slope of -40 dB/decade. In the vicinity of ω_0, the gain curve is highly dependent on the damping factor; a small damping factor produces a pronounced peak or resonance in the frequency response; increasing the damping factor reduces the peak. The phase Bode plot of Figure 3.9 is -90° at the undamped natural frequency; approximately 0° one decade below ω_0 and approximately -180° one decade above the cut-off frequency. The exact nature of the curve is again dependent on the damping factor.

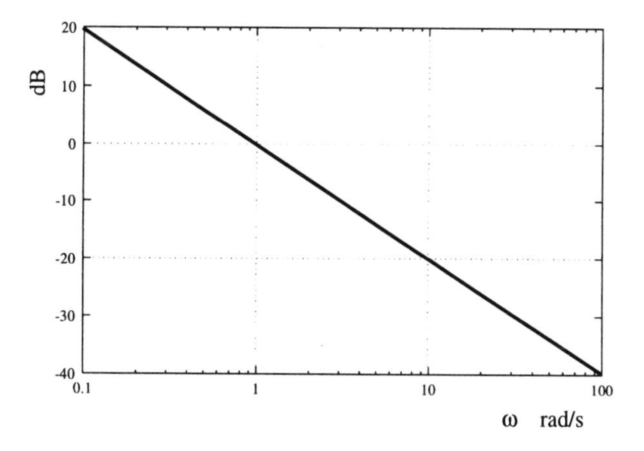

Figure 3.7 *Gain Bode diagram for a single pole at the origin.*

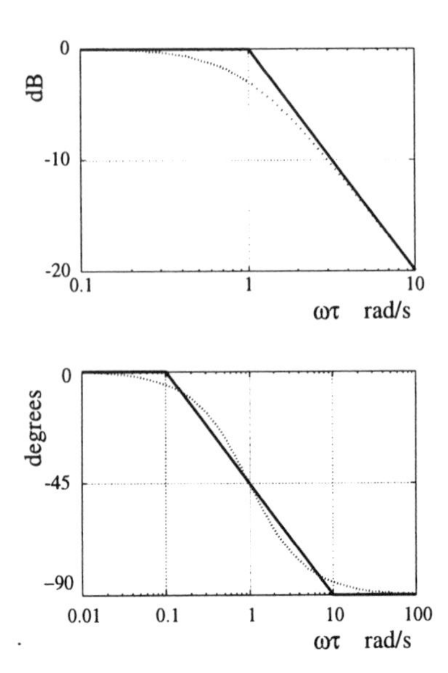

Figure 3.8 *Bode diagram for a real pole term: (a) gain plot – continuous line is asymptotic curve; (b) phase plot – continuous line is linear approximation.*

The complete gain or phase Bode plot for a system is simply the addition of these individual terms, as illustrated in the following example.

EXAMPLE 3.4

Sketch the Bode plot of a system with the following transfer function:

$$H(s) = \frac{s + 20}{s + 2000}$$

The frequency response is obtained by replacing s with $j\omega$:

$$H(\omega) = \frac{j\omega + 20}{j\omega + 2000}$$

Then normalise the numerator and denominator to obtain recognisable terms:

$$H(\omega) = \frac{j\omega/20 + 1}{(j\omega/2000 + 1)100}$$

The individual terms are plotted as dashed lines in Figure 3.10: the constant gain term with a gain of $20\log_{10}(1/100) = -40$ dB and a phase shift of zero; the zero term $(j\omega/20 + 1)$ with a cut-in frequency of 20 rad/s; the pole term $(j\omega/2000 + 1)$ with a cut-off frequency of 2000 rad/s. The total response (asymptotic) is formed by adding the individual responses. The actual Bode plots evaluated numerically are also shown for comparison.

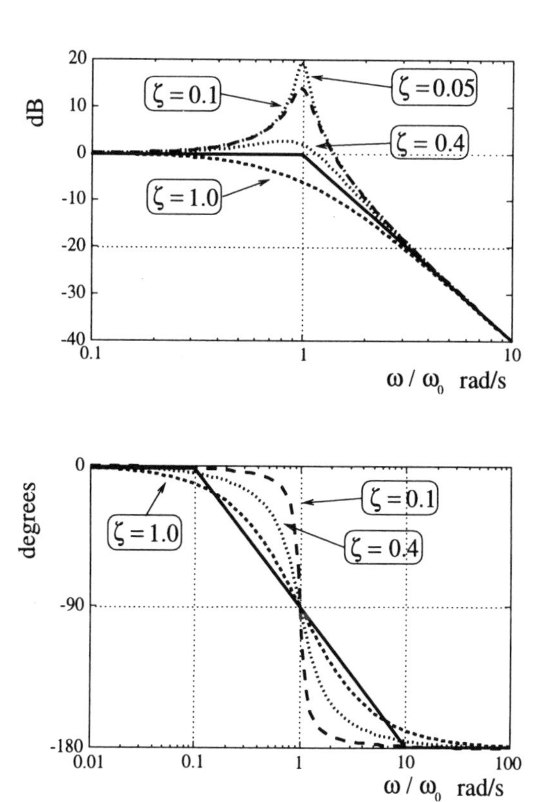

Figure 3.9 *Bode diagrams for a complex conjugate pole pair of poles at a range of ζ values: (a) gain plot (continuous line is the asymptotic curve); (b) phase plot (continuous line is linear approximation).*

Self assessment question 3.4: Do the above example again for the following transfer function: $H(s) = (10s + 200)/(s + 200)$.

3.3.5 Fourier and Laplace

In historical terms, the Fourier series appeared first but could only be used for periodic signals. The Fourier transform was then developed to extend these concepts to non-periodic signals. There are, however, certain technical problems with the Fourier transform that we do not have time to explore. In particular, certain signals of practical importance cannot be represented in terms of the Fourier transform (at least not easily). The Laplace transform was developed to overcome these problems and is thus the most general of the three.

Why do we need all three? There are several reasons: (i) the Fourier series is still used because it is the easiest to understand and because periodic signals are useful in practice; (ii) the Fourier transform tends to be used in the analysis of signals and in applications where the signal is of prime importance, e.g. communications and radar; it

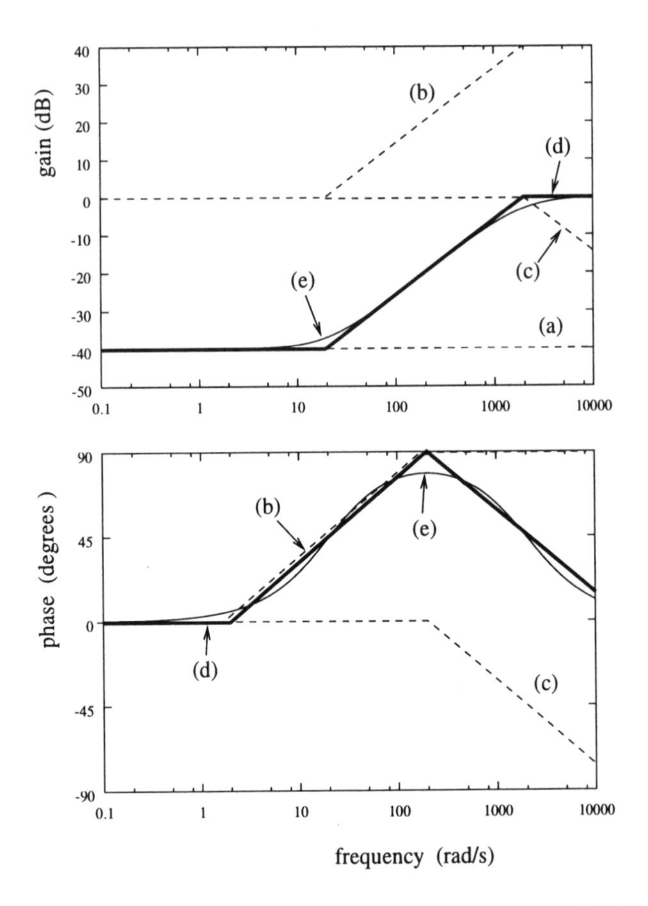

Figure 3.10 *Bode plot example: (a) constant gain term; (b) zero on the real axis; (c) pole on the real axis; (d) asymptotic response; (e) actual response.*

is possible to calculate the Fourier transform numerically using a very efficient algorithm called the fast Fourier transform (FFT) – this has led to the development of spectrum analysers which can be used in a similar manner to an oscilloscope to display the Fourier transform of any signal you connect to them (see Chapter 10); (iii) the Laplace transform is used in the design of systems and in applications where the system is of prime importance, e.g. control engineering and filter design; by manipulating the Laplace transfer function algebraically we can effectively manipulate a filter and examine alternative ways of building it. These are not hard and fast rules. Fourier methods can, for example, be used to design control systems, and Laplace transforms are used to design filters for communications systems.

3.4 Transfer function and impulse response

In the previous section the effect of the position of the poles and zeros on the frequency response of the system was considered. In this section the effect of the position of the poles on the time-domain response will be examined – starting with the impulse response. Consider a transfer function $H(s)$ which has a pair of complex conjugate poles at:

$$s = p_1 = \sigma_a + j\omega_a$$

$$s = p_2 = \sigma_a - j\omega_a$$

To keep track of the effect of each pole we label terms with the subscripts p_1 and p_2. If the system is assumed to have no zeros its transfer function is:

$$H(s) = \frac{1}{(s - \sigma_a - j\omega_a)_{p_1}\,(s - \sigma_a + j\omega_a)_{p_2}}$$

To find the impulse response $h(t)$ take the inverse Laplace transform. One simple way of doing this is to take a partial fraction expansion (PFE):

$$H(s) = \frac{A_1}{(s - \sigma_a - j\omega_a)_{p_1}} + \frac{A_2}{(s - \sigma_a + j\omega_a)_{p_2}}$$

Note that the poles define the denominators in the partial fraction expansion. The next step is to evaluate the constants A_1 and A_2 in the PFE. A_1 and A_2 have the following values:

$$A_1 = \frac{1}{2j\omega_a}\ ;\ A_2 = \frac{-1}{2j\omega_a}$$

Note also that $A_1 = A_2^*$, i.e. they are complex conjugate. So it is only necessary to calculate A_1, then A_2 can be determined from the symmetry. Thus the transfer function is:

$$H(s) = \frac{\frac{1}{2j\omega_a}}{(s - \sigma_a - j\omega_a)_{p_1}} + \frac{\frac{-1}{2j\omega_a}}{(s - \sigma_a + j\omega_a)_{p_2}}$$

The impulse response is obtained by taking inverse Laplace transforms using the tables in Appendix B. Each pole gives rise to a separate complex exponential term:

$$\begin{aligned} h(t) &= \left[\frac{1}{2j\omega_a}\exp((\sigma_a + j\omega_a)t)\right]_{p_1} - \left[\frac{1}{2j\omega_a}\exp((\sigma_a - j\omega_a)t)\right]_{p_2} \\ &= \frac{1}{2j\omega_a}\exp(\sigma_a t)\,[\exp(j\omega_a t) - \exp(-j\omega_a t)] \\ &= \frac{1}{\omega_a}\exp(\sigma_a t)\sin(\omega_a t) \end{aligned}$$

The real part of the pole, σ_a, controls the decay rate of the impulse response, and the imaginary part, ω_a, controls the frequency of the oscillation. Thus the nature of the impulse response is strongly related to the position of the poles. Figure 3.11 shows impulse response plots for various pole locations in the s-plane. Notice in particular what would happen if the real part of the pole was kept constant and the imaginary part was increased from zero to infinity. The envelope of the impulse response waveform stays constant while the frequency of the oscillation increases. On the other hand, if the imaginary part of the pole ω_a is held constant and the real part σ_a increased from zero to minus infinity, the frequency of the oscillation remains constant while the rate of decay increases. Notice also that if the pole lies on the real axis (i.e. $\omega_a = 0$) then there is no oscillation associated with the impulse response. Further, if the pole lies on the imaginary axis (i.e. $\sigma_a = 0$) then there is no decay or growth associated with the impulse response – a pure oscillation. This relationship between the position of the pole in the s-plane and the nature of the impulse response also provides clues about the stability of the system. If a system has a pole which has a positive real part, i.e. $\sigma_a > 0$, applying an impulse to the input of such a system will produce an output which grows and grows exponentially. Thus a brief disturbance or 'kick' to the system will produce an output which just gets bigger and bigger. Such a system is *unstable*. In a similar manner, if a system has a pole which has a negative real part, i.e. $\sigma_a < 0$, applying an impulse to the input of such a system will produce an output which decays exponentially. Thus a brief disturbance or 'kick' to the system will produce an output which gets smaller. Such a system is *stable*.

In general, a system will have a more complicated transfer function with several poles and zeros. Assuming that all poles are simple ones, the transfer function is:

$$H(s) = \frac{A\,(s - z_1)(s - z_2)\cdots(s - z_m)}{(s - p_1)(s - p_2)\cdots(s - p_n)}$$

The system has m zeros and n poles and $m < n$, i.e. $H(s)$ is a proper fraction. A is a constant. Expressing this as a PFE as before, with constants $A_1, A_2, \cdots, A_n$, gives:

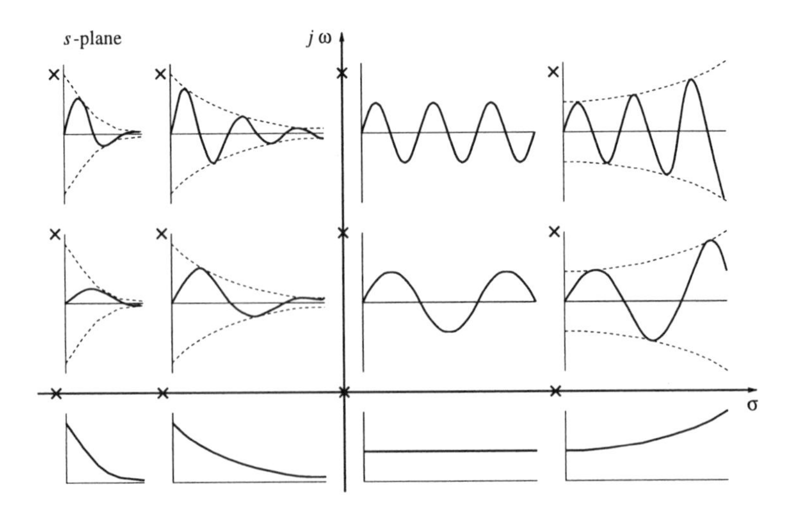

Figure 3.11 *Impulse response for various pole positions in the s-plane.* □

$$H(s) = \frac{A_1}{(s - p_1)} + \frac{A_2}{(s - p_2)} + \cdots + \frac{A_n}{(s - p_n)}$$

Taking the inverse Laplace transform yields the impulse response:

$$h(t) = A_1\, e^{p_1 t} + A_2\, e^{p_2 t} + \cdots A_n\, e^{p_n t}$$

Thus the impulse response is the sum of complex phasors such as $e^{p_n t}$. These phasors are often called the *modes* of the system. The nature of each mode depends on the position of the associated pole in the s-plane. Since the impulse response of the general system is just the weighted sum of the modes, it takes only one unstable mode to make the whole system unstable.

- *A necessary and sufficient condition that a system is stable is that all the poles of the transfer function have negative real parts, i.e. all the poles lie in the left half of the s-plane.*

It is worth while at this point recalling the discussion in section 3.3 on a simple method for obtaining the Fourier transform from the Laplace transform. If the system is stable, all the poles lie in the left half of the s-plane, and simple substitution of s with $j\omega$ is effective.

Self assessment question 3.5: Is the transfer function $H(s) = 1/(s^2 + 0.2s + 1.01)$ stable or unstable? If it is stable, make a sketch of its impulse response, clearly indicating the time constant and the period of oscillation.

3.5 Time-domain response of first and second order systems

Steps and ramps are useful test signals in assessing a system's time-domain or transient response. The unit step $u(t)$ is illustrated in Figure 3.12. In applying a unit step to the input of a system we are examining how the output responds to sudden changes. The Laplace transform of a unit step is:

$$L[u(t)] = \frac{1}{s}$$

The unit step is important because it is easy to generate and because of the close

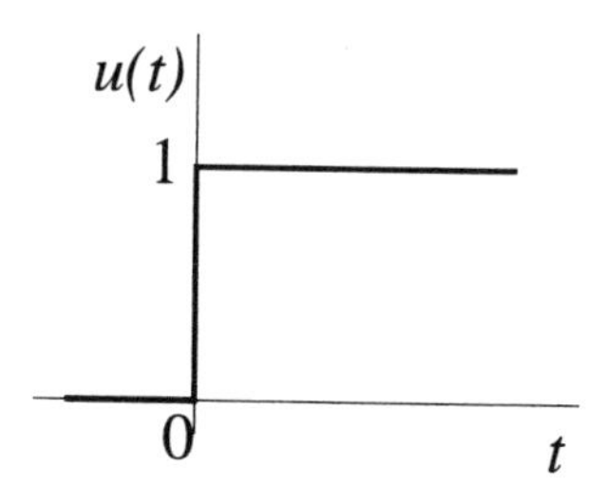

Figure 3.12 *Unit step function.*

relationship between the *impulse response* and the *step response*. To see this relationship, apply a unit step to a system and observe its response $y(t)$ as illustrated in Figure 3.13. Using transform analysis to develop an expression for the output gives:

$$Y(s) = H(s)\,\frac{1}{s}$$

Bring the s over to the left-hand side:

$$s\,Y(s) = H(s)$$

Take inverse Laplace of both sides:

$$\frac{dy}{dt} = h(t)$$

In words, this means that the impulse response is the derivative of the step response. In practice, we could measure the step response (since steps are easy to generate in the lab – unlike impulses), and differentiate it to find the impulse response.

3.5.1 First order systems

Consider once more the circuit illustrated in Figure 1.12 as an example of a first order system. As before, the transfer function is:

$$H(s) = \frac{1}{(1 + sRC)}$$

The highest power of s is s^1 and hence it is a first order system. To find the poles, set the denominator to zero and hence there is one real pole at $s = -1/RC$. The transfer function can be used to calculate the step response:

$$\begin{aligned}
Y(s) &= H(s)\,X(s) \\
&= \frac{1}{(1 + sRC)}\,\frac{1}{s} \\
&= \frac{\frac{1}{RC}}{\left(s + \frac{1}{RC}\right)}\,\frac{1}{s} \\
&= \frac{1}{s} - \frac{1}{\left(s + \frac{1}{RC}\right)} \qquad \text{(PFE)}
\end{aligned}$$

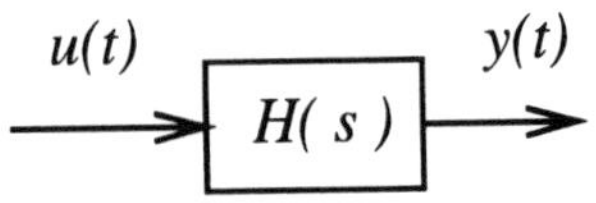

$u(t)$ → $H(\,s\,)$ → $y(t)$

Figure 3.13 *Measurement of step response.*

Take inverse Laplace transforms:

$$y(t) = 1 - \exp\left(-\frac{t}{RC}\right) \quad t \geq 0.$$

The step input and the response are illustrated in Figure 3.14.

3.5.2 Second order systems

As an example of a second order system, we consider the circuit of Figure 3.14. Using the impedance analysis of section 1.7, an expression for the transfer function can be developed:

$$Y(s) = \frac{\dfrac{1}{sC}}{R + sL + \dfrac{1}{sC}} X(s)$$

Therefore:

$$\begin{aligned} H(s) &= \frac{1}{s^2LC + sCR + 1} \\ &= \frac{\dfrac{1}{LC}}{s^2 + \dfrac{R}{L}s + \dfrac{1}{LC}} \end{aligned}$$

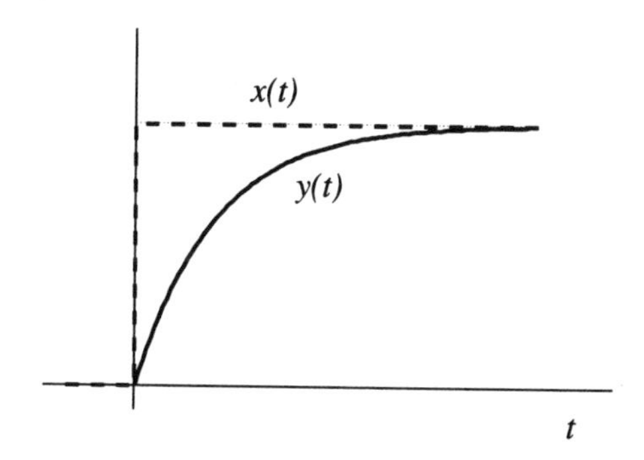

Figure 3.14 *Step response $y(t)$ of a first order system to a unit step function $x(t)$.*

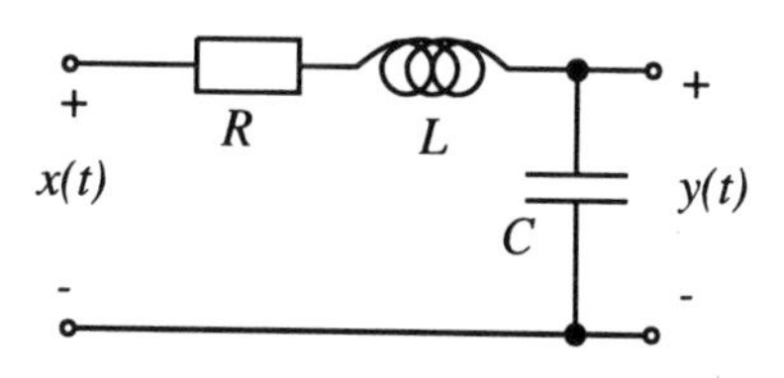

Figure 3.15 *RLC circuit.*

As in the case of Bode diagrams, it is usual to normalise such second order systems in terms of the undamped resonant frequency, ω_0, and the damping factor, ζ. Thus:

$$H(s) = \frac{\omega_0^2}{s^2 + 2\zeta\omega_0 s + \omega_0^2}$$

For this particular system:

$$\omega_0^2 = \frac{1}{LC} \qquad \text{(omega)}$$

$$\zeta = \frac{R}{2\sqrt{\frac{L}{C}}} \qquad \text{(zeta)}$$

and the poles of $H(s)$ are at:

$$s_1 = -\zeta\omega_0 + \omega_0\sqrt{\zeta^2 - 1}$$

$$s_2 = -\zeta\omega_0 - \omega_0\sqrt{\zeta^2 - 1}$$

There are three possibilities for the poles which are dependent on the value of the damping factor:

$\zeta > 1$ => two real roots (overdamped);
$\zeta = 1$ => two real roots at $s = -\omega_0$ (critical damping);
$\zeta < 1$ => two complex roots (underdamped).

All possible pole positions for a second order system are illustrated in the s-plane diagram of Figure 3.16. If the damping factor is zero we have two complex poles at $s = \pm j\omega_0$. As the damping factor is increased, the poles move along a circle of radius ω_0. When the damping factor is 1, the poles are coincident on the real axis at $s = -\omega_0$. If the damping factor is increased further, one pole moves along the real axis towards the origin while the other moves along the real axis towards $-\infty$.

Having examined the pole position for a second order system, it is natural to investigate the nature of the impulse response as the damping factor and undamped natural frequency are changed. Consider first the underdamped case where the damping factor $\zeta < 1$. The poles are complex conjugate:

$$p_1, p_2 = -\zeta\omega_0 \pm j\omega_0\sqrt{1 - \zeta^2}$$

The impulse response can be shown to be:

$$h(t) = \frac{\omega_0}{\sqrt{1 - \zeta^2}} e^{-\zeta\omega_0 t} \sin(\sqrt{1 - \zeta^2}\,\omega_0 t)$$

The impulse response is obtained by taking the inverse Laplace transform of the transfer function. Note that the real part of the poles determines the time constant associated with the exponential decay of the impulse response, while the imaginary part of the poles determines the frequency of oscillation.

The second case is one of a critically damped system with damping factor $\zeta = 1$. This has two identical real poles at:

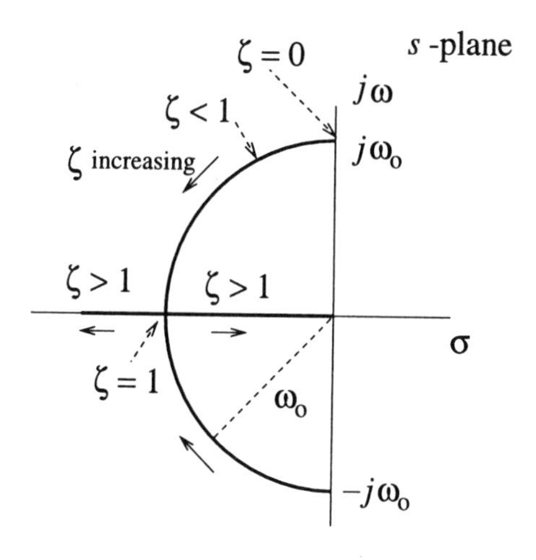

Figure 3.16 *Root locus of a second order system, showing how the poles of a simple second order system move as the damping factor is increased.*

$$p_1 = p_2 = -\,\omega_0$$

The impulse response is:

$$h(t) = \omega_0\ t\ e^{-\omega_0 t}$$

Because the poles are real there is no oscillatory term in the impulse response. The t term in the impulse response is because there are two poles in the same place. Try proving this result. Typical impulse responses for the underdamped and critically damped cases are illustrated in Figure 3.17.

Finally there is the overdamped case. This case has already been examined – it is equivalent to two first order systems with two different real poles. The damping factor $\zeta > 1$. The two real poles:

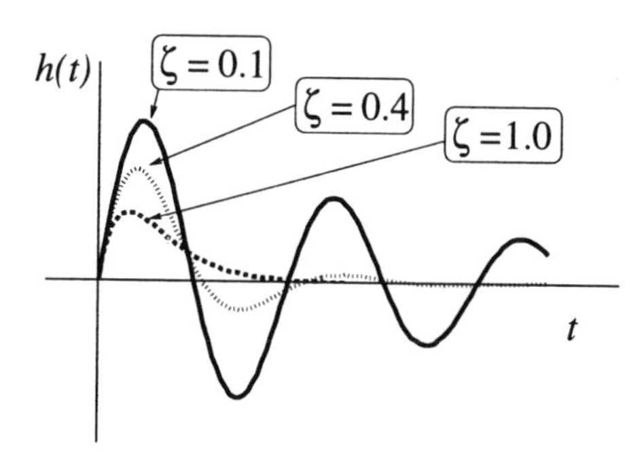

Figure 3.17 *Impulse responses for underdamped and critically damped second order system over a range of damping factors.*

$$p_1, p_2 = -\zeta\omega_0 \pm \omega_0\sqrt{\zeta^2 - 1}$$

The impulse response is:

$$h(t) = \frac{\omega_0}{\sqrt{\zeta^2 - 1}}\, e^{-\zeta\omega_0 t}\, \sinh(\sqrt{\zeta^2 - 1}\,\omega_0\, t)$$

Note that a hyperbolic sin is simply formed from the difference of two real exponential terms. Thus there is no oscillation present in the impulse response as the two poles are real.

The step response of a second order system can also be evaluated using transform analysis techniques:

$$Y(s) = H(s)\, L[u(t)]$$

$$= \frac{\omega_0}{(\,s^2 + 2\zeta\omega_0 s + \omega_0^2\,)}\,\frac{1}{s}$$

Take inverse transforms to find the step response:

$$y(t) = L^{-1}[Y(s)]$$

$$= 1 - \frac{\exp(\,-\zeta\omega_0 t\,)}{\sqrt{1-\zeta^2}}\sin\left(\sqrt{1-\zeta^2}\,\omega_0\, t + \theta\right)$$

where

$$\theta = \tan^{-1}\left(\frac{\sqrt{1-\zeta^2}}{\zeta}\right)$$

This result only applies to the underdamped case. Typical step responses for various damping factors are illustrated in Figure 3.18.

Self assessment question 3.6: What is the damping factor and undamped natural frequency of a system with transfer function: $H(s) = 1/(10s^2 + 20s + 40)$?

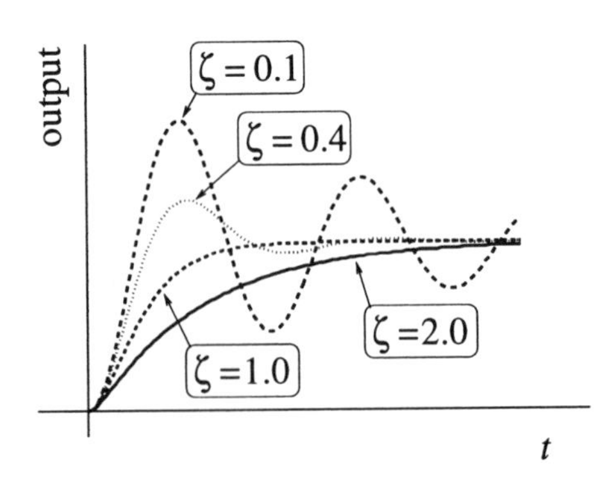

Figure 3.18 *Typical step responses of underdamped, critically damped and overdamped second order systems.*

3.6 Rise time and bandwidth

The speed of response of a system is usually measured by examining its step response. Figure 3.19 illustrates the step response of a general system which exhibits overshoot and oscillation. The rise time T_r is the time taken to get to within 90% of the final value F. The peak time T_p is the time taken to reach the first peak which has a value P. Note that not all systems will have a peak in the step response, e.g. overdamped second order systems have no peak in the step response. The overshoot is normally described as a percentage:

$$\text{percentage overshoot} = 100\,\frac{(P-F)}{F}$$

Finally the settling time T_s is the time taken to get to within 2% of the final value.

Most systems have a contiguous range of frequencies over which the gain $|H(\omega)|$ remains approximately constant. This property gives rise to the concept of bandwidth. More formally, the half power bandwidth or passband (ω_B) is defined as the interval of positive frequencies over which the gain does not vary by more than 3 dB. Two typical cases are illustrated in Figure 3.20. Applying this definition, the bandwidth of the system whose frequency response is illustrated in Figure 3.20(a) is $\omega_B = \omega_1$ and the bandwidth of the system whose frequency response is illustrated in Figure 3.20(b) is $\omega_B = \omega_2 - \omega_1$. In general, systems may have several such passbands.

In section 3.5 the step response of a first order system was calculated, i.e.:

$$y(t) = 1 - \exp\left(-\frac{t}{RC}\right)$$

Since the final value is $F = y(\infty) = 1$, the rise time can be found by solving:

$$0.9F = 1 - \exp\left(-\frac{T_r}{RC}\right)$$

Thus

$$T_r = RC\,\ln(10)\ \text{s}$$

The frequency response of a first order system was considered in section 3.2 under

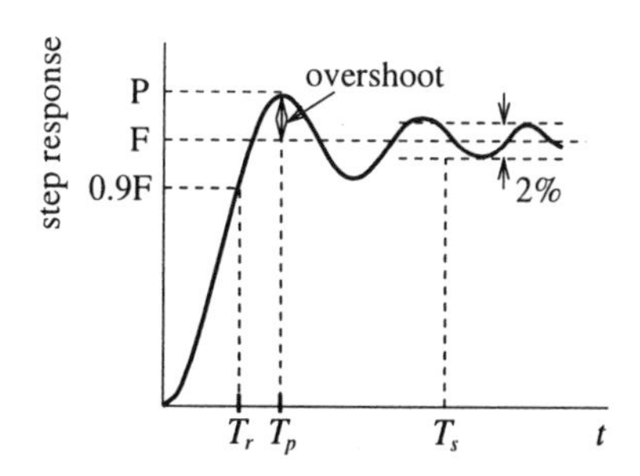

Figure 3.19 *Time-domain performance specification.*

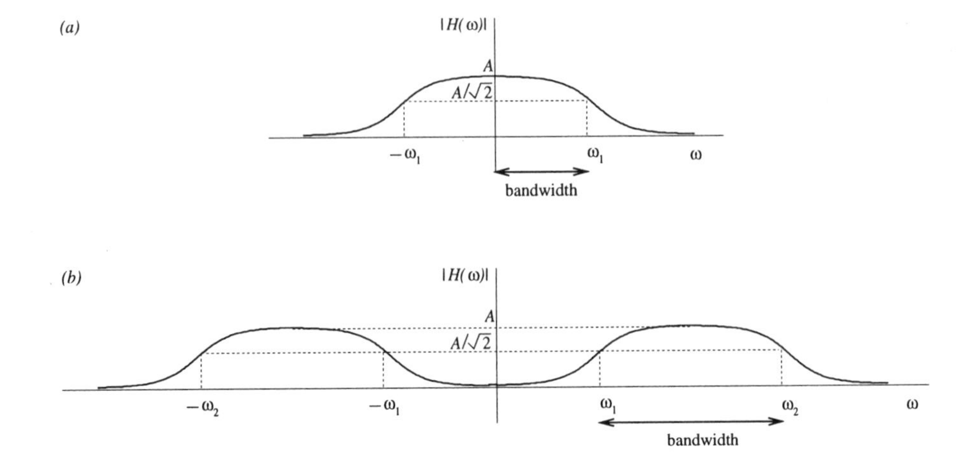

Figure 3.20 *Definition of bandwidth: (a) low-pass system; (b) bandpass-system.*

Bode plots and is illustrated in Figure 3.8(a). Its bandwidth is $1/\tau$ which in this case is $1/RC$ rad/s. Thus the relationship between rise time and bandwidth for a first order system is a simple one, i.e. the rise time is the reciprocal of the bandwidth. If a faster response (a smaller rise time) is required, more bandwidth is needed, and vice versa. This is a useful intuitive relationship. However, in the real world more complicated systems exist, and things may not be quite that simple.

Second order systems exhibit other characteristics which are more typical of the higher order systems that are encountered in practice. However, unlike higher order systems, there are some simple relationships which aid in understanding the trade-offs which are possible in a second order system. For an underdamped second order system, the peak time is a function of the undamped natural frequency and the damping factor:

$$T_p = \frac{\pi}{\omega_0 \sqrt{1-\zeta^2}} \text{ s}$$

Similarly the percentage overshoot (PO) is given by:

$$\text{PO} = 100 \exp\left(\frac{-\zeta \pi}{\sqrt{1-\zeta^2}} \right)$$

Figure 3.21 illustrates the general form of the frequency response of a second order system. As before, there is the undamped natural frequency ω_0 and the resonant frequency ω_R which is associated with the peak in the frequency response. The resonant frequency is less than the undamped naturally frequency. Finally, there is the bandwidth ω_B whose exact position will depend on the damping factor. The following heuristic argument illustrates two important trade-offs which are possible in a second order system:

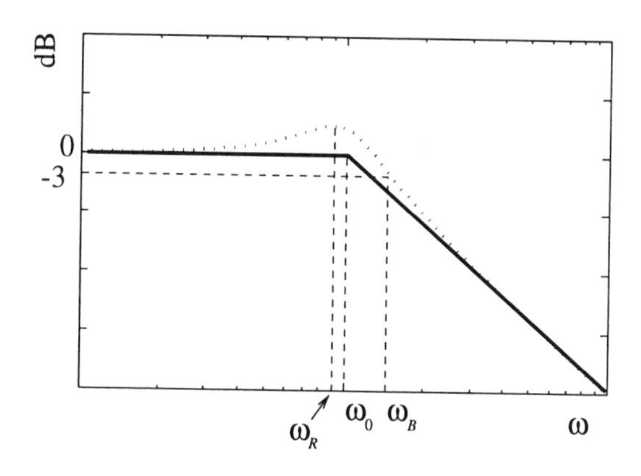

Figure 3.21 *Frequency response of second order system.*

- *Case (i), constant damping.* As a first step, consider the case where the damping factor ζ is held constant. If we increase bandwidth, ω_B rad/s, we also increase the natural frequency, ω_0 rad/s. In increasing the natural frequency we reduce the peak time. In reducing the peak time we reduce the rise time, T_r seconds. So for second order systems with damping factor held constant we get the same results as for first order systems, i.e. the rise time is inversely proportional to the bandwidth. To get a faster response you need more bandwidth.
- *Case (ii), constant bandwidth.* If we wished to hold the bandwidth, ω_B, constant we could do this (approximately) by holding the undamped natural frequency ω_0 constant. In the underdamped condition we could then reduce the peak time, T_p, by reducing the damping factor, ζ. Reducing the peak time would also reduce the rise time, T_r. The net effect is that you can speed up the response without increasing the bandwidth by reducing the damping factor. The disadvantages are more overshoot in the time domain and a bigger resonant peak in the frequency domain.

Thus, first order systems indicate that bandwidth is inversely proportional to rise time. For higher order system this relationship also holds but we can also achieve a faster response within the same bandwidth. The price to be paid for the latter is more overshoot and oscillation in the step response.

3.7 Summary

The theme of this chapter has been connections and relationships between the alternative descriptions of a linear continuous time system. Figure 3.22 summarises the interrelationships between four alternative descriptions of a linear system.

Many linear systems can be described by differential equations. Taking the Laplace transform of such equations and performing some simple algebraic manipulation yields the system transfer function $H(s)$. If this transfer function is expressed as a ratio of polynomials in s, the roots of the numerator polynomial are the zeros and the roots of

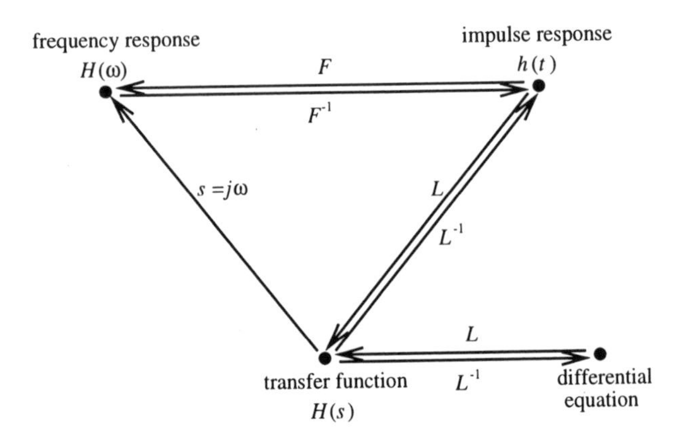

Figure 3.22 *Hitchhiker's guide to system response.*

the denominator polynomial are the poles. The similarity between the Fourier and Laplace transforms leads to a straightforward way of obtaining the frequency response $H(\omega)$ from the transfer function $H(s)$, i.e. replace s with $j\omega$. This substitution, combined with the properties of complex numbers, makes it possible to infer the nature of the frequency response from a pole/zero diagram and are also the foundations of the Bode diagram. Both forms of graphical representation are useful in providing insight into how the position of the poles and zeros affects the frequency response of a linear system.

The poles of a linear system define the denominators in a partial fraction expansion of the transfer function and, as Figure 3.22 suggests, are closely related to the impulse response of the system. Hence a system will be stable provided all of its poles lie in the left half of the s-plane.

The speed of response of a linear system to sudden change is usually quantified through its step response in terms of rise time, percentage overshoot and settling time. Both first and second order systems demonstrate the general trend that rise time is the reciprocal of bandwidth. However, in second and higher order systems, other trade-offs are possible. In particular, it is possible to reduce rise time while maintaining bandwidth. This is achieved, as has been demonstrated for second order systems, at the expense of greater percentage overshoot.

3.8 Problems

3.1 With reference to problem 1.7, draw a pole/zero map for each circuit.

3.2 Draw the pole/zero maps and hence sketch the frequency responses of a system with the following transfer functions:

$$H(s) = \frac{s^2 - 0.4s + 400.04}{(s^2 + 2s + 101)(s^2 + 2s + 901)}$$

3.3 If a 1 kHz sine wave is applied to the circuit of problem 2.3, what is the amplitude and phase of the output with respect to the input after all transients have died away. [Amplitude 5/6; phase 180°]

3.4 A radar waveform is generated using the mechanism shown below. Sketch the time-domain waveforms and the magnitude Fourier transform of the waveforms. The frequency of the sine wave (carrier frequency) is 10 kHz and the pulse duration $\tau = 1$ ms. The pulse repetition period T used in the system is 10 ms.

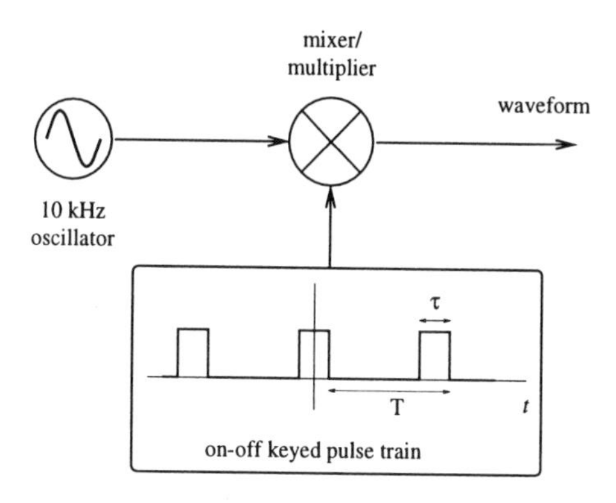

3.5 The network shown in (a) of the figure below is known as a *shunt peaking* network. Show that the impedance has the form

$$Z(s) = \frac{V(s)}{I(s)} = \frac{K(s - z_1)}{(s - p_1)(s - p_2)}$$

and determine z_1, p_1 and p_2 in terms of R, L and C. If the poles and zero of $Z(s)$ have the locations shown in (b) of the figure with $Z(0) = 1$, find the values for R, L and C. [1 Ω; 1/3 H; 1/10 F]

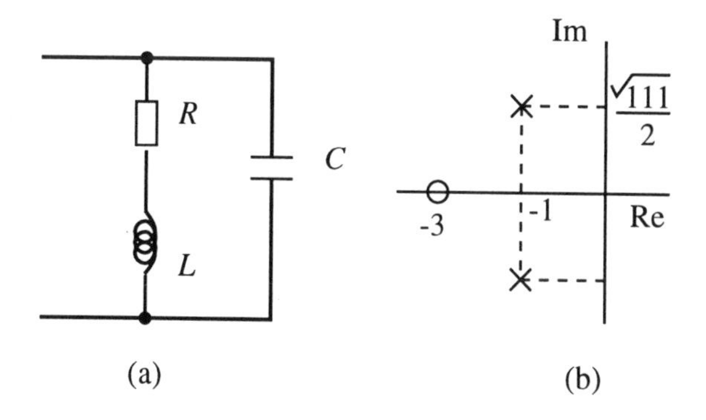

3.6 The transfer function of a linear analogue system is given by the following expression:

$$H(s) = \frac{2500\,(s + 10)}{(s^2 + 2s)\,(s^2 + 30s + 2500)}$$

Draw the magnitude and phase Bode plots for the system.

3.7 For the circuit shown below, calculate the rise time and determine the relationship between the rise time and the bandwidth for circuits of this type.

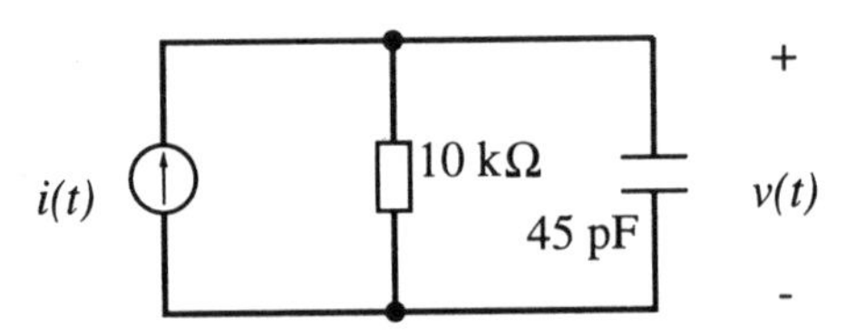

[Rise time = 1.04 μs]
Find the rise time of the 'shunt peaking' circuit shown below for values of $L = 0.5$ mH and $L = 2.0$ mH.

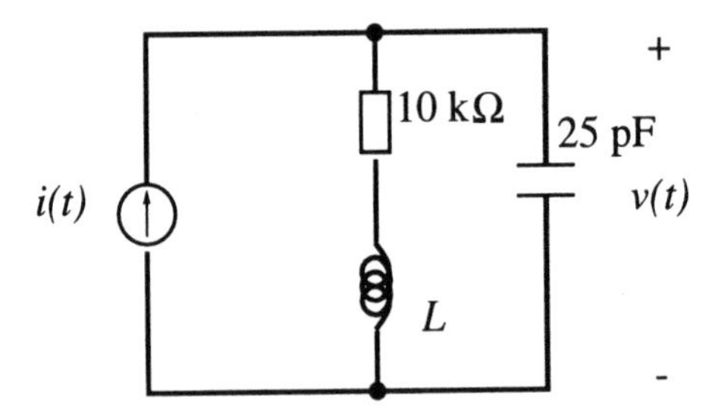

[Rise time = 0.46 μs ($L = 0.5$ mH); rise time = 0.27 μs ($L = 2.0$ mH); – both approximate]

CHAPTER 4

Sampled data systems and the z-transform

4.1 Introduction

Many naturally occurring signals of interest are analogue or continuous-time. Analogue filters can be used to process or extract information from such signals. As shown in Figure 4.1(a), a continuous-time input signal, $x(t)$, is applied to a linear filter to produce a continuous-time output signal, $y(t)$. The filter is completely characterised by its transfer function (as in Chapter 1) or equivalently by its impulse response (as in Chapter 2).

There are two major problems associated with analogue filters: (i) the values of the components, such as the resistors and capacitors, can drift with temperature and age – thus a piece of equipment may work well when it is first constructed but may become

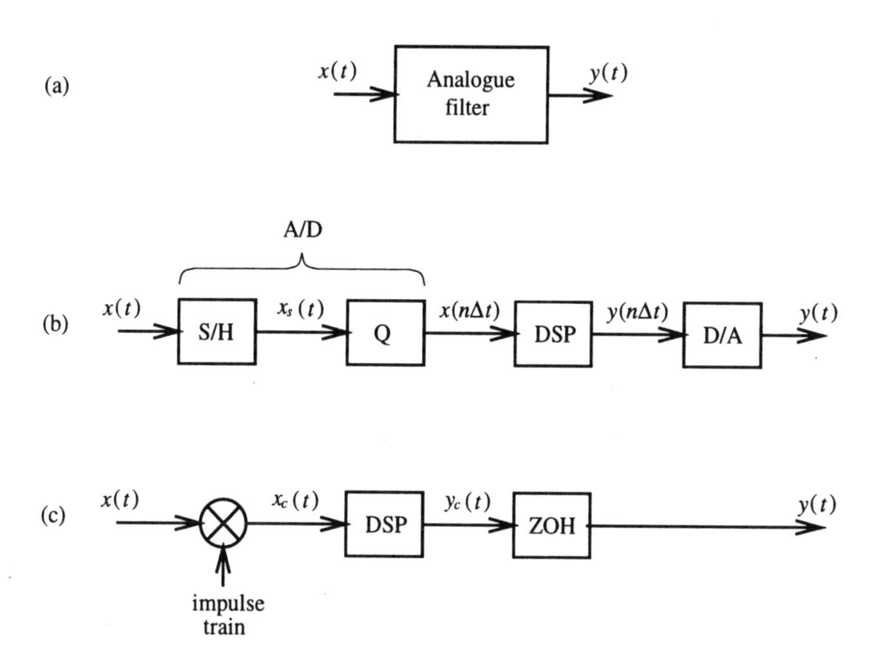

Figure 4.1 *From analogue to digital: (a) analogue filter; (b) sampled data equivalent; (c) mathematical model of sampled data system.*

less than perfect after years of use or harsh conditions; (ii) the functions that the designer can specify are limited to those which map directly to analogue circuit building blocks – thus high-pass and low-pass filters are readily available but adaptive filtering (discussed in Chapter 10) is extremely difficult to realise because of the lack of an analogue memory unit. It is these difficulties, plus the availability of large scale integration techniques, which make digital or (more precisely) discrete-time signal processing attractive. Figure 4.1(b) illustrates a sampled data system which might be used to replace the purely analogue system of Figure 4.1(a). Again the input is the continuous-time signal, $x(t)$, and the output is the continuous-time signal, $y(t)$. A typical example of the input is illustrated in Figure 4.2(a).

Before the continuous-time input can be processed digitally it must be converted to a sequence of binary numbers. This is usually a two stage process; (i) sampling; (ii) quantisation. Sampling may be performed by a sample-and-hold (S/H) device which takes a snapshot of the analogue signal every Δt seconds and then holds that value constant for Δt seconds until the next snapshot is obtained. This produces the characteristic 'staircase' waveform illustrated in Figure 4.2(b). If the original signal is $x(t)$ then the 'heights' of the 'steps' on the 'staircase' are obtained by simply setting $t = n\Delta t$ where $n = 0, 1, 2, \ldots$ etc. This produces a sequence of numbers $x(n\Delta t)$ where $\{.\}$ denotes a sequence, i.e.:

$$\{x(n\Delta t)\} = \{x(0),\, x(\Delta t),\, x(2\Delta t),\, \cdots\, x(n\Delta t)\, \cdots\}$$

During the Δt seconds between samples, a second device known as a quantiser (Q in Figure 4.1(b)) calculates the nearest binary number which can approximate the particular analogue voltage level $x(n\Delta t)$. An input–output characteristic for a 4-bit quantiser is illustrated in Figure 4.3. The quantisation is, by its very nature, an approximation and introduces errors or noise to the signal as the quantised values must fall on the staircase in Figure 4.3. The error between the closest staircase value and the original input $x(n\Delta t)$ is unavoidable but it can be reduced by employing more steps to the staircase, i.e. more bits to the quantiser. A more detailed discussion of quantisation is left to Chapter 7. For the rest of this chapter, the effects of quantisation will be ignored.

The S/H plus the quantiser together form the analogue-to-digital (A/D) converter, whose input is $x(t)$ and whose output is the sequence of numbers $x(n\Delta t)$. The heights of these numbers, as illustrated in Figure 4.2(c), are proportional to the values of the signal at times $t = n\Delta t$. The digital signal processing (DSP) unit reads in each number $x(n\Delta t)$, does some calculations based on it and other variables stored in memory, and then writes the number $y(n\Delta t)$ out to the digital-to-analogue (D/A) converter. It repeats this process every Δt seconds and thus generates an output sequence $y(n\Delta t)$, as illustrated in Figure 4.2(d). The DSP can take many forms: a general purpose microprocessor; a specialised microprocessor; a dedicated piece of VLSI circuitry. The numbers in the sequence $y(n\Delta t)$ are written in turn to the D/A converter every Δt seconds. The D/A produces an analogue voltage at its output which is proportional to a particular sample $y(n\Delta t)$ and holds the output at that value until it receives the next number in the sequence. It thus produces another 'staircase' waveform, as illustrated in Figure 4.2(e).

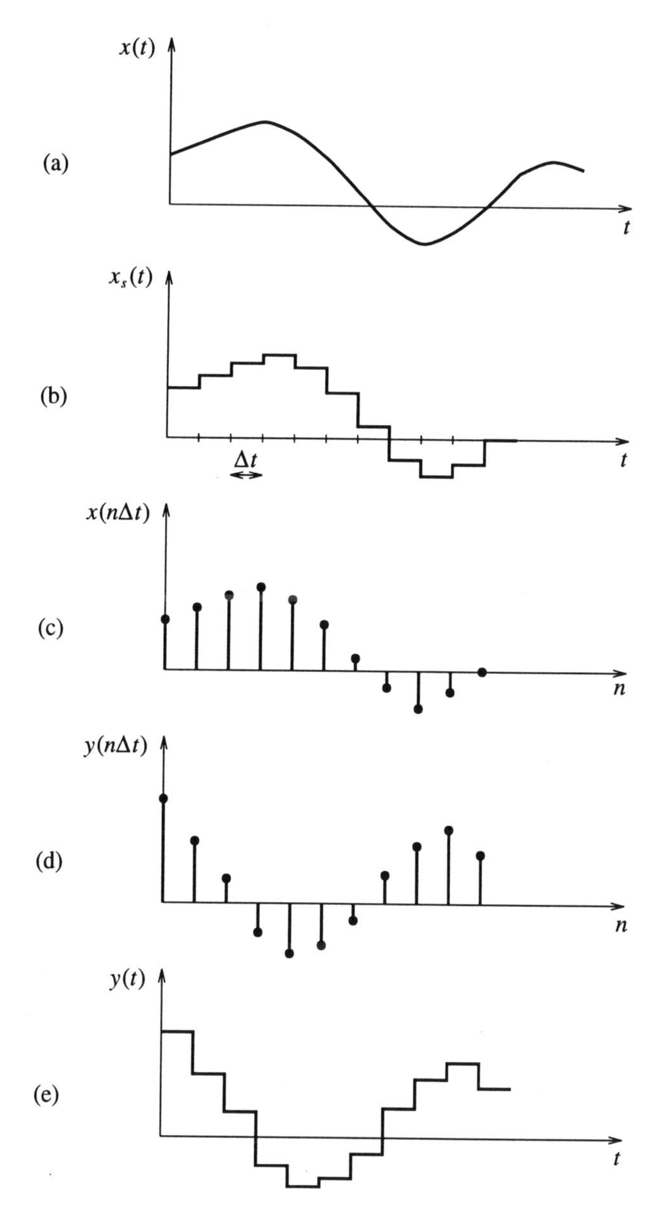

Figure 4.2 *Signals present in sampled data system: (a) analogue input $x(t)$; (b) output from S/H; (c) input data sequence $\{x(n\Delta t)\}$; (d) output data sequence $\{y(n\Delta t)\}$; (e) analogue output from D/A converter.*

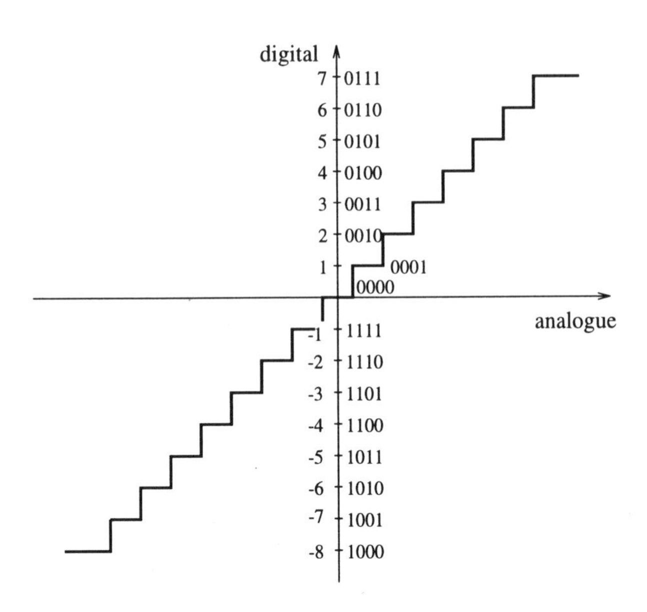

Figure 4.3 *Input–output characteristic of 4-bit quantiser (two's complement notation).*

While Figure 4.1(b) and the associated waveforms of Figures 4.2 are not characteristic of every sampled data system, they are characteristic of many and this provides the basis for discussion of the key concepts associated with such systems. In this chapter, initial consideration will be given to the sampling process and the effects it has on signal fidelity. This is followed by the development of the z-transform in section 4.3. The importance of the z-transform is that it plays the same role in digital filtering as the Laplace transform plays in analogue filtering. This leads directly to a detailed discussion of digital filters and discrete convolution in section 4.4. Many concepts used to describe continuous-time signals and systems have a direct parallel in discrete-time signal processing. Thus there is no necessity to learn a completely new set of techniques when discrete-time systems are met for the first time. In section 4.5 it is demonstrated that the poles of a digital filter are as critical in assessing the stability of a digital filter as they were for the stability of an analogue one – the only difference being that in the former it is the poles of the z-transform transfer function rather than the Laplace transfer which are being considered. In section 4.6 the relationship between the z-transform transfer function and the frequency response of a digital filter is explored. Again parallels can be drawn with the discussion in section 3.3. Finally, a complete example of the analysis of a simple practical system is presented in section 4.7.

4.2 Sampled data systems and aliasing

While the block diagram of Figure 4.1(b) is a fairly accurate description of a practical system, it is not very tractable when an analysis of the system's effect on the input signal is required or design tools are needed. From the perspective of analysis and design of the DSP unit, the mathematical model of Figure 4.1(c) has proven to be much more useful for analysis and design. The sampling process of Figure 4.1(b) is replaced by a multiplication process in Figure 4.1(c). The input signal, $x(t)$, is multiplied by a periodic impulse train of period $\delta\Delta t$. Thus the impulse train $\delta_T(t)$ is defined as:

$$\begin{aligned}\delta_T(t) &= \cdots + \delta(t+2\Delta t) + \delta(t+\Delta t) \\ &\qquad + \delta(t) + \delta(t-\Delta t) + \delta(t-2\Delta t) + \cdots \\ &= \sum_{n=-\infty}^{\infty} \delta(t-n\Delta t)\end{aligned}$$

This is illustrated in Figures 4.4(a) to (c). When the continuous-time input is multiplied by the impulse train, the 'sampled signal', $x_c(t)$, is obtained:

$$\begin{aligned}x_c(t) &= x(t)\ \delta_T(t) \\ &= x(t)\ [\delta(t) + \delta(t-\Delta t) + \delta(t-2\Delta t) + \cdots +] \\ &= x(t)\ \delta(t)\ +\ x(t)\ \delta(t-\Delta t) + x(t)\ \delta(t-2\Delta t) \cdots \\ &= x(0)\ \delta(t)\ +\ x(\Delta t)\ \delta(t-\Delta t) + x(2\Delta t)\ \delta(t-2\Delta t) \cdots \\ &= \sum_n x(n\Delta t)\ \delta(t-n\Delta t)\end{aligned}$$

Thus $x_c(t)$ is a train of impulses whose strength is proportional to the signal at $t = n\Delta t$. This is illustrated in Figure 4.4(c) by using the height of the impulse to indicate its strength. Compare this with Figure 4.2(c) which illustrates the sequence of numbers in the actual system. The output of the DSP is also assumed to be an impulse train (Figure 4.4(d)) whose strengths are proportional to the size of the numbers at the output of the actual DSP, as illustrated in Figure 4.2(d). To produce the overall output $y(t)$ from the impulse train $y_c(t)$, all that is required is to apply $y_c(t)$ to a simple analogue filter known as a zero-order-hold (ZOH) whose impulse response $h_{ZOH}(t)$ is defined as:

$$h_{ZOH}(t) = \begin{cases} 1, & 0 \leq t < \Delta t \\ 0, & \text{otherwise} \end{cases} \tag{4.1}$$

This is illustrated in Figure 4.5. Since the ZOH is a simple filter it is straightforward to calculate its frequency response by taking the Fourier transform of its impulse response, i.e.:

$$\begin{aligned}H_{ZOH}(\omega) &= F[h_{ZOH}(t)] \\ &= \Delta t \exp\left(-\frac{j\omega\Delta t}{2}\right) \mathrm{sa}\left(\frac{\omega\Delta t}{2}\right)\end{aligned} \tag{4.2}$$

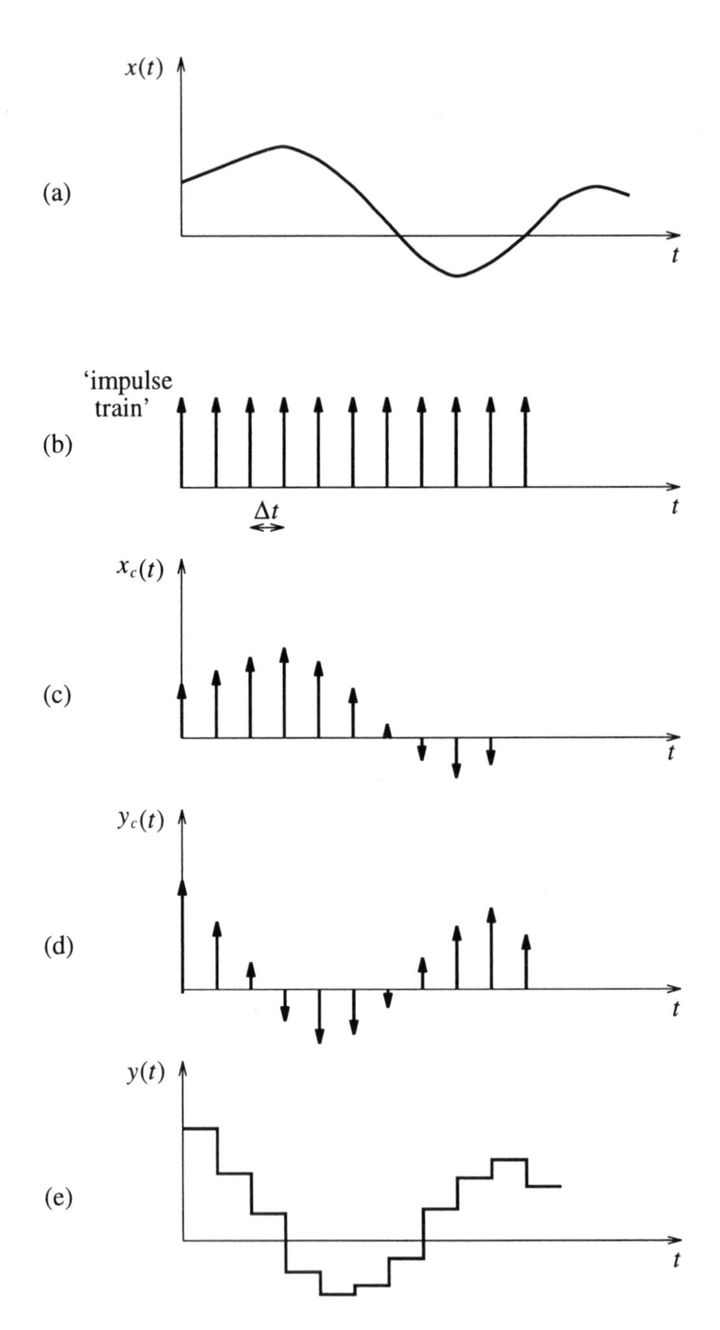

Figure 4.4 *Signals present in mathematical model of sampled data system: (a) analogue input, $x(t)$; (b) periodic impulse train, $\delta_T(t)$; (c) sequence of impulses at input to DSP, $x_c(t)$; (d) sequences of impulses at output of DSP, $y_c(t)$; (e) analogue output from zero-order-hold (ZOH).*

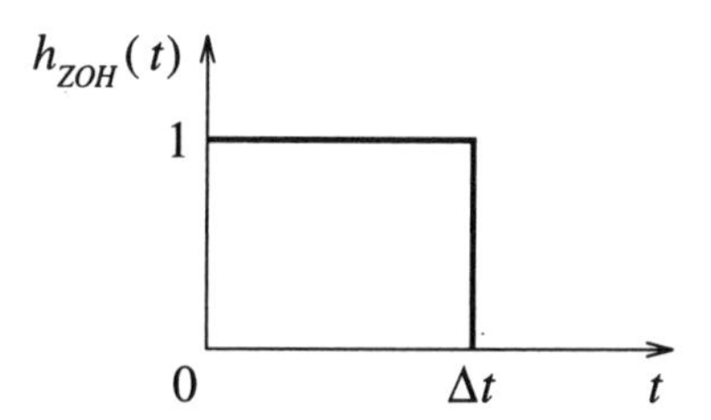

Figure 4.5 *Impulse response of ZOH.*

The amplitude and phase response of the ZOH are illustrated in Figure 4.6. Further, the ZOH is a linear filter and hence the output $y(t)$ is the convolution of $y_c(t)$ with $h_{ZOH}(t)$, i.e.:

$$
\begin{aligned}
y(t) &= h_{ZOH}(t) * y_c(t) \\
&= h_{ZOH}(t) * \sum_n y(n\Delta t)\ \delta(t - n\Delta t) \\
&= \sum_n y(n\Delta t)\ [h_{ZOH}(t) * \delta(t - n\Delta t)] \\
&= \sum_n y(n\Delta t)\ h_{ZOH}(t - n\Delta t)
\end{aligned}
$$

This is the staircase waveform illustrated in Figure 4.4(e).

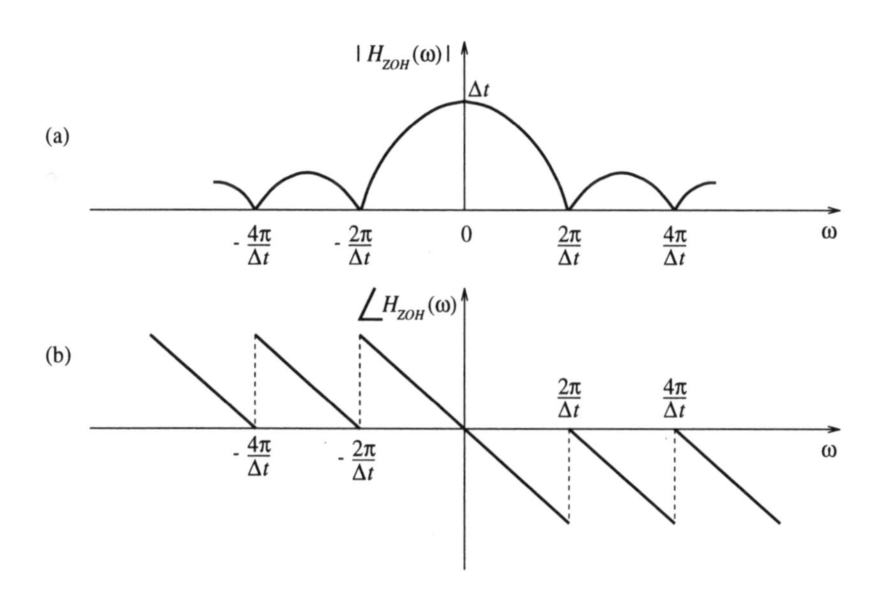

Figure 4.6 *Frequency response of ZOH: (a) amplitude; (b) phase.*

Comparison of Figures 4.2 and 4.4 shows that the input and output waveforms $x(t)$ and $y(t)$ are identical. Thus, although the mathematical model of Figure 4.1(c) is very different in structure from Figure 4.1(b), it has an identical effect on the input signal. The advantage of the mathematical model is that all its elementary functions are defined in terms of analogue signals and thus the techniques defined in Chapters 1 to 3 can be used to analyse them. In fact, apart from the multiplication operation required to model the sampling processing, all other operations can be defined in terms of linear filtering functions. This even includes the operation of the D/A which is modelled by the zero-order-hold (ZOH) filter. The mathematical model can be used to address the effect of the sampling process on signal fidelity. To do this, a simple example is considered where the DSP does not alter the signal samples but merely reads signal samples from the A/D and writes them directly to the D/A. Under what conditions is no information lost or corrupted in this processing of the input? To answer this question, a frequency-domain approach will be adopted.

The sampled signal is obtained by multiplication in the time domain, i.e.:

$$x_c(t) = x(t)\, \delta_T(t) \tag{4.3}$$

Thus the Fourier transform of the sampled signal can be obtained by convolution in the frequency domain using equation (2.15):

$$\begin{aligned} X_c(\omega) &= F[x_c(t)] \\ &= \frac{1}{2\pi}\, X(\omega) * F[\delta_T(t)] \end{aligned} \tag{4.4}$$

The second term on the right-hand side is the Fourier transform of an impulse train $\delta_T(t)$ which is defined as:

$$F[\delta_T(t)] = \frac{2\pi}{\Delta t} \sum_n \delta\left(\omega - \frac{2\pi n}{\Delta t}\right)$$

i.e. a train of impulses in the frequency domain. This is illustrated in Figure 4.7. Note that reducing the value of Δt moves the impulses closer together in the time domain but increases the separation in the frequency domain. Using the property from Chapter 2 that multiplication in the time domain is equivalent to convolution in the frequency domain gives:

$$\begin{aligned} X_c(\omega) &= \frac{1}{\Delta t}\, X(\omega) * \sum_n \delta\left(\omega - \frac{2\pi n}{\Delta t}\right) \\ &= \frac{1}{\Delta t} \sum_n X(\omega)\ \delta\left(\omega - \frac{2\pi n}{\Delta t}\right) \\ &= \frac{1}{\Delta t} \sum_n X\left(\omega - \frac{2\pi n}{\Delta t}\right) \end{aligned}$$

A simple example of this process is illustrated in Figure 4.8. For this example, the

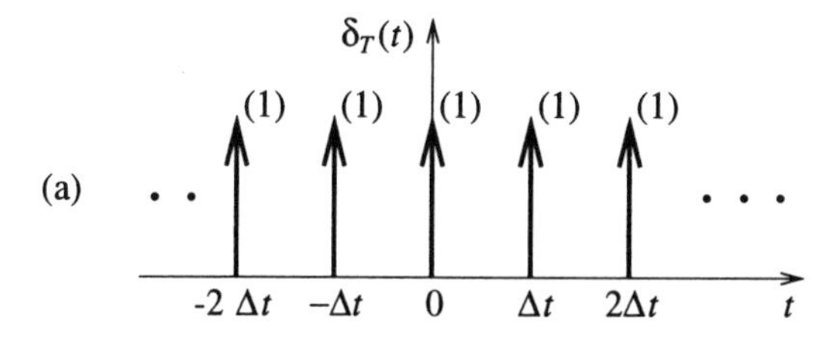

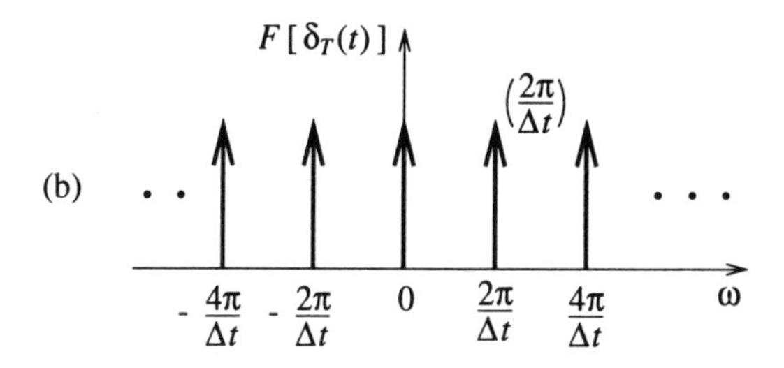

Figure 4.7 *Periodic impulse train (a) and its Fourier transform (b).*

Fourier transform of the analogue signal is assumed to be triangular in shape with a maximum frequency component at ω_m rad/s. To obtain the Fourier transform of the sampled signal we convolve $X(\omega)$ in turn with each of the impulses in $F[\delta_T(t)]$. First $X(\omega)$ is convolved with the impulse at $\omega = 0$ to form the primary component at $\omega = 0$ in $X_c(\omega)$. Then $X(\omega)$ is convolved with the impulse at the sampling frequency, i.e. at $\omega = 2\pi/\Delta t$ rad/s, to produce the secondary component at $\omega = 2\pi/\Delta t$ rad/s. The process is repeated for all impulses at positive and negative frequencies to produce all the complementary components. Note in particular that $X_c(\omega)$ is a periodic function of ω with period $2\pi/\Delta t$. This is a general property of the Fourier transform of sampled sequences. For this particular example, the sampling rate in rad/s $(2\pi/\Delta t)$ has been chosen to be much larger than the highest frequency present, i.e.:

$$\frac{2\pi}{\Delta t} \gg \omega_m$$

or equivalently in Hz:

$$\frac{1}{\Delta t} \gg \frac{\omega_m}{2\pi}$$

The obvious question to ask is: 'is it possible to reconstruct the original signal, $x(t)$, from the sampled signal, $x_c(t)$?'. If the answer is affirmative then nothing has been lost in the sampling process, i.e. all the information in $x(t)$ is also present in $x_c(t)$. Figure 4.9 illustrates how this can be done. To recover the original signal from the sampled signal all that is required is that $x_c(t)$ should be applied to an ideal low-pass filter with frequency response $H_L(\omega)$, i.e.:

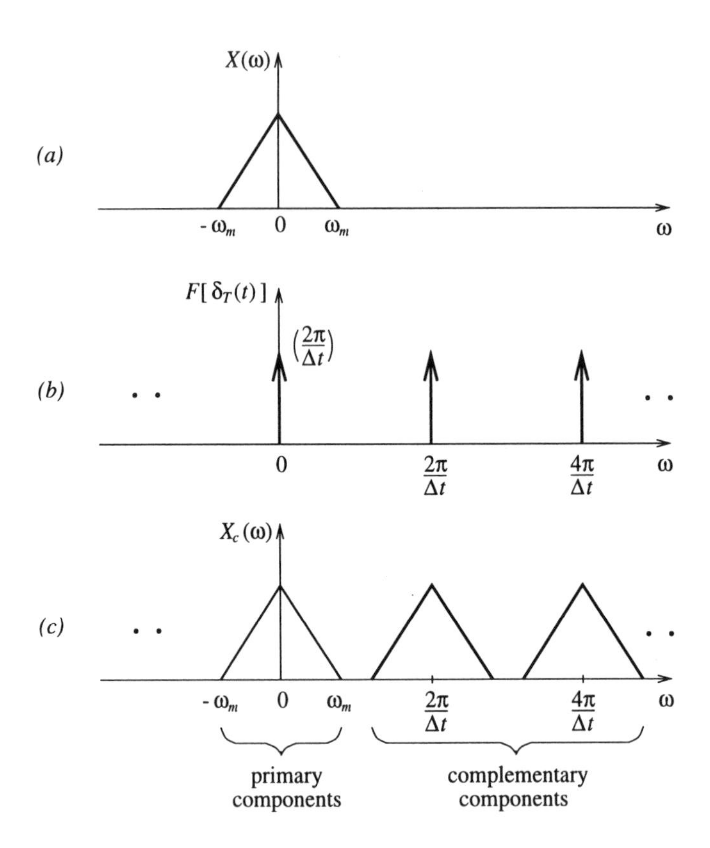

Figure 4.8 *Frequency content of a signal sampled at an adequate rate: (a) Fourier transform of analogue input; (b) Fourier transform of impulse train; (c) Fourier transform of sampled signal.*

$$X(\omega) = H_L(\omega)\ X_c(\omega)$$

Figure 4.9(b) illustrates the Fourier transform of the sampled signal. To obtain the Fourier transform of the filter output we simply multiply the frequency response of the LPF by $X_c(\omega)$. This ideal LPF with bandwidth ω_m removes the complementary components and isolates the primary component. Thus, in theory, we can recover the original signal from the sampled signal provided the sampling rate is much higher than the highest frequency present in the original signal.

In Figure 4.10 the case of a 'just adequate' sampling rate is considered. As before, $X(\omega)$ is convolved with $F[\delta_T(t)]$ to obtain $X_c(\omega)$. This time however we are considering the special case where the sampling rate is exactly equal to twice the highest frequency, i.e.:

$$\frac{2\pi}{\Delta t} = 2\omega_m$$

As before, $X_c(\omega)$ contains a set of images of $X(\omega)$ repeated at $\omega = 0$, $\omega = 2\pi/\Delta t$, $\omega = 4\pi/\Delta t$, etc. In this case the edges of these images just touch at $\omega = \pi/\Delta t$,

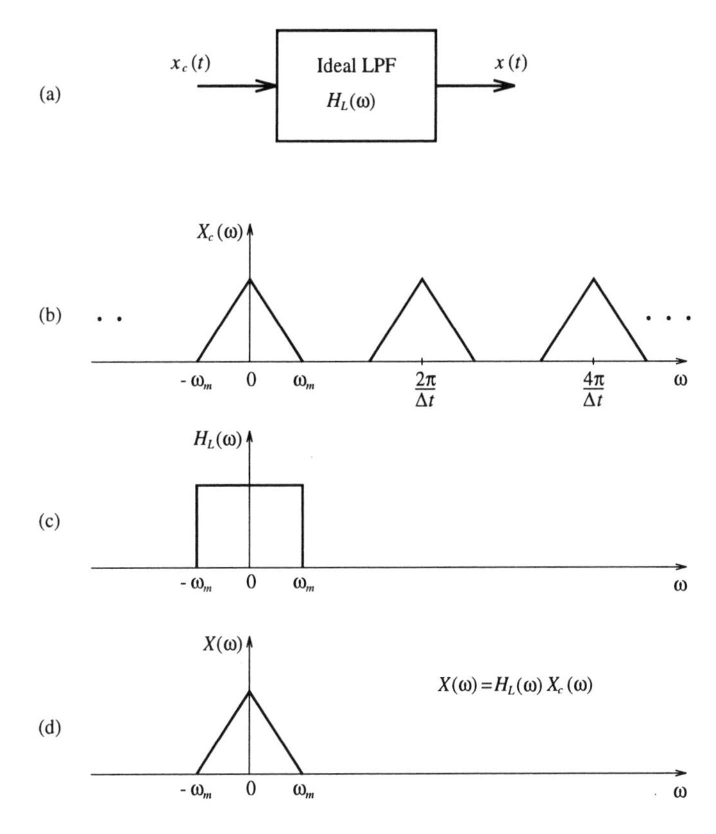

Figure 4.9 *Analogue signal reconstruction from a sampled signal: (a) low-pass filter; (b) frequency content of sampled signal; (c) frequency response of ideal low-pass filter; (d) frequency content of output from low-pass filter – perfect signal reconstruction.*

$\omega = 3\pi/\Delta t$, etc. It is still theoretically possible to reconstruct the original signal by applying an ideal low-pass filter to the sampled signal.

Figure 4.11 illustrates the case where the sampling rate is too low to adequately represent $x(t)$. As in the previous two examples, $X(\omega)$ is convolved in turn with each of the impulses in $F[\delta_T(t)]$. This time however the images of $X(\omega)$ in $X_c(\omega)$ overlap. Since convolution is a linear operation, the images will simply add when they overlap. Note however that since $X(\omega)$ will in general be complex, the addition will be the addition of complex numbers. When $X(\omega)$ is applied to a low-pass filter, the result is a distorted version of $X(\omega)$. This *aliasing* in the frequency domain is caused by not sampling fast enough (undersampling) in the time domain. The maximum frequency present in the analogue signal is greater than half the sampling frequency.

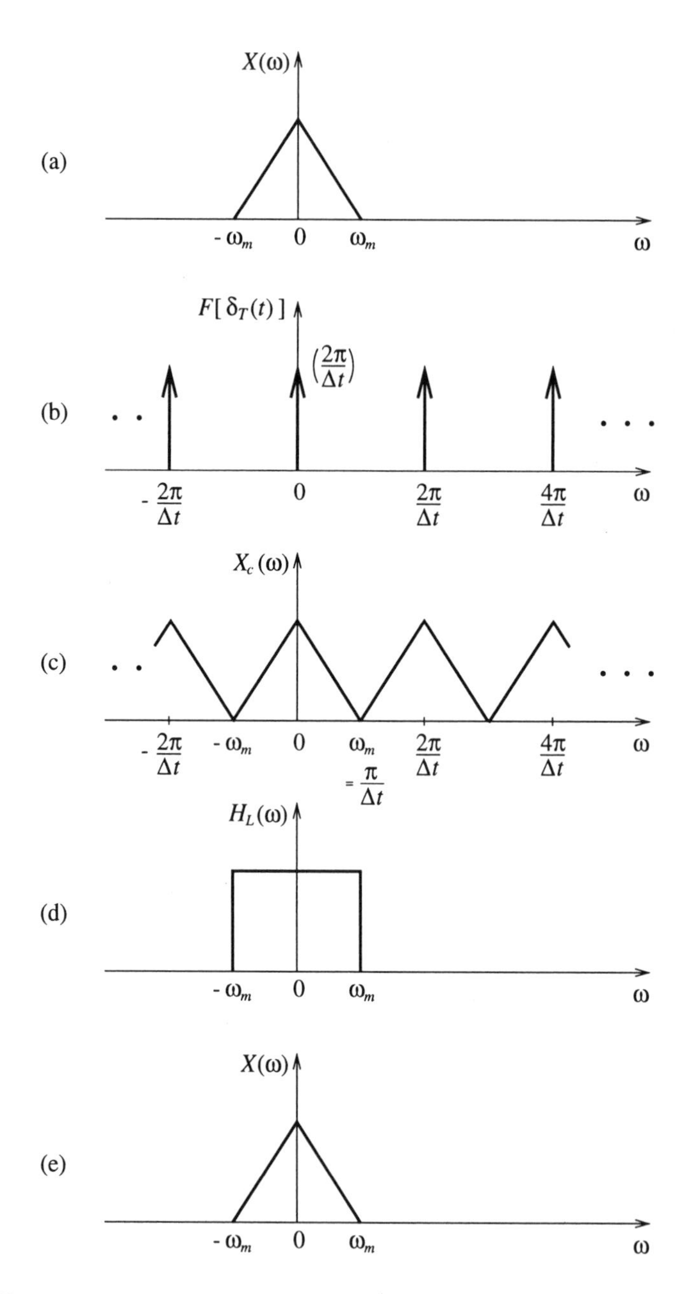

Figure 4.10 *Frequency content of a signal sampled at barely adequate rate: (a) Fourier transform of analogue input; (b) Fourier transform of impulse train; (c) Fourier transform of sampled signal; (d) frequency response of ideal low-pass filter; (e) frequency content of output from low-pass filter.*

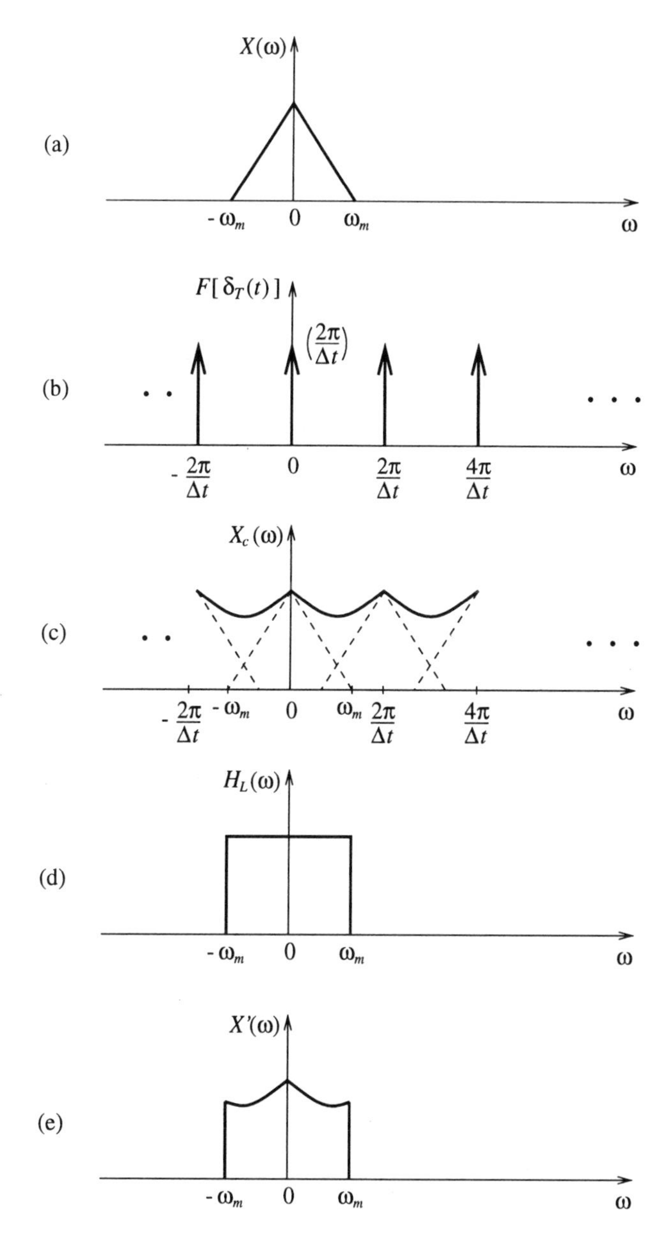

Figure 4.11 *Aliasing – frequency content of a signal sampled at an inadequate rate: (a) Fourier transform of analogue input; (b) Fourier transform of impulse train; (c) Fourier transform of sampled signal, (d) frequency response of ideal low-pass filter; (e) frequency content of output from low-pass filter – poor reconstruction.*

4.2.1 Sampling theorem – the Nyquist criterion

Thus, in order to prevent aliasing distortion, sampling must occur at a frequency greater than twice the highest frequency present. The sampling theorem can be stated more formally as:

- to prevent aliasing distortion, a bandlimited signal of bandwidth ω_B rad/s must be sampled at a rate of at least $2\omega_B$ rad/s.

where 'a bandlimited signal of bandwidth ω_B' is a signal whose Fourier transform is zero at all values of ω except in a region of width ω_B for positive values of ω (and a corresponding region for negative values of ω).

4.2.2 Practical sampled data systems

Provided that the sampling rate is fast enough to meet the Nyquist criterion, it is theoretically possible to reconstruct the original analogue signal from the sampled signal using an ideal low-pass filter. In practice, this perfect reconstruction is not possible for two reasons: (i) as will be seen in Chapter 5, it is impossible to construct an ideal LPF from a finite number of components; (ii) in a real system where an analogue output is required, a D/A converter is essential – this D/A converter is modelled as a ZOH whose frequency response, $H_{ZOH}(\omega)$ as illustrated in Figure 4.6, is far from that of the ideal LPF of Figure 4.10(d). Thus in the complete block diagram of Figure 4.12, the D/A converter is followed by an analogue low-pass reconstruction filter with a nominal cut-off of $1/2\Delta t$ Hz. These filters are usually designed using standard analogue techniques and are frequently Butterworth designs (described in Chapter 5).

When designing a sampled data system it is not always possible to ensure that the Nyquist criterion is met. Often it may be difficult to ascertain what precisely is the highest frequency present in the analogue signal. The obvious solution is to place an analogue LPF prior to the S/H whose function is to remove or attenuate all frequencies above half the sampling frequency. This is illustrated in Figure 4.12 and is discussed again in Chapter 12. Practical analogue filter designs will not remove the frequency components above $1/2\Delta t$ completely, rather they will attenuate or strongly attenuate them depending upon the complexity of the analogue filter used. Thus some degree of aliasing is to be expected in practical systems.

Self assessment question 4.1: A 7 kHz sine wave is sampled at 10 kHz. What frequencies are present at the output of the A/D?

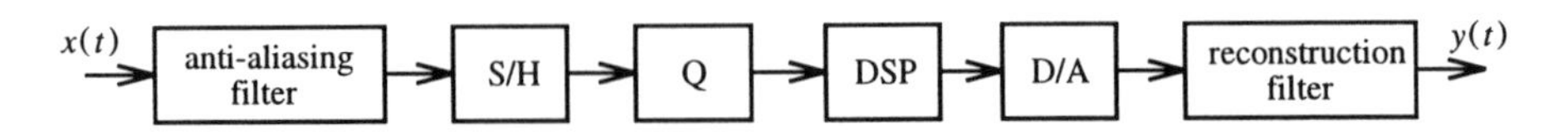

Figure 4.12 *Block diagram of practical sampled data system.*

4.3 The z-transform

In this section the z-transform method, which is an essential tool when analysing and designing the DSP element in a sampled data system, is developed. In the analysis and design of continous-time analogue systems, Laplace and Fourier transforms play a key role. In the analysis and design of discrete-time digital systems, the Laplace transform is replaced by the z-transform. In attempting to apply Laplace techniques to sampled data systems, problems are encountered. Solving these problems leads to the z-transform. The manipulation of the z-transform is very similar to the manipulation of the s-transform in analogue systems.

Start with the sampled analogue signal of section 4.2 and evaluate its Laplace transform. A sampled analogue signal has the form:

$$x_c(t) = \sum_{n=0}^{\infty} x(n\Delta t)\,\delta(t - n\Delta t)$$

Take Laplace transforms:

$$X_c(s) = L[x_c(t)]$$

$$= \sum_{n=0}^{\infty} x(n\Delta t)\exp(-n\Delta ts)$$

The problem is that $X_c(s)$ is periodic in that

$$X_c(s) = X_c\left(s + \frac{2\pi j}{\Delta t}\right)$$

This periodic nature causes problems when evaluating the inverse Laplace transform. Usually a partial fraction expansion is performed based on a finite number of poles – roughly one term per pole. The next step is to look up the inverse of each term. Unfortunately $X_c(s)$ cannot have a finite number of poles. If it has a pole at A it will also have poles at B, C, D, etc. – an infinite number, as illustrated in Figure 4.13. Recall that the Fourier transform of $x_c(t)$ is also periodic in frequency (repeating at the sampling frequency). The solution to the problem is to make a simple substitution, i.e. let:

$$z = \exp(\Delta t\ s)$$

This can be viewed as either a mapping of all points in the s-plane to points in a z-plane, as illustrated in Figure 4.14, or as a convenient short-hand to save writing down the exp term many times. This defines the z-transform:

$$X(z) = X_c(s)\,|_{\exp(\Delta t\ s) = z}$$

Thus:

$$X(z) = \sum_{n=0}^{\infty} x(n\Delta t)\,z^{-n} \qquad (4.5)$$

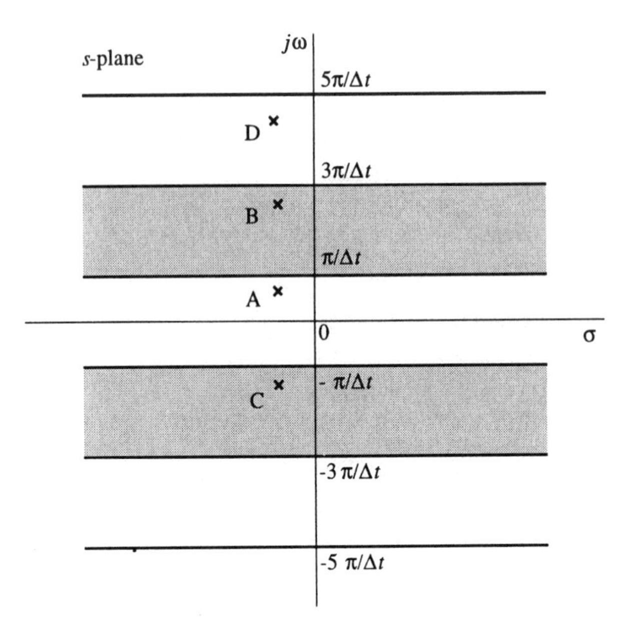

Figure 4.13 *Regions of periodicity of $X_c(s)$.*

It can be summarised as follows:

(i) The analogue signal, $x(t)$, is sampled by multiplying it by the impulse train, $\delta_T(t)$, to give $x_c(t)$:

$$x_c(t) = x(t)\ \delta_T(t)$$

(ii) Take the Laplace transform of $x_c(t)$ to give $X_c(s)$:

$$X_c(s) = L[x_c(t)]$$

(iii) Replace $\exp(\Delta t\ s)$ by z in $X_c(s)$ to get $X(z)$:

$$X(z) = X_c(s)\ |_{\exp(\Delta t\ s) = z}$$

This particular interpretation of the z-transform is very important when dealing with systems which have mixtures of analogue and digital processing. Many people (in particular control engineers) talk about the z-transform of an analogue signal such as $x(t)$. What they actually mean are the above three steps.

Alternatively if only a sequence of numbers $\{x(n\Delta t)\}$ exists where $n = 0, 1, 2$, etc., the z-transform is defined as:

$$X(z) = Z[\{x(n\Delta t)\}]$$

$$= \sum_{n=0}^{\infty} x(n\Delta t)\ z^{-n} \tag{4.6}$$

With the same defining equation for the z-transform we can provide an alternative

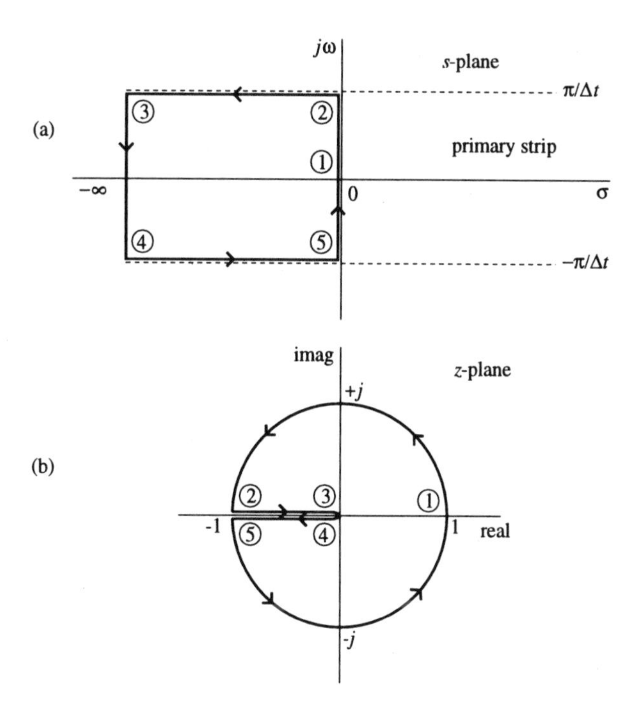

Figure 4.14 *Mapping of the primary strip in the left half of the s-plane: (a) into the circle in the z-plane; (b) by the z-transform.*

interpretation. In many applications the discrete signal or sequence of samples does not arise from sampling an analogue signal. For example, it may be necessary to process a sequence of numbers where each number represents the hours of daylight on a particular day, e.g. $x(n)$ represents the number of hours of daylight on day number n – this signal does not have an analogue equivalent. Both interpretations are useful in practice. The following are examples of calculating the z-transform of simple sequences.

EXAMPLE 4.1

A unit pulse $\delta(n\Delta t)$: The simplest example of a discrete sequence is a single sample of value one at time zero – a unit pulse illustrated in Figure 4.15. By definition:

$$x(n\Delta t) = \begin{cases} 1, & n = 0 \\ 0, & n \neq 0 \end{cases}$$

Take the z-transform as defined in equation (4.6):

$$\begin{aligned} X(z) &= \sum_{n=0}^{\infty} x(n\Delta t)\, z^{-n} \\ &= x(0)\, z^{-0} \\ &= 1\,(1) = 1 \end{aligned}$$

Figure 4.15 *A unit pulse.*

The unit pulse can also be called a discrete impulse or just an impulse (which can cause confusion with the analogue impulse).

EXAMPLE 4.2

A unit step: A discrete step (illustrated in Figure 4.16) is a sampled version of the analogue or continuous step:

$$x(n\Delta t) = \begin{cases} 1, & n \geq 0 \\ 0, & n < 0 \end{cases}$$

Take the z-transform:

$$X(z) = \sum_{n=0}^{\infty} x(n\Delta t)\, z^{-n}$$

$$= \sum_{n=0}^{\infty} (1)\, z^{-n}$$

$$= \sum_{n=0}^{\infty} z^{-n}$$

An important result when dealing with the z-transform – the sum of geometric series is:

$$\sum_{n=0}^{\infty} c^n = \frac{1}{1-c} \quad \text{provided} \quad |c| < 1$$

Using this properties of a geometric series, gives:

$$X(z) = \sum_{n=0}^{\infty} (z^{-1})^n = \frac{1}{1-(z^{-1})}$$

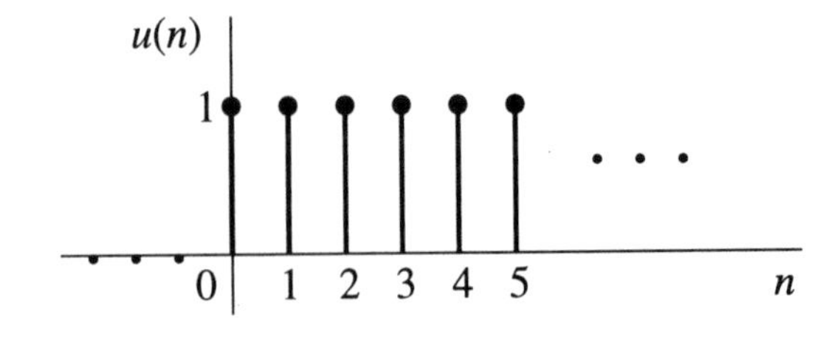

Figure 4.16 *A unit step.*

EXAMPLE 4.3
A sampled exponential: An important discrete signal is the sampled exponential illustrated in Figure 4.17:

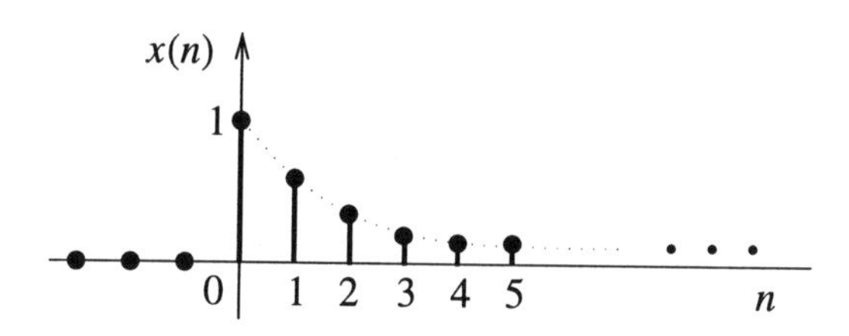

Figure 4.17 *A sampled exponential.*

$$x(n\Delta t) = \exp(-\alpha n \Delta t)$$

Take the z-transform:

$$\begin{aligned} X(z) &= \sum_{n=0}^{\infty} \exp(-\alpha n \Delta t)\, z^{-n} \\ &= \sum_{n=0}^{\infty} (\exp(-\alpha \Delta t)\, z^{-1})^n \\ &= \frac{1}{1 - \exp(-\alpha \Delta t)\, z^{-1}} \\ &= \frac{z}{z - \exp(-\alpha \Delta t)} \end{aligned}$$

Note that the constant α could be complex with magnitude less than one which would give rise to a discrete decaying sine wave.

Self assessment question 4.2: Using equation (4.6), develop an expression for the z-transform of the infinite sequence $\{1, \alpha, \alpha^2, \alpha^3, \cdots \alpha^n \cdots\}$.

4.3.1 The inverse z-transform

The z-transform of a data sequence $\{x(n)\}$ can be written in short-hand form in the same manner as the Laplace transform:

$$\begin{aligned} X(z) &= Z[\{x(n)\}] \\ &= \sum_{n=0}^{\infty} x(n)\, z^{-n} \end{aligned}$$

Note in particular that the term Δt has been removed to simplify the notation. The summation of equation (4.6) is not dependent on Δt and hence it has no effect on the z-transform. However it is sometimes useful to re-introduce it when dealing with systems which contain both continuous-time and discrete-time processing to emphasise the link between the two. The inverse z-transform operates on $X(z)$ to produce the time-domain sequence:

$$\{x(n)\} = Z^{-1}[X(z)]$$

Like the Laplace transform, the inverse transform can be evaluated by performing a partial fraction expansion (PFE) and then looking up tables. There are some subtle differences between the Laplace and the z-transform with respect to the PFE which are best examined using an example.

EXAMPLE 4.4

A typical example of the type of transform which might be encountered in analysing or designing a sampled data system is given below:

$$X(z) = \frac{3 - \frac{5}{2} z^{-1}}{1 - \frac{3}{2} z^{-1} + \frac{z^{-2}}{2}} = \frac{3z^2 - \frac{5}{2} z}{\left(z^2 - \frac{3}{2} z + ½\right)}$$

This is not a proper fraction because the highest power in the numerator is equal to the highest power in the denominator. It is more convenient to form a PFE of $X(z)/z$ rather than $X(z)$. The first step is to divide both sides by z:

$$\frac{X(z)}{z} = \frac{3z - \frac{5}{2}}{\left(z^2 - \frac{3}{2} z + ½\right)} = \frac{3z - \frac{5}{2}}{(z - ½)\,(z - 1)}$$

$$= \frac{A}{(z - ½)} + \frac{B}{(z - 1)}$$

Hence:

$$\left(3z - \frac{5}{2}\right) = A(z - 1) + B(z - ½)$$

Equating terms in z on the left-hand and right-hand sides:

$$3 = A + B$$

and equating constants:

$$-\frac{5}{2} = -A - \frac{B}{2}$$

giving $A = 2$ and $B = 1$. Substitute for the values of A and B:

$$\frac{X(z)}{z} = \frac{2}{(z - ½)} + \frac{1}{(z - 1)}$$

and multiply both sides by z:

$$X(z) = \frac{2z}{(z - ½)} + \frac{z}{(z - 1)}$$

From tables we can work out the inverse z-transform of each of these terms:

$$Z^{-1}\left[\frac{z}{(z-1)}\right] = 1$$

$$Z^{-1}\left[\frac{z}{(z-½)}\right] = (½)^n$$

The complete sampled sequence is then obtained by combining the two:

$$x(n) = \frac{1}{2^{n-1}} + 1, \qquad n \geq 0$$

If the sequence has a finite number of terms, the z-transform is easy to evaluate.

EXAMPLE 4.5

If we have a sequence of three samples, $x(0) = 1$, $x(1) = 2$ and $x(2) = 3$, the sequence is written:

$$\{x(n)\} = 1, 2, 3$$

The z-transform of the sequence is:

$$\begin{aligned} X(z) &= \sum_{n=0}^{\infty} x(n)\, z^{-n} \\ &= x(0)\, z^{-0} + x(1)\, z^{-1} + x(2)\, z^{-2} \\ &= 1 + 2z^{-1} + 3z^{-2} \end{aligned}$$

This argument can be operated in reverse. If a z-transform is given, such as

$$X(z) = 1 + 2z^{-1} + 3z^{-2}$$

then by inspection the sequence of samples is:

$$\{x(n)\} = \{1, 2, 3\}$$

Self assessment question 4.3: Evaluate the inverse z-transform of $(z^2 - z)/(z^2 - 5z + 6)$.

4.3.2 Delay theorem

A very important property of the z-transform is the delay theorem. If we know the z-transform of a sequence, e.g. if

$$Z[\{x(n)\}] = X(z)$$

then the z-transform of the same sequence delayed by one sample is

$$Z[\{x(n-1)\}] = z^{-1}\, X(z) \tag{4.7}$$

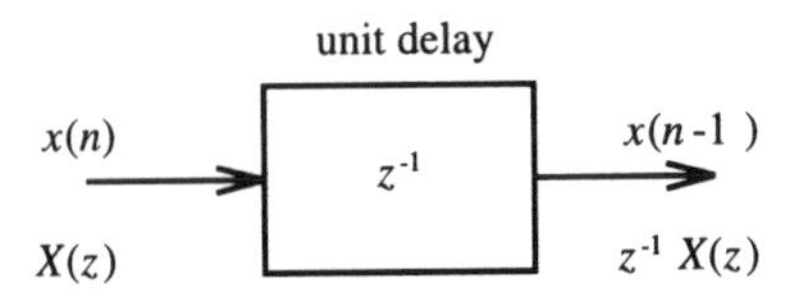

Figure 4.18 *A unit delay.*

A delay of one sample is very easy to implement in a microprocessor system where there is a clock with period Δt seconds. On one clock edge, write the data sample into a particular memory address. On the next clock edge, read the same sample out of memory. The sample will have been delayed by one clock period. For this reason the term z^{-1} is often called a 'delay'. The idea can also be extended. A delay of two samples or clock periods is represented by z^{-2}:

$$Z[\{x(n-2)\}] = z^{-2}\ X(z)$$

The block diagram representation of a unit delay is illustrated in Figure 4.18.

4.4 Digital filters and discrete convolution

Digital filtering is one significant function that can be implemented in the DSP unit. A digital filter can be described in three equally valid ways – all are important. This simple example illustrates the process. First it can be described as a mathematical algorithm – a difference equation:

- *mathematical algorithm (difference equation)*

  $$y(n) = a\ y(n-1) + x(n)$$

  The output at the present time $y(n)$ is calculated from the previous output $y(n-1)$ and the present input $x(n)$. For this example the properties of the filter will be entirely described by the constant a – in the same way that the values of R and C define the properties of an analogue filter.

Second it can be described in block diagram form as in Figure 4.19:

- *block diagram* – this might be separate hardware units, e.g. Σ is an adder, $\times$ is a multiplier and z^{-1} is a memory unit; the output of the memory unit is $y(n-1)$.

Finally the filter can be described with some lines of pseudo-code:

- *software*

```
1    read x
     y = a * y + x
     write y
     go to line 1
```

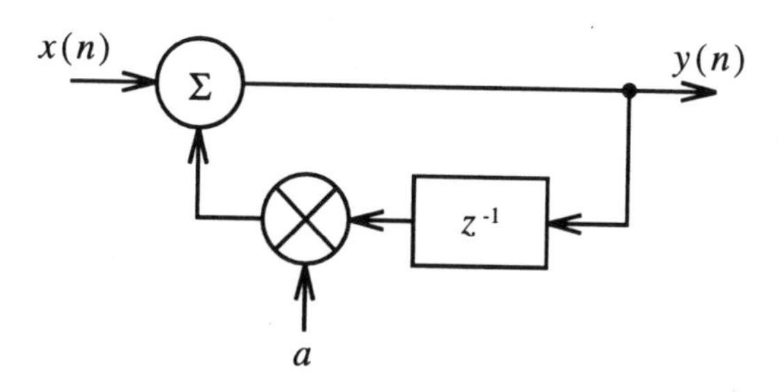

Figure 4.19 *Block diagram of a simple digital filter.*

Read in the next sample x from the A/D. Take the previous output from memory (y), multiply it by a, add the product to the value of x and overwrite y with the result. Then write the value of y to the D/A. Go back to line 1 and wait for a new sample from the A/D.

A slightly more complicated example of a difference equation can be used to illustrate the use of the z-transform in the analysis of digital filters, and leads to the concept of the transfer function of such filters:

$$y(n) = a_0\, x(n) + a_1\, x(n-1) + a_2\, x(n-2) + b_1\, y(n-1) + b_2\, y(n-2)$$

Take the z-transform of both sides – using the delay theorem:

$$\begin{aligned} Y(z) &= Z[\{y(n)\}] \\ &= a_0\, X(z) + a_1\, z^{-1}\, X(z) + a_2\, z^{-2}\, X(z) \\ &\quad + b_1\, z^{-1}\, Y(z) + b_2\, z^{-2}\, Y(z) \end{aligned}$$

Collecting terms in $Y(z)$ and $X(z)$ on either side of the equation:

$$Y(z)\,[1 - b_1\, z^{-1} - b_2\, z^{-2}] = X(z)\,[a_0 + a_1\, z^{-1} + a_2\, z^{-2}]$$

The transfer function is then the ratio of the output transform over the input transform:

$$H(z) = \frac{Y(z)}{X(z)} = \frac{a_0 + a_1\, z^{-1} + a_2\, z^{-2}}{1 - b_1\, z^{-1} - b_2\, z^{-2}}$$

The block diagram of a general digital filter is illustrated in Figure 4.20. This structure is known as a direct realisation of a digital filter and it has two main sections: feedforward and feedback. The input signal is applied directly to the feedforward section – characterised by the coefficients $a_0, a_1, \cdots, a_{N-1}$. The shift register formed from the delay units holds 'older' input signal samples, e.g. $x(n-1)$, $x(n-2)$, etc. Each input signal sample is multiplied by a corresponding weight or coefficient. The N products are added together (or accumulated) to form the output from the feedforward section. The overall output $y(n)$ is formed by adding together the outputs from the feedforward and feedback sections. In the feedback section, 'older' values of the output, such as $y(n-1)$, $y(n-2)$, are held in a shift register. These are multiplied by a set of coefficients, $b_1, b_2, \cdots, b_M$, and the products accumulated to form the output of the feedback section. The general form of the difference equation is given by:

$$y(n) = \sum_{i=0}^{N-1} a_i\, x(n-i) + \sum_{i=1}^{M} b_i\, y(n-i) \tag{4.8}$$

Taking z-transforms of both sides leads to the general form of the transfer function:

$$H(z) = \frac{\sum_{i=0}^{N-1} a_i\, z^{-i}}{1 - \sum_{i=1}^{M} b_i\, z^{-i}} \tag{4.9}$$

The feedforward and feedback sections have characteristic effects on the impulse response. Consider the case where all the feedback coefficients are zero – i.e. there is no feedback. Apply a single sample of value 1 to the filter – all samples after it are zero. This is a discrete-time impulse. At time zero the output will simply be '1' multiplied by a_0. We then clock the filter which causes the single sample to shift one memory unit to the right. The output of the filter is now 1 multiplied by a_1. We could keep clocking the filter until the single sample reaches the end of the feedforward section, in which case the output will be a_{N-1}. If we clock the filter once more, the output will be zero. We have applied a unit impulse to the filter and the output is the sequence $\{a_0, a_1, \cdots, a_{N-1}\}$. The sequence is finite and hence the filter is the finite impulse response (FIR). Note also that for a FIR filter the impulse response is defined directly by the coefficients.

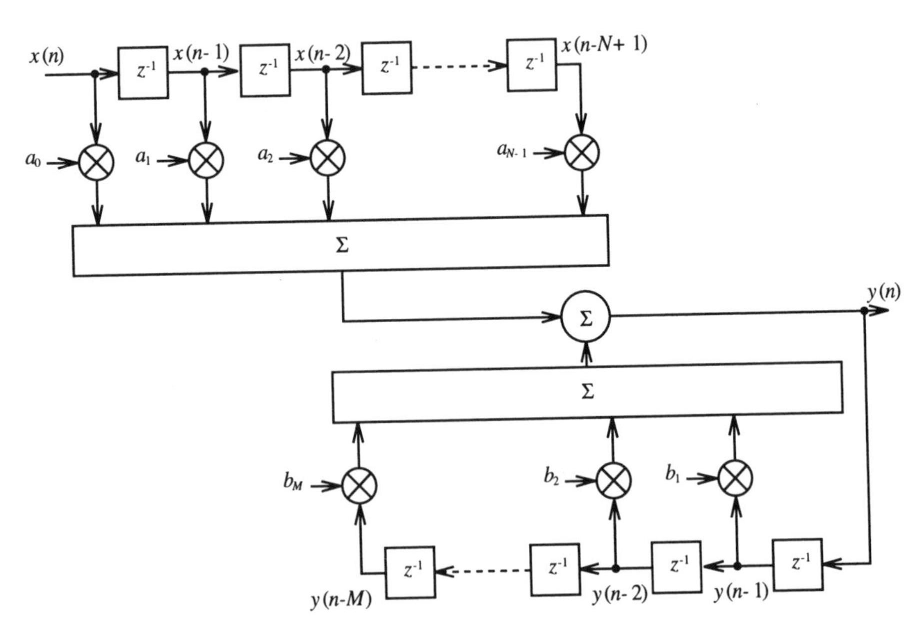

Figure 4.20 *Direct realisation of a digital filter.*

Alternatively, consider the case where there is no feedforward, i.e. all the a_i coefficients are zero except a_0 which is 1. Apply a unit pulse to the filter. The first output will be 1. We then clock the filter. Although the new input is zero, the previous output value of 1 is sitting at the output of the first delay in the feedback section. Thus the second output will be b_1. If we continue to clock the filter we will continue to get an output as the data circulate round and round the feedback section. If the filter is stable, the output will decay towards zero but never quite get there. The impulse response is infinite and the filter is the infinite impulse response (IIR). In summary:

- *if* $b_i = 0$ *for all values of i, the filter is the finite impulse response (FIR);*
- if any $b_i \neq 0$, the filter is the infinite impulse response (IIR).

Like the analogue case discussed in Chapter 1, the transfer function can be used to calculate the output response of a digital filter to an input sequence $\{x(n)\}$. This involves the following steps:

(i) Take the z-transform of the input sequence:

$$X(z) = Z[\{x(n)\}]$$

(ii) Multiply by the transfer function:

$$Y(z) = H(z)\,X(z)$$

(iii) Take the inverse z-transform:

$$\{y(n)\} = Z^{-1}[Y(z)]$$

4.4.1 Discrete convolution

The time-domain description of a discrete-time linear system (digital filter) has two equivalent forms:

$$y(n) = \sum_{m=0}^{\infty} x(m)\,h(n-m) \tag{4.10}$$

$$y(n) = \sum_{m=0}^{\infty} h(m)\,x(n-m) \tag{4.11}$$

This *discrete-time convolution* is often denoted with an asterisk:

$$\{y(n)\} = \{h(n)\} * \{x(n)\}$$

Thus, in an analogous manner to the continuous-time case, convolution in the time domain is equivalent to multiplication in the z-domain.

Equations (4.10) and (4.11) are similar in form to equations (2.8) and (2.9) respectively. For example, in equation (4.10) the impulse response sequence is time-reversed and moved forward sample by sample. At each point in time n, the product $x(m)h(n-m)$ is evaluated and a summation is performed over the index m. Similarly, in equation (2.8), the impulse response $h(t)$ is time-reversed and moved forward in time. At each time instant t, the product is evaluated and integrated with respect to the variable τ.

Whilst discrete convolution has its own intrinsic value as a tool for the analysis and design of digital filters, it also offers insight into the continuous-time convolution of equations (2.8) and (2.9). If we rewrite the approximation to the continuous convolution provided by equation (2.7):

$$\hat{y}(t) = \sum_{m=-\infty}^{\infty} x(m\Delta t)\, h'(t - m\Delta t)\, \Delta t$$

it is similar in form to (4.10). If we further chose to evaluate $\hat{y}(t)$ at discrete points in time, i.e. at $t = n\Delta t$, we can view the discrete convolution as an approximation to continuous-time convolution:

$$\hat{y}(n\Delta t) = \sum_{m=0}^{\infty} x(m\Delta t)\, h'(n\Delta t - m\Delta t)\, \Delta t$$

The summation starts at $m = 0$ if we assume that $\{x(m\Delta t)\}$ is a causal sequence, i.e.: $x(m\Delta t) = 0$ for $m < 0$. Continuous-time and discrete-time convolution are thus very similar processes. However the discrete form is conceptually simpler to grasp since we are dealing with summation operations rather than integration. Further, the finite impulse response (FIR) digital filter can be viewed as a convolution machine. The following example illustrates some of these points.

EXAMPLE 4.6

A sequence $\{1, 2, 3\}$ is applied to a FIR filter with transfer function $1 + 0.5z^{-1} + 0.25z^{-2}$. Use discrete convolution to evaluate the output.

Solution

Using the methods employed in example 4.5, the impulse response sequence is $\{1,\ 0.5,\ 0.25\}$. This is illustrated at the bottom of Figure 4.21. Because the sequence is finite there is a simple one-to-one relationship between it and the coefficients of a FIR filter. In this case the three elements of the sequence are the coefficients of a three-tap FIR filter as in Figure 4.21. For this example we will use equation (4.11) which simplifies to the following form because the filter is FIR with three taps:

$$y(n) = \sum_{m=0}^{2} h(m)\, x(n-m)$$

The input sequence $\{x(n)\}$ is illustrated at the top of Figure 4.21. If this sequence is applied to the filter, the sample of value 1 reaches the filter first, followed by the sample of value 2 and then the sample of value 3. In order to draw the input sequence in a manner which is consistent with the block diagram, we must time-reverse the sequence. This is illustrated in the plot of $x(-m)$ against m. In accordance with the above

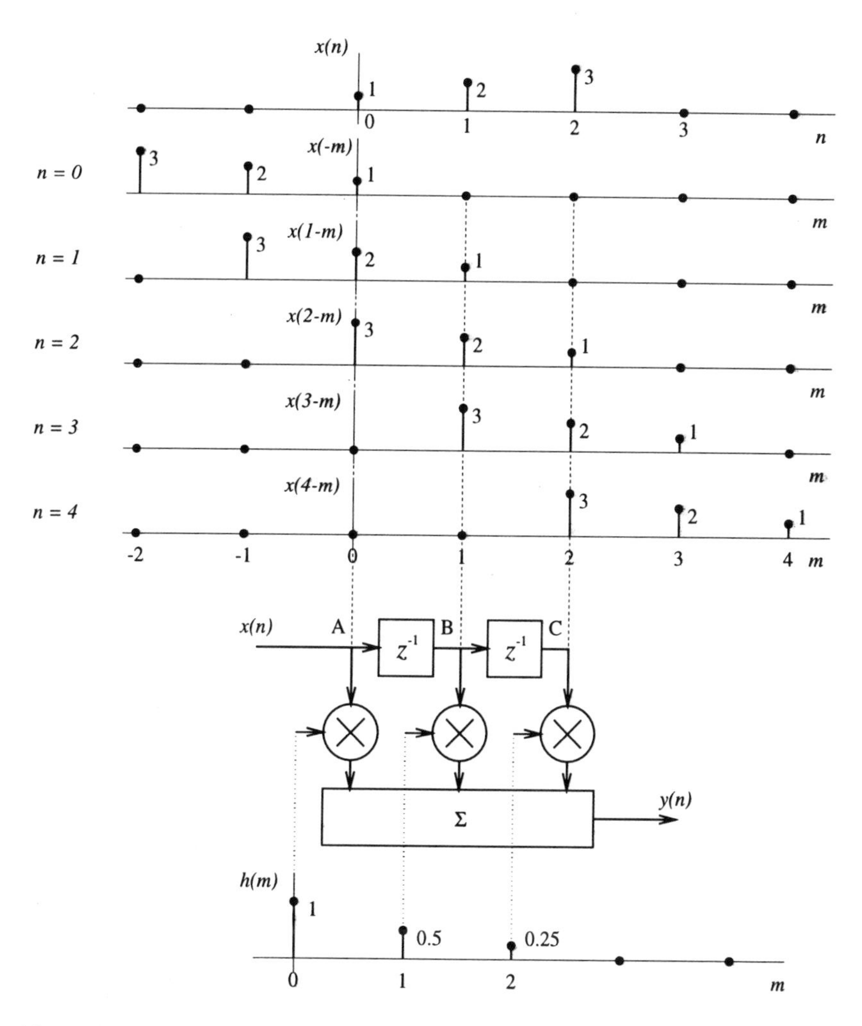

Figure 4.21 *Example of discrete convolution for input sequence* $\{1, 2, 3\}$ *and impulse response sequence* $\{1, 0.5, 0.25\}$.

equation:

$$y(0) = \sum_{m=0}^{2} h(m)\, x(-m)$$

At $n = 0$ the sample of value 1 sits at point A on the filter. The samples at points B and C are both zero. Hence the filter output is:

$$y(0) = h(0)\, x(0) = 1(1) = 1$$

At $n = 1$ the convolution equation becomes:

$$y(1) = \sum_{m=0}^{2} h(m)\, x(1 - m)$$

The time-reversed input sequence has moved forward one sample in time (or, equivalently, shifted to the right by one sample in the FIR filter). The sample at point C in the filter is still zero and hence the output is:

$$y(1) = h(0)\ x(1-0) + h(1)\ x(1-1) = 1(2) + 0.5(1) = 2.5$$

The calculation proceeds in a similar manner for $n = 2$ and 3. The last non-zero output from the filter occurs at $n = 4$. At this point:

$$\begin{aligned} y(4) &= \sum_{m=0}^{2} h(m)\ x(4-m) \\ &= h(2)\ x(4-2) = 0.25(3) = 0.75 \end{aligned}$$

The complete output sequence is: $\{1,\ 2.5,\ 4.25,\ 2,\ 0.75\}$.

Self assessment question 4.4: For example 4.6, evaluate $y(2)$ and $y(3)$.

4.5 Poles and stability

The poles and zeros of a digital filter are defined in a similar manner to that for an analogue filter. The position of the poles is also an indicator of stability. Consider a system, $H(z)$, with two complex conjugate poles in the z-plane:

$$\begin{aligned} p_1 &= r\ \exp(j\phi) \\ p_2 &= r\ \exp(-j\phi) \end{aligned}$$

A typical transfer function might be:

$$H(z) = \frac{z}{(z - r\ \exp(j\phi))\ (z - r\ \exp(-j\phi))} \tag{4.12}$$

To find the impulse response, make a PFE and look up tables. In accordance with example 4.4, move a z term to the right-hand side before starting:

$$\begin{aligned} \frac{H(z)}{z} &= \frac{1}{(z - r\ \exp(j\phi))\ (z - r\ \exp(-j\phi))} \\ &= \frac{A}{(z - r\ \exp(j\phi))} + \frac{B}{(z - r\ \exp(-j\phi))} \qquad \dagger \end{aligned}$$

Moving the z back to the right-hand side gives terms which can be found in tables:

$$H(z) = \frac{Az}{(z - r\ \exp(j\phi))} + \frac{Bz}{(z - r\ \exp(-j\phi))}$$

Take inverse z-transforms to find the impulse response:

$$\begin{aligned} h(n) &= A\ (r\ \exp(j\phi))^n + B\ (r\ \exp(-j\phi))^n \\ &= \frac{r^n\ \exp(j\phi n)}{2jr\ \sin(\phi)} - \frac{r^n\ \exp(-j\phi n)}{2jr\ \sin(\phi)} \end{aligned}$$

† $A = 1/(2jr\ \sin(\phi))$; $B = -\,1/(2jr\ \sin(\phi))$.

$$= \frac{r^{n-1}}{\sin(\phi)} \left\{ \frac{\exp(j\phi n) - \exp(-j\phi n)}{2j} \right\}$$

$$= \frac{r^{n-1}}{\sin(\phi)} \sin(\phi n)$$

This is the unit pulse response or impulse response of the digital filter. We note that the impulse response will decay away to zero provided r is less than one. Recall that r is also the distance from the origin in the z-plane to the poles p_1 or p_2, so the system will be stable if the poles in the z-plane lie inside a circle of radius one – the unit circle. Figure 4.22 illustrates the relationship between the position of a pole and the impulse response for both purely real poles and complex conjugate pairs of poles. The angle of the pole ϕ determines the frequency of oscillation and the radius r determines the rate of decay or growth. Thus:

- *a system will be stable if all its poles lie inside the unit circle in the z-plane.*

4.6 Frequency response of a digital filter

Like analogue filters, we are interested in the frequency response of digital filters. Recall that the relationship between the Laplace transform of a sampled signal and the z-transform is simply a matter of interchanging $\exp(\Delta t\ s)$ with z:

$$X_c(s) = X(z)|_{z = \exp(\Delta t\ s)}$$

Further, it is possible to obtain the Fourier transform from the Laplace transform by replacing s with $j\omega$:

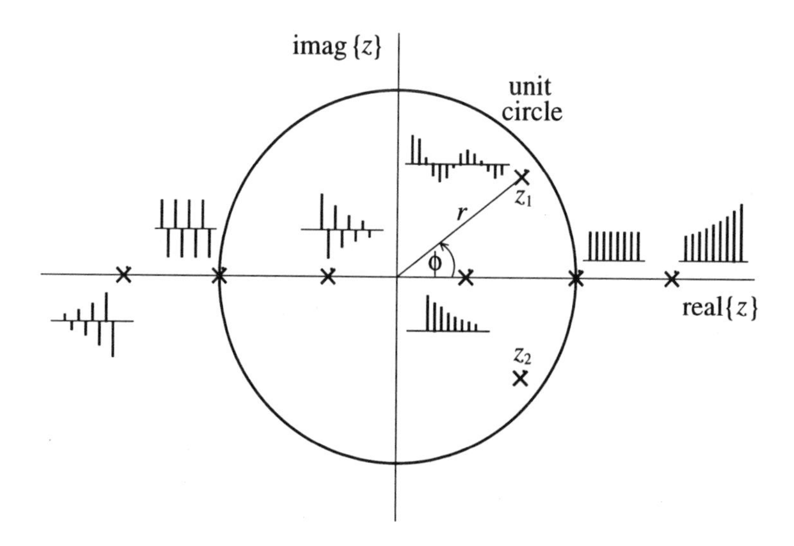

Figure 4.22 *Positions of poles and corresponding impulse response.* □

$$X_c(\omega) = X_c(s)|_{s = j\omega}$$

If we wish to obtain the Fourier transform from the z-transform, the two steps can be conveniently combined:

$$X(\omega) = X(z)\,|_{z = \exp(\Delta t\ j\omega)}$$

If we have the transfer function of a digital filter, $H(z)$, this also provides a convenient mechanism for obtaining the frequency response, $H(\omega)$:

$$H(\omega) = H(z)\,|_{z = \exp(\Delta t\ j\omega)} \tag{4.13}$$

EXAMPLE 4.7

If the transfer function of a digital filter is

$$H(z) = \frac{z}{z - a}$$

then

$$H(\omega) = \frac{\exp(j\omega\Delta t)}{\exp(j\omega\Delta t) - a}$$

The gain and phase response are $|H(\omega)|$ and $\angle H(\omega)$ respectively.

Self assessment question 4.5: Calculate the gain, at a quarter of the sampling frequency, of a digital filter with transfer function $H(z) = 2 + z^{-1}$.

The frequency response of a digital filter is a very similar concept to the frequency response of an analogue filter. If we apply a sampled cosine wave $x(n)$ of frequency ω_0 rad/s, where

$$x(n) = \cos(\omega_0\ \Delta t\ n)$$

to a digital filter with frequency response $H(\omega)$, then the steady-state output $y(n)$ will be

$$y(n) = |H(\omega_0)|\ \cos\,(\omega_0\ \Delta t\ n + \angle H(\omega_0))$$

It is not convenient to draw Bode plots for discrete time systems but the positions of the poles and zeros give important indicators to the shape of the frequency response. Example 4.6 has a pole at $z = a$ and a zero at the origin, $z = 0$, as illustrated in Figure 4.23. In this example the pole gives rise to a term $(\exp(j\omega\Delta t) - a)$ in the denominator of the frequency response. The magnitude of the frequency response or amplitude response is

$$|H(\omega)| = \frac{1}{|\ \exp(j\omega\Delta t) - a\ |}$$

and the phase response is

$$\angle H(\omega) = \omega\Delta t - \angle(\ \exp(j\omega\Delta t) - a)$$

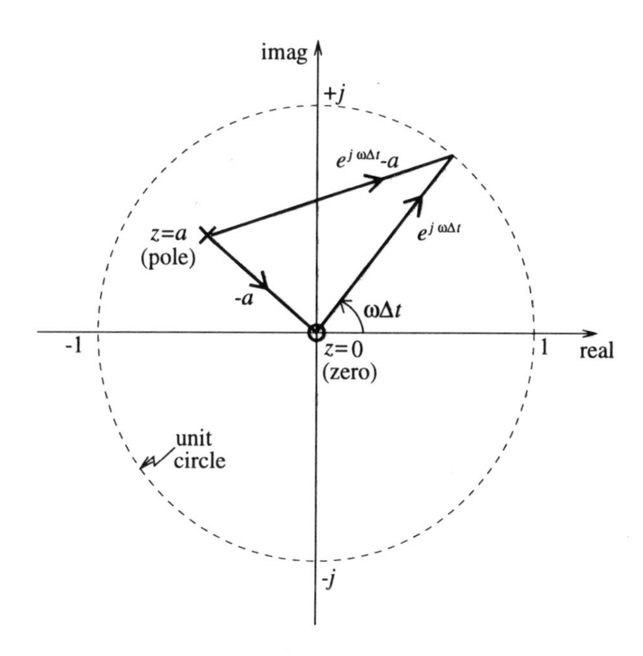

Figure 4.23 *Pole/zero maps and frequency response.* □

Thus in examining the amplitude and phase response of the filter we are interested in how the magnitude and phase of the complex variable $(\exp(j\omega\Delta t) - a)$ changes as we change the frequency ω. Like the analogue case, a vector interpretation of the complex numbers is useful. Figure 4.23 illustrates $(\exp(j\omega\Delta t) - a)$ as a vector from the pole to a point on the unit circle $\exp(j\omega\Delta t)$. The vector from the origin to this point has length one and angle $\omega\Delta t$ radians. When the frequency ω is 0, the vector $(\exp(j\omega\Delta t) - a)$ is a line starting at the pole and ending on the real axis $1 + j0$. As we increase the frequency, the end of the vector $(\exp(j\omega\Delta t) - a)$ moves around the unit circle. When the angle $\omega\Delta t$ is equal to π radians, the end of the vector is at $-1 + j0$. This corresponds to half the sampling frequency since:

$$\omega = \frac{\pi}{\Delta t}$$

Clearly the vector $(\exp(j\omega\Delta t) - a)$ will be shortest when the angle $\omega\Delta t = \angle a$, i.e.:

$$\omega = \frac{\angle a}{\Delta t}$$

At that frequency the gain will be a maximum since $(\exp(j\omega\Delta t) - a)$ is on the bottom line of the expression for gain.

This approach can be extended to the situation where $H(z)$ has m zeros and n poles, i.e.:

$$H(z) = \frac{(z - z_1)(z - z_2) \cdots (z - z_m)}{(z - p_1)(z - p_2) \cdots (z - p_n)}$$

If we are asked for the frequency response, we simply replace z with $e^{j\omega\Delta t}$ thus:

$$H(\omega) = \frac{A(e^{j\omega\Delta t} - z_1)(e^{j\omega\Delta t} - z_2)\cdots(e^{j\omega\Delta t} - z_m)}{(e^{j\omega\Delta t} - p_1)(e^{j\omega\Delta t} - p_2)\cdots(e^{j\omega\Delta t} - p_n)}$$

To obtain the amplitude response of the system, we take the magnitude of $H(\omega)$, which because of the properties of complex numbers simplifies to:

$$|\,H(\omega)\,| = \frac{A|e^{j\omega\Delta t} - z_1||e^{j\omega\Delta t} - z_2|\cdots|e^{j\omega\Delta t} - z_m|}{|e^{j\omega\Delta t} - p_1||e^{j\omega\Delta t} - p_2|\cdots|e^{j\omega\Delta t} - p_n|} \tag{4.14}$$

In a similar way we can obtain the phase response of the system by taking the argument of $H(\omega)$. Again, because of the properties of complex numbers, this simplifies to:

$$\begin{aligned}\angle H(\omega) = {} & \angle(e^{j\omega\Delta t} - z_1) + \angle(e^{j\omega\Delta t} - z_2) + \cdots + \angle(e^{j\omega\Delta t} - z_m) \\ & - \angle(e^{j\omega\Delta t} - p_1) - \angle(e^{j\omega\Delta t} - p_2)\cdots - \angle(e^{j\omega\Delta t} - p_n)\end{aligned} \tag{4.15}$$

In this general case the amplitude response is obtained by taking the products of the lengths of the various vectors drawn from the zeros to the point on the unit circle $e^{j\omega\Delta t}$ associated with the frequency ω, and dividing this by the product of the lengths of the vectors from the poles – as indicated by equation (4.14). The phase is found by summing the phases of the individual vectors from the zeros minus the sum of the phases of the individual vectors from the poles – as indicated by equation (4.15). This is illustrated in Figure 4.24 which shows the pole/zero plot for a transfer function with a complex conjugate pair of poles and a zero at the origin. The figure explicitly shows the lengths a, b and c associated with calculating the gain when the angle $\omega\Delta t$ is equal to $\pi/2$. This angle corresponds to a quarter of the sampling frequency ω_s rad/s since:

$$\begin{aligned}\omega\Delta t &= \frac{\pi}{2} \\ \omega &= \frac{\pi}{2\Delta t} \\ &= \left(\frac{2\pi}{\Delta t}\right)\frac{1}{4} \\ &= \frac{\omega_s}{4}\end{aligned}$$

The gain at $\omega = \omega_s/4$ is indicated on the amplitude response of Figure 4.24 as $a/(bc)$. The phase shift at this frequency is $\alpha - \beta - \gamma$ degrees. The process is repeated at every angle $\omega\Delta t$ on Figure 4.24(a) to obtain the amplitude and phase response of the system. The value of this technique is not as a precise tool for obtaining the frequency response but rather that it gives an idea of what the frequency response might look like, Figure 4.24(b). In this example we would expect a peak in the frequency response at a frequency of around $\omega_s/8$ because the complex conjugate pair of poles lie at angles of $\pi/4$ and $-\pi/4$ radians. This technique can provide further insights at the design stage. For instance, we could make the peak in the frequency response more pronounced by keeping the poles at $\pm\pi/4$ radians and moving them closer to the unit circle.

Self assessment question 4.6: Sketch the frequency response of a digital filter with transfer function $H(z) = z/(z+1)$.

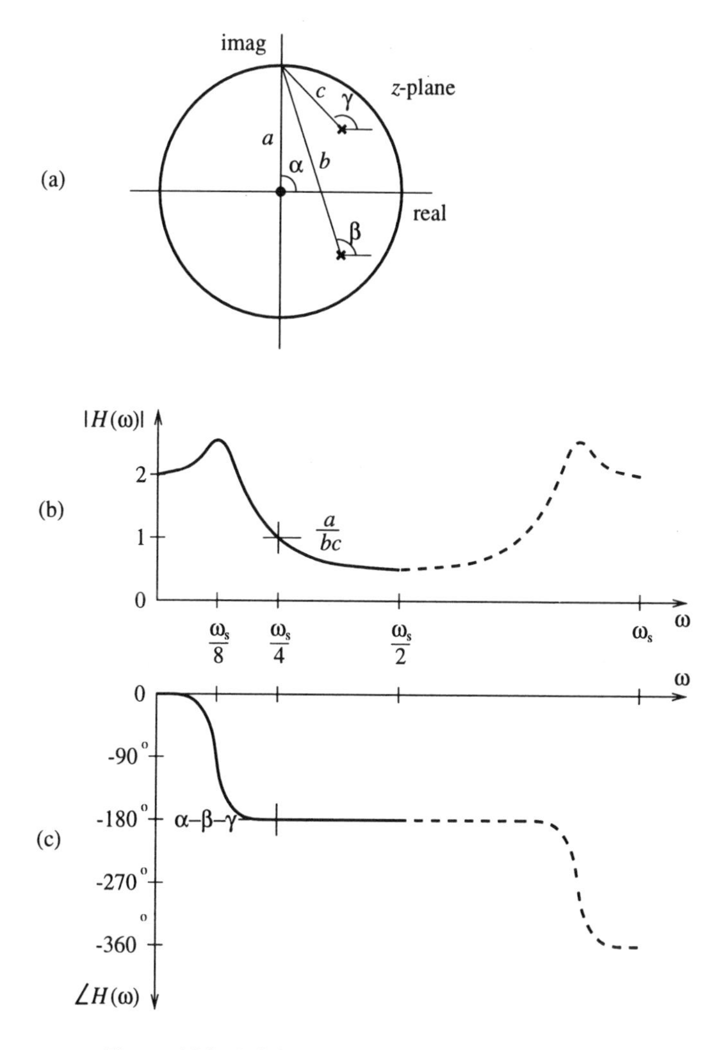

Figure 4.24 *Pole/zero maps and frequency response.* □

4.7 Example of a complete system

The following worked example illustrates how the steady-state response of a complete sampled data system can be analysed using some of the techniques described in this chapter.

EXAMPLE 4.8

An analogue signal, $x(t)$, is applied to the sampled data system of Figure 4.25 which has a sampling rate of 10 kHz. The sample and hold (S/H) has a high impedance input stage and hence does not draw any current from the RC circuit ($R = 21\ \text{k}\Omega$ and $C = 3\ \text{nF}$). There are two sine waves present in the analogue signal: one has an amplitude of A and frequency of 1 kHz, and the other has an amplitude of $A/2$ and a frequency of 7.5 kHz. What frequencies are present at the output of the analogue-to-digital (A/D) converter and what are their amplitudes? The digital signal processing (DSP) unit implements a digital filter which is illustrated in Figure 4.26. What frequencies are present at the output of the digital signal processor and what are their amplitudes? It may be assumed that the low-pass filter (LPF) is ideal and has a cut-off frequency of 5 kHz.

What frequencies are present at the output of the system and what are their amplitudes?

Solution

The input $x(t)$ contains two sine waves. From Figure 3.6, the Fourier transform of a sine wave or cosine wave consists of two impulses. For the 1 kHz component, the impulses are at $\pm 2\pi \times 10^3$ rad/s, each with strength $A\pi$. For the 7.5 kHz component, the impulses are at $\pm 2\pi(7.5 \times 10^3)$ rad/s, each with strength $A\pi/2$. Since we are not asked to consider phase shift in this question, there is no need to distinguish between the transform of a cosine wave and a sine wave – the former, being easier to work with, is adopted here. This Fourier transform is illustrated in Figure 4.27(a). Note: for ease of representation, the frequency axis in Figure 4.27 is labelled in kHz rather than rad/s.

The gain of the RC filter is:

$$|H_{RC}(\omega)| = \frac{1}{|1 + j\omega RC|}$$

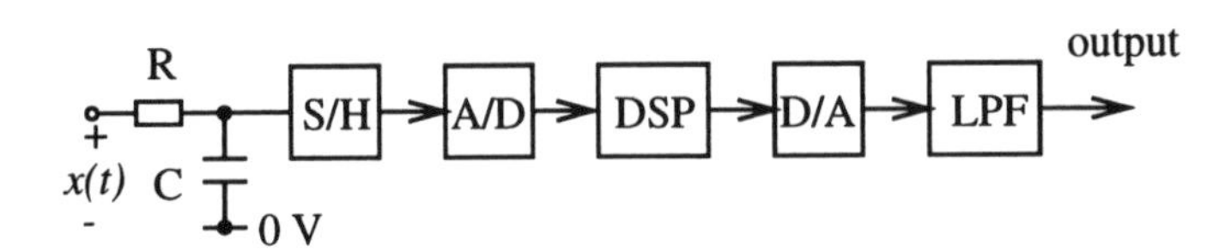

Figure 4.25 *A sampled data system.*

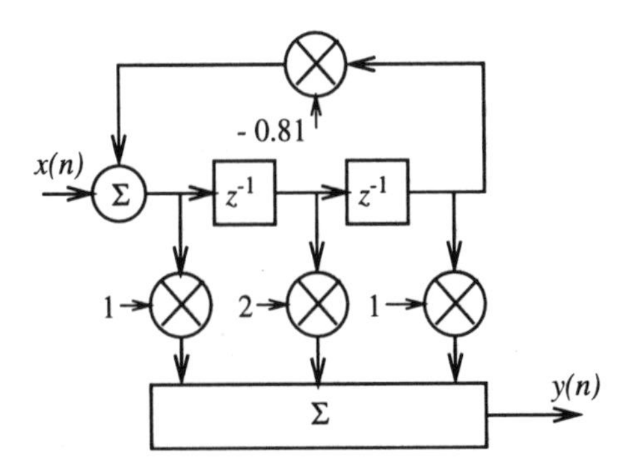

Figure 4.26 *Digital filter block diagram.*

This is illustrated approximately in Figure 4.27(b). It is worth noting that for the particular values of R and C, the -3 dB bandwidth of this filter is 5 kHz – half the sampling frequency. At 1 kHz, the gain of this filter is 0.9298 and at 7.5 kHz the gain is 0.3192. While the latter is clearly above the cut-off frequency of the filter, the gain is not negligible. In no sense could it be considered that the low-pass filter has removed the component at 7.5 kHz. The amplitude of the 1 kHz component at the output of the LPF is $0.9298A$ and the amplitude of the 7.5 kHz component is $0.3192 \times A/2 = 0.1596A$. For the 1 kHz component, the Fourier transform consists of impulses at $\pm 2\pi \times 10^3$ rad/s, each with strength $0.9298A\pi$. For the 7.5 kHz component, the Fourier transform consists of impulses at $\pm 2\pi(7.5 \times 10^3)$ rad/s, each with strength $0.1596A\pi$. This Fourier transform is illustrated in Figure 4.27(c). Effectively the Fourier transform of Figure 4.27(c) is obtained by multiplying together the Fourier transform of the input, Figure 4.27(a), and the frequency response of the filter, Figure 4.27(b).

As in section 4.2, the action of the sampler is modelled as multiplication in the time domain with the impulse train of Figure 4.7(a). Hence, using equation (2.15), the Fourier transform of the signal at the output of the sampler is formed by convolving the Fourier transform of Figure 4.27(c) with the Fourier transform of 4.7(b). The 1 kHz component at the output of the LPF is not aliased by the sampling operation and gives rise to impulses at $\pm 2\pi \times 10^3$, $\pm 2\pi(9 \times 10^3)$, $\pm 2\pi(11 \times 10^3)$ etc. rad/s in the Fourier transform of the signal which is present at the output of the sampler. Each impulse has strength $0.9298A\pi/\Delta t$. The 7.5 kHz component at the output of the LPF is aliased by the sampling operation and gives rise to impulses at $\pm 2\pi(2.5 \times 10^3)$, $\pm 2\pi(7.5 \times 10^3)$, $\pm 2\pi(12.5 \times 10^3)$ etc. rad/s, each with strength $0.1596A\pi/\Delta t$. This Fourier transform is illustrated in Figure 4.27(d).

To find the frequency response of the digital filter we require first to calculate the transfer function of the filter. It is convenient to label the signal at the left-most summer of Figure 4.26 as $x_1(n)$. We can then write a difference equation to relate $x(n)$ and $x_1(n)$:

$$x_1(n) = x(n) - 0.81x_1(n-2)$$

Taking z-transforms of both sides yields:

$$X_1(z) = \frac{X(z)}{(1 + 0.81z^{-2})} \tag{4.16}$$

The difference equation which relates the output $y(n)$ to the intermediate signal $x_1(n)$ is

$$y(n) = x_1(n) + 2x_1(n-1) + x_1(n-2)$$

Again taking z-transforms yields:

$$Y(z) = X_1(z)\left(1 + 2z^{-1} + z^{-2}\right) \tag{4.17}$$

Substituting equation (4.16) into (4.17) for $X_1(z)$ leads to the required transfer function:

$$H(z) = \frac{Y(z)}{X(z)} = \frac{1 + 2z^{-1} + z^{-2}}{1 + 0.81z^{-2}}$$

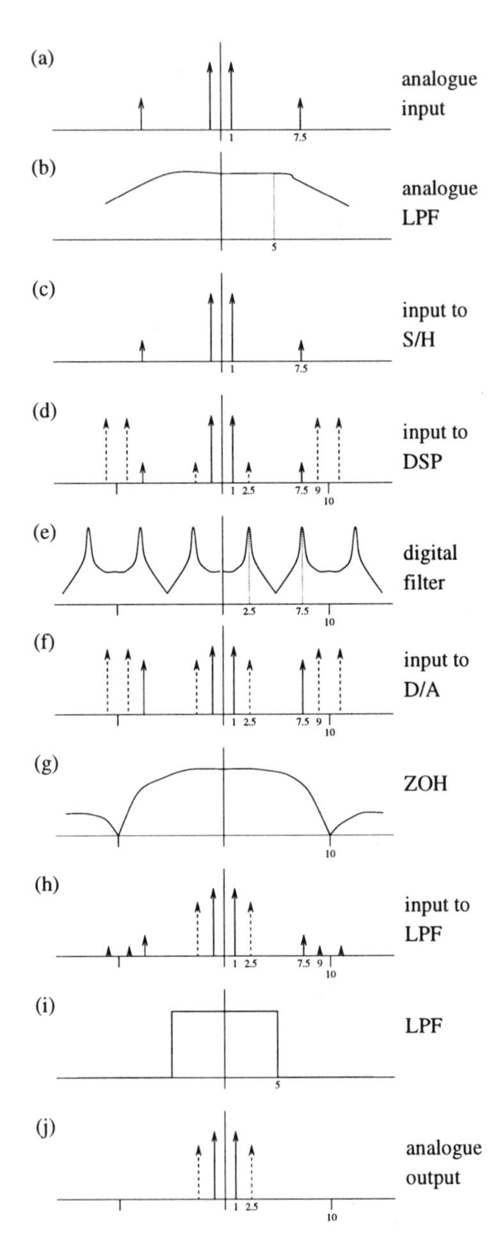

Figure 4.27 *Sampled data system: (a) frequency content of analogue signal; (b) frequency response of anti-aliasing LPF; (c) frequency content of analogue signal at input to S/H; (d) frequency content of discrete signal at input to DSP; (e) frequency response of digital filter; (f) frequency content of discrete signal at input to D/A; (g) frequency response of ZOH; (h) frequency content of analogue signal at input to reconstruction LPF; (i) frequency response of ideal reconstruction filter; (j) frequency content at output.*

$$= \frac{(1+z^{-1})^2}{1+0.81z^{-2}}$$

As in section 4.2, the frequency response is obtained by substituting $\exp(j\omega\Delta t)$ for z in the expression for the transfer function. Thus:

$$|H(\omega)| = \frac{|1+\exp(-j\omega\Delta t)|^2}{|1+0.81\exp(-2j\omega\Delta t)|}$$

At 1 kHz, $\omega = 2\pi \times 10^3$, $\omega\Delta t = \pi/5$ and $|H(\omega)| = 2.464$. Because of the properties of digital filters, this will also be the gain at 9, 11, 19, 21 etc. kHz. At 2.5 kHz, $\omega = 2\pi(2.5 \times 10^3)$ rad/s, $\omega\Delta t = \pi/2$ and $|H(\omega)| = 10.526$. Because of the properties of digital filters this will also be the gain at 7.5, 12.5, 17.5 etc. kHz. Although it is not asked for directly in this problem, it is useful to locate the poles and zeros and sketch the frequency response of this digital filter. The result is illustrated in Figure 4.27(d). It is worth noting that this filter is particularly good at amplifying the aliased component at 2.5 kHz.

The frequencies present at the output of the digital filter are 1, 9, 11, 19, 21 etc. kHz – each with an amplitude of

$$2.464\,\frac{0.9298A}{\Delta t} = \frac{2.291A}{\Delta t}$$

and 2.5, 7.5, 12.5, 17.5, 22.5 etc. kHz – each with an amplitude of

$$10.526\,\frac{0.1596A}{\Delta t} = \frac{1.68A}{\Delta t}$$

The Fourier transform of the signal at the output of the digital filter is illustrated in Figure 4.27(f). It has impulses at $\pm 2\pi \times 10^3$, $\pm 2\pi(9 \times 10^3)$, $\pm 2\pi(11 \times 10^3)$ etc. rad/s, each with strength $\pi 2.291A/\Delta t$ and impulses at $\pm 2\pi(2.5 \times 10^3)$, $\pm 2\pi(7.5 \times 10^3)$, $\pm 2\pi(12.5 \times 10^3)$ etc. rad/s, each with strength $\pi 1.68A/\Delta t$. Like any other filter, the Fourier transform of the signal at the output of the digital filter is simply obtained by multiplying the Fourier transform of the input by the frequency response.

The action of the D/A converter is modelled with a ZOH as in section 4.2. From equation (4.2) the frequency response of the ZOH is:

$$|H_{ZOH}(\omega)| = \Delta t \left| \mathrm{sa}\left(\frac{\omega\Delta t}{2}\right) \right|$$

This is illustrated in Figure 4.27(g). As can also be seen from Figure 4.27, the action of the ZOH is to reduce components at 7.5, 9, 11, 12.5 etc. kHz – however it does not completely remove them. In the particular problem, the reconstruction filter is assumed to be an ideal LPF with a cut-off frequency of 5 kHz, Figure 4.27(i). Hence the components at 7.5, 9, 11, 12.5 etc. kHz will be completely removed and it is only necessary to calculate the gain of the ZOH for the 1 and 2.5 kHz components, as these will be the only ones present at the output of the system. At 1 kHz, $\omega\Delta t = \pi/5$ and $|H_{ZOH}(\omega)| = 0.984\Delta t$. The amplitude of the 1 kHz sine wave at the output of the reconstruction filter is thus:

$$0.984\Delta t\,\frac{2.291A}{\Delta t} = 2.254A$$

At 2.5 kHz, $\omega\Delta t = \pi/2$ and $|H_{ZOH}(\omega)| = 0.9\Delta t$. The amplitude of 1 kHz sine wave at the output of the reconstruction filter is thus:

$$0.9\Delta t \frac{1.68A}{\Delta t} = 1.512A$$

This is shown in Figure 4.27(j). It is worth noting that the scale factor $1/\Delta t$ on the amplitude of the sampled signals is an arbitrary artefact of the modelling process. It is effectively cancelled by the scale factor Δt present in the gain of the ZOH.

4.8 Summary

Sampled data systems offer an alternative to conventional continuous-time analogue processing. The key elements in a sampled data system are: the sample and hold, the quantiser, a DSP unit and a D/A converter. Such systems can be analysed and designed by modelling the sampling process by multiplication with an impulse train and the D/A by a ZOH. Using this model, the effects of undersampling and aliasing can be examined and the sampling theorem can be verified. Practical systems often require two additional components, namely an anti-aliasing filter and a reconstruction filter.

The z-transform is a modified version of the Laplace transform and provides a convenient mechanism for the design of discrete-time systems. Its application is very similar to the Laplace transform and many of the techniques associated with the Laplace transform have direct counterparts in the z-transform.

Figure 4.28 illustrates the interrelationships between four alternative descriptions of a discrete-time linear system. Such systems can be described by difference rather than differential equations. Taking the z-transform of the difference equation and performing some simple algebraic manipulation yields the system transfer function $H(z)$. If this transfer function is expressed as a ratio of polynomials in z, the roots of the numerator polynomial are the zeros and the roots of the denominator polynomial are

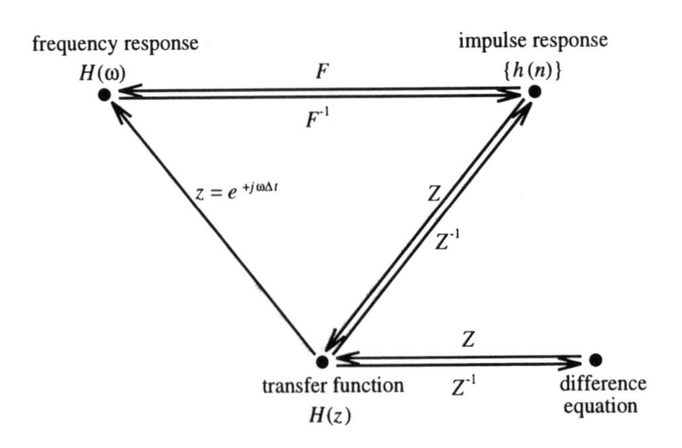

Figure 4.28 *Hitchhiker's guide to discrete-time linear systems.*

the poles. The similarity between the Fourier, Laplace and z-transforms leads to a straightforward way of obtaining the frequency response $H(\omega)$ from the transfer function $H(z)$, i.e. replace z with $\exp(j\omega\Delta t)$. This substitution, combined with the properties of complex numbers, makes it possible to infer the nature of the frequency response from a pole/zero diagram, providing insight into how the position of the poles and zeros affects the frequency response.

The poles of a discrete-time linear system define the denominators in a partial fraction expansion of the transfer function, and are closely related to the impulse response of the system. Hence a system will be stable provided all of its poles lie inside the unit circle in the z-plane.

4.9 Problems

4.1 An S/H operates at a sampling frequency of 10 kHz. Determine and sketch the Fourier transform of the sampled signal when: (a) a 3 kHz cosine wave is applied; (b) an 8 kHz cosine wave is applied; (c) a 12 kHz cosine wave is applied; (d) a 118 kHz cosine wave is applied. If the sampled signal is applied to a perfect low-pass filter with a cut-off frequency of 5 kHz, what frequency will be present at its output in the above four cases?
[(a) 3 kHz; (b) 2 kHz; (c) 2 kHz; (d) 2 kHz]

4.2 In recording studios it is possible to enhance a singer's performance by constructing a digital delay and adding the delayed signal to the original signal. A delay of T seconds is achieved by sampling the signal, storing each sample in random access memory (RAM), and reading the sample out of RAM T seconds later. If the highest frequency to be recorded is 15.8 kHz, what is the minimum sampling frequency which can be used? How many memory locations would be required to produce a delay of 0.5 s. [31.6 kHz; 15 800 locations]
A cheaper delay unit can be manufactured for the guitar because the highest note that the guitar produces has a fundamental frequency of 2 kHz and the only significant harmonic is the third. What is the minimum sampling frequency in this case? How many memory locations would be required to produce a delay of 0.5 s? If the fourth harmonic of the highest note is present on some makes of guitar, at what frequency will it appear at the output of the cheaper unit? Will this be pleasing to the ear?
[12 kHz; 6000 locations; 4 kHz, in this case yes]

4.3 Derive the impulse response, transfer function, magnitude and frequency response of a ZOH. Sketch your results.

4.4 A signal, $x(t)$, has the following magnitude Fourier transform: The signal is applied to a S/H–A/D combination with a sampling rate of 8 kHz. The samples are read in by a digital filter implemented on a TMS32010 microprocessor with the following algorithm:

$$y(n) = x(n) + x(n-3)$$

Finally there is a D/A to produce an analogue output $y(t)$ from the sampled signal produced by the digital filter. Sketch the magnitude Fourier transform of the output.

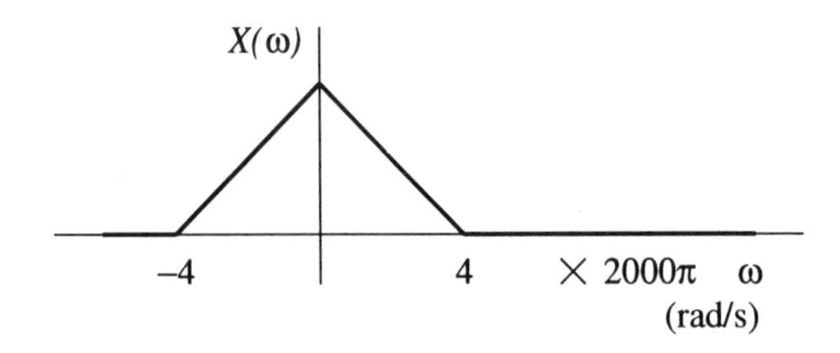

4.5 Given the following unit sample response sequences, calculate $H(z)$ without using z-transform tables. Sketch the pole/zero pattern in the z-plane and the filter block diagram. (a) $h(0) = 1$, $h(1) = 1$, $h(n) = 0$ otherwise; (b) $h(0) = 1$, $h(1) = -1$, $h(n) = 0$ otherwise; (c) $h(0) = 1$, $h(1) = -2$, $h(2) = 1$, $h(n) = 0$ otherwise; (d) $h(n) = r^n \sin(\omega_0 n)$ (let $\omega_0 = \pi/8$ and $r = 0.9$).
[(a) $H(z) = 1 + z^{-1}$; (b) $H(z) = 1 - z^{-1}$; (c) $H(z) = 1 - 2z^{-1} + z^{-2}$; (d) $H(z) = 0.344z^{-1}/(1 - 1.663z^{-1} + 0.81z^{-2})$]

4.6 Determine the transfer function and the pole/zero map for the following digital filter:

$$y(n) = 1.6\ y(n-1) - 0.8\ y(n-2) + x(n)\ .$$

Determine the first eight terms of the unit pulse response $\{h(n)\}$ using the above algorithm. Derive a general expression for the unit pulse response using z-transform methods. Use discrete convolution to calculate the first eight samples of the output when the following sequence is applied to the filter:

$$0,\ 0.25,\ 0.5,\ 0.75,\ 1.0,\ 0,\ 0,\ \cdots,\ 0$$

[1.0, 1.6, 1.76, 1.54, 1.05, 0.45, –0.12, –0.55, $\cdots$;
$h(n) = 2.236(0.894)^n \sin(0.4636(n+1))$; 0, 0.25, 0.9, 1.99, 3.46, 3.95, 3.55, 2.52, $\cdots$]

4.7 For the transfer function

$$H(z) = \frac{z^2 + 2z + 1}{z^3}$$

make sketches of the magnitude and phase frequency response based on the position of the z-plane poles and zeros.

4.8 A digital filter is described by the following transfer function:

$$H(z) = \frac{(z+1)(z^2 - 0.6489z + 1.1025)}{(z^2 + 1.3435z + 0.9025)}$$

Is the filter stable? Sketch the frequency response. At what frequency is the maximum gain? Without using graphical methods, calculate the maximum gain in dB.
[Yes; maximum gain at 3/8th of the sampling frequency; gain $= 27.5$ dB]

4.9 The music signal from an electronic synthesiser is a periodic waveform with a fundamental frequency of 1 kHz. All the power in the signal is contained in the first three harmonics. The amplitude of the second harmonic is half the amplitude of the fundamental and the amplitude of the third harmonic is half the amplitude of the second. The signal is applied to a digital signal processing system which includes an anti-aliasing filter with frequency response $2000\pi/(2000\pi + j\omega)$ and an analogue-to-digital converter which operates at a sampling frequency of 5.4 kHz. The sampled signal is processed using a digital filter which is defined by the following difference equation:

$$y(n) = x(n) - 0.7922x(n-1) + x(n-2)$$

What frequencies are present at the output of the digital filter and what are their relative powers? [2 kHz and 2.4 kHz with relative powers of 0.12 and 0.02 respectively]

CHAPTER 5

Infinite impulse response digital filters

5.1 Introduction

This chapter reviews the realisation of analogue low-pass filters using Butterworth and Chebyshev polynomials to control the pole/zero placement positions and consequent frequency response. It investigates the differences in the passband shape and transition band roll-off rate for these particular prototype filter designs. Digital filter design is then introduced via the difference equation of Chapter 4 and the various filter structures are explored. This chapter concentrates on the design approaches for realising infinite impulse response (IIR) digital filters, showing how the Butterworth and Chebyshev prototype designs are developed into practical digital filters via the bilinear z-transform. Finally there is a brief discussion of some of the limitations imposed on the resulting frequency response by finite precision arithmetic effects in these digital filters.

This chapter thus concentrates exclusively on the recursive digital filters which are developed directly from Figure 4.20. There are four major reasons for the use of digital filters as opposed to passive or active analogue filters:

(i) It is extremely difficult to realise *RLC* filters with low cut-off frequencies. However, in the case of digital filters, cut-off frequency is determined by the clock frequency and the filter coefficient values. It is, therefore, possible to realise digital filters with very low cut-off frequencies which achieve high filter performance, i.e. provide good stopband rejection.

(ii) Digital filters are programmable. They may be multiplexed with different coefficient values to realise distinct characteristics. For instance, one signal can experience a low-pass filter response and another signal a bandpass response, simply by time-division multiplexing the input signal and accessing the stored frequency characteristics using sets of coefficients stored in random access memory (RAM). Alternatively a desired filter design can be split into stages and the input signal progressively filtered in these multiplexed stages within a single filter structure.

(iii) Digital filters do not suffer from the aging distortion of analogue filters.

(iv) There is also no need for impedance matching with digital filters.

The disadvantages of digital filters are few. The most significant of these has, historically, been the relatively high cost of digital filters. However, the increasing use of custom VLSI circuits and fast DSP microprocessor architectures, □, has displaced analogue filters to only low cost and high frequency applications.

Before introducing the design of digital filters, it is necessary to review the analogue filter prototypes that are widely used as the basis for specifying the frequency response of a digital filter. This chapter subsequently investigates recursive digital filters whose design is based on these analogue prototype filters.

5.2 Analogue prototype filters

5.2.1 Introduction

Most required frequency filters fall into four categories, i.e. low-pass, high-pass, bandpass and bandstop. The usual technique used to design these filters is to specify a prototype low-pass filter function. This is usually normalised to provide a cut-off frequency at 1 rad/s and so transformations are applied to achieve first the actual desired cut-off frequencies and second to alter the filter type to a bandpass or high-pass design. In this section some common approximations used for the design of prototype low-pass filters are discussed. The transformations used to realise the practical digital filter functions are then presented in section 5.4.

The ideal, unity gain, low-pass filter prototype with a 'brick wall' response has a unit amplitude frequency response from DC to 1 rad/s, with the response then dropping to zero (see Figure 5.1). [Papoulis] defines this in terms of the squared magnitude transfer function:

$$H(\omega)\,H(-\omega) = |H(\omega)|^2 = \frac{1}{1 + F(\omega^2)} \tag{5.1}$$

where:

$$F(\omega^2) = \begin{cases} 0, & 0 < \omega < 1 \\ \infty, & \omega > 1 \end{cases} \tag{5.2}$$

Since $F(\omega^2)$ is a polynomial in ω^2:

$$F(\omega^2) = \lim_{n \to \infty} \omega^{2n}$$

to satisfy the lower condition in equation (5.2). It is impractical to attempt to form a

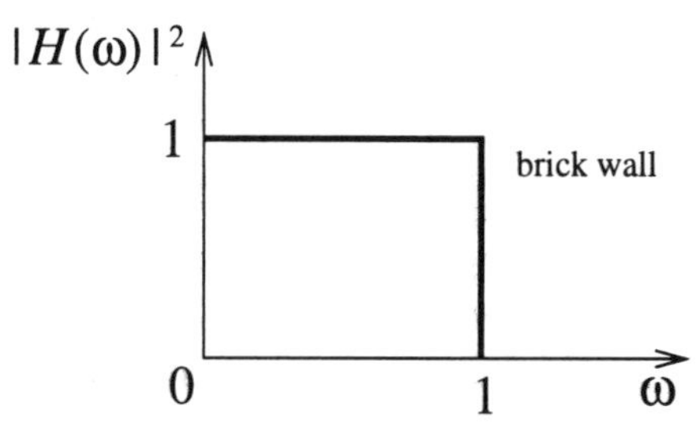

Figure 5.1 *Ideal 'brick wall' low-pass filter.*

polynomial with infinite order, since an infinite number of real reactive components are required. Therefore approximations to $F(\omega^2)$ must be made.

Self assessment question 5.1: What overall shape will the envelope of the impulse response possess for the 'brick wall' frequency response of Figure 5.1?

5.2.2 Butterworth prototype polynomials

The approximation for $F(\omega^2)$ due to Butterworth is:

$$F(\omega^2) = \omega^{2n} \tag{5.3}$$

This provides the following squared magnitude transfer function:

$$|H(\omega)|^2 = \frac{1}{1+\omega^{2n}} \tag{5.4}$$

which leads to:

$$H(\omega) = \frac{1}{\sqrt{(1+\omega^{2n})}} \tag{5.5}$$

where n is a positive integer. To localise the poles of this function, let $p = j\omega$. Therefore (5.5) becomes:

$$H(p) = \begin{cases} \dfrac{1}{(1+p^{2n})^{1/2}}, & \text{for } n \text{ even} \\ \dfrac{1}{(1-p^{2n})^{1/2}}, & \text{for } n \text{ odd} \end{cases} \tag{5.6}$$

It can be shown that for n even, the poles are given by $p = \exp(\pm j\pi/2n)$, while for n odd the poles are given by $p = \exp(\pm j\pi/n)$.

For this Butterworth filter the poles all lie in the left-hand s-plane on a trajectory comprising a circle of unit radius whose centre is the s-plane origin. (The poles for the $|H(-\omega)|$ filter lie in the right-hand s-plane.) For the case $n = 3$, these are shown in the outer curve of Figure 5.2, which is derived from Figure 3.1. Note that the poles are spaced apart by π/n radians or 60° in Figure 5.2 for the 3rd order filter. Odd order filters have a negative real pole, while even order filters have only complex conjugate pole pairs.

The Butterworth filter transfer function possesses a magnitude frequency response which is called 'maximally flat' as there are *no* significant deviations in the passband. The parameter, n, determines how closely we approximate to the ideal 'brick wall' frequency response. Plots of magnitude frequency response $|H(\omega)|$ versus gain in dB, for three values of filter order, $n = 3, 5, 7$, are shown in Figure 5.3(a), while the phase responses are provided in Figure 5.3(b). These are normalised, with respect to a unity −3 dB cut-off frequency. In the derivation of a Butterworth transfer function, $H(s)$ becomes more complicated as n increases beyond $n = 3$. Tables of the polynomial coefficients:

$$H(s) = \frac{1}{b_n s^n + b_{n-1}s^{n-1} + \cdots + b_1 s + 1} \tag{5.7}$$

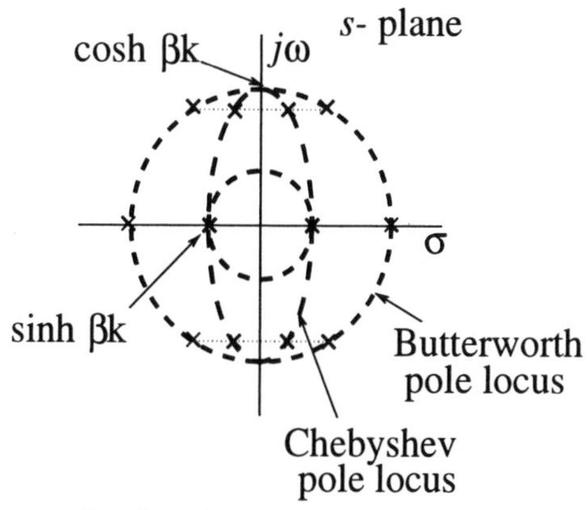

Figure 5.2 *Butterworth and Chebyshev pole positions for order 3 filters.*

are found in many texts [Zverev]. In Figure 5.3(a) the filter roll-off rate into the stopband is approximately $6n$ dB per octave. Thus higher-order filters produce much more rapid roll-off and hence attenuation of unwanted components.

It is more common to realise high-order filters as a cascade of 1st and 2nd order filter sections. It is then required to factor the denominator polynomial of $H(s)$ and again tables of these factored forms may be found in numerous texts. The first 5 orders are listed in Table 5.1.

Self assessment question 5.2: For a 4th order Butterworth filter with a −3 dB cut-off frequency of 1 kHz, how many poles are there in the filter design?

Self assessment question 5.3: For a 4th order Butterworth filter with a −3 dB cut-off frequency of 1 kHz, what is the theoretical filter attenuation at 2, 10 and 100 kHz?

Self assessment question 5.4: You are required to design a filter with a −3 dB cut-off frequency of 3 kHz and −60 dB stopband rejection at 12 kHz. What order of filter would you select?

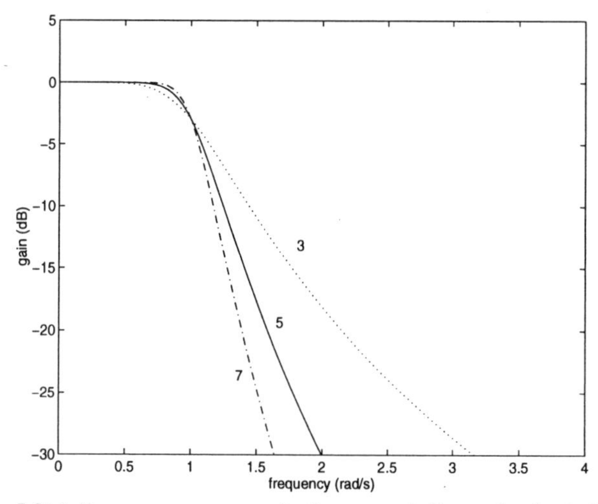

Figure 5.3(a) *Frequency responses for Butterworth filters of order 3, 5 and 7.*

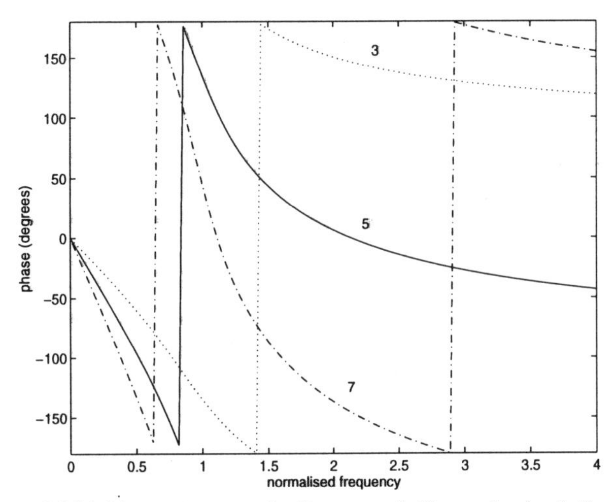

Figure 5.3(b) *Phase responses for Butterworth filters of order 3, 5 and 7.*

5.2.3 Chebyshev prototype polynomials

This polynomial for $F(\omega^2)$ is the so-called equal or 'equiripple' approximation, which has a passband ripple of constant amplitude with maximum band edge roll-off rate. Plots of Chebyshev filter magnitude response $|H(\omega)|$, in dB, and phase, in degrees, are shown in Figure 5.4, for a 2 dB passband ripple design. Note that the number of ripples in the passband is approximately half the filter order. The magnitude roll-off rate in Figure 5.4(a) is much faster than in Figure 5.3(a), but the rate of phase change with frequency is degraded in Figure 5.4(b) compared with Figure 5.3(b). The transfer function, $H(\omega)$, for the Chebyshev filter is given by:

$$H(\omega) = \frac{1}{(1 + \varepsilon^2\, C_n^2(\omega))^{1/2}} \tag{5.8}$$

where $C_n(\omega)$ is the Chebyshev polynomial of order n which is defined by:

$$C_n(\omega) = \begin{cases} \cos\,(n \cos^{-1} \omega), & |\omega| \leq 1 \\ \cosh\,(n \cosh^{-1} \omega), & |\omega| > 1 \end{cases} \tag{5.9}$$

Table 5.1 *Butterworth polynomials in factored form.*

n	*Butterworth polynomials*
1	$s + 1$
2	$s^2 + \sqrt{2}s + 1$
3	$(s^2 + s + 1)\,(s + 1)$
4	$(s^2 + 0.76536s + 1)\,(s^2 + 1.84776s + 1)$
5	$(s + 1)\,(s^2 + 0.6180s + 1)\,(s^2 + 1.6180s + 1)$
6	$(s^2 + 0.518s + 1)\,(s^2 + 1.414s + 1)\,(s^2 + 1.932s + 1)$

The magnitude of the passband ripple is controlled by the design parameter ε. One can solve:

$$\delta = 1 - \frac{1}{(1+\varepsilon^2)^{1/2}} \tag{5.10}$$

for a given value of ripple, δ, to set the appropriate value for the parameter ε. The definition of the Chebyshev polynomial in equation (5.9) means that rather than having poles with a circular locus on the s-plane, a Chebyshev filter has poles with an elliptical locus (see Figure 5.2). The actual values of the polynomials, $C_n(\omega)$, may be found in standard filter design texts and an example set are provided in Table 5.2, again factored into second order sections, for the more widely applied $\delta = ½$ dB ripple design rather

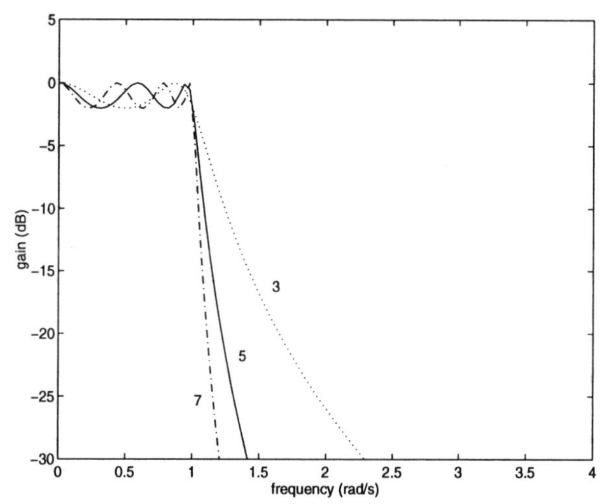

Figure 5.4(a) *Frequency responses for 2 dB ripple Chebyshev filters of order 3, 5 and 7.*

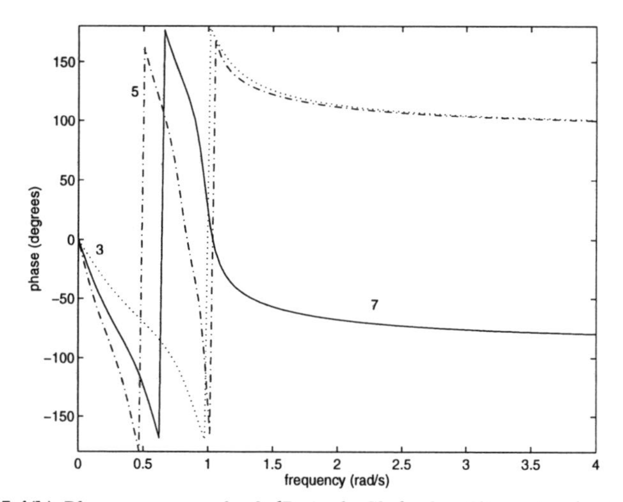

Figure 5.4(b) *Phase responses for 2 dB ripple Chebyshev filters of order 3, 5 and 7.*

than the relatively high $\delta = 2$ dB ripple employed for display purposes in Figure 5.4.

Table 5.2 *Chebyshev polynomials for ½ dB ripple filter design.*

n	*Chebyshev polynomials*
1	$s + 2.863$
2	$s^2 + 1.425s + 1.516$
3	$(s^2 + 0.626s + 1.142)\,(s + 0.626)$
4	$(s^2 + 0.351s + 1.064)\,(s^2 + 0.845s + 0.356)$

A comparison of Butterworth and Chebyshev filter responses for several different complexities or orders is shown in Figures 5.3 and 5.4. Many other types of filter response approximation are available, e.g. linear phase, approximations based on Bessel polynomials. Bessel filters provide least phase distortion (for a simple design) but their roll-off rate into the stopband is inferior to that of Butterworth designs.

Self assessment question 5.5: What are the principal advantages and disadvantages of Chebyshev prototype filters compared with Butterworth filters?

5.3 Digital filter structures

5.3.1 Introduction

The emphasis in Chapter 4 was on the *analysis* of digital filters, e.g. given the difference equation, what is the corresponding transfer function and frequency response. In this chapter the emphasis is on filter *design* so that, given a desired frequency response, it is shown how we determine the required transfer function and hence arrive at the required difference equation to construct the digital filter. The z-transform eases the manipulation of the difference equation and hence can lead to alternative realisations for a given digital filter. These alternative realisations are important when considerations such as hardware or software capabilities are included in the design process.

One of the most important procedures is translating a given transfer function into a difference equation. This is usually straightforward and is illustrated by the following example.

EXAMPLE 5.1
For the z-transform equation:

$$H(z) = \frac{z^2 - 0.2z - 0.08}{z^2 + 0.5}$$

derive the difference equation for the corresponding digital filter.

Solution
This must be normalised into negative powers of z and, using equation (4.9):

$$H(z) = \frac{1 - 0.2z^{-1} - 0.08z^{-2}}{1 + 0.5z^{-2}}$$

$$Y(z)\,(1 + 0.5z^{-2}) = X(z)\,(1 - 0.2z^{-1} - 0.08z^{-2})$$

The inverse z-transform then yields the difference equation:

$$y(n) + 0.5y(n-2) = x(n) - 0.2x(n-1) - 0.08x(n-2)$$

or:

$$y(n) = x(n) - 0.2x(n-1) - 0.08x(n-2) - 0.5y(n-2)$$

This gives $y(n)$ as a function of present and past input $x(n-k)$ and output $y(n-k)$ sample values. The difference equation is very important in the sense that it illustrates the fact that the present sampled data output $y(n)$ from a digital filter is a combination of the weighted present and past inputs to the system as well as weighted past outputs.

Self assessment question 5.6: For the difference equation $y(n) = x(n) + 10x(n-2) + 5y(n-2)$ draw the block diagram for the direct form of this structure with delay and multiply symbols.

5.3.2 The canonical form

The direct form considered in Chapter 4 is the simplest realisation or architecture for a digital filter. When implementing a filter in hardware or software, alternative realisations may be considered to minimise the propagation of numerical errors or to minimise memory requirements. The canonical form of a digital filter is the realisation which requires the *minimum* memory to implement a particular transfer function.

Figure 5.5 shows how the canonical form can be developed from the direct form. At first sight it is not obvious that Figure 5.5(a) and Figure 5.5(c) have the same transfer function. They have the same coefficients yet their structure is quite different. The key to understanding the relationship between them is linearity. Figure 5.5(a) (which repeats Figure 4.20) shows the direct form which is essentially two filters connected in series (cascade). The input signal $x(n)$ is applied to filter 1 to form an intermediate output $x'(n)$. This signal in turn is the input to a second filter (filter 2) whose output is $y(n)$. Since both filter operations are linear it does not matter in which order they are applied. So if we use filter 2 first, followed by filter 1, as in Figure 5.5(b), the overall output will still be $y(n)$.

Examination of Figure 5.5(b) suggests that there is a lot of redundancy in terms of data storage within this structure. The signal sample at point A' in filter 2 will obviously be the same as the signal at point A in filter 1. More importantly, the signal sample at point B' in filter 2 will be the same as the signal at point B in filter 1, since the signal at B is the signal at A delayed by one sample, and the signal at B' is the signal at A' delayed by one sample. By a similar argument, the signal sample at C is identical to the signal at C'. Hence the delay line structure in filter 2 is identical to the delay line structure in filter 1. When this redundancy is removed it leads to the canonical form structure of Figure 5.5(c). In this final design it is normal for M to equal N.

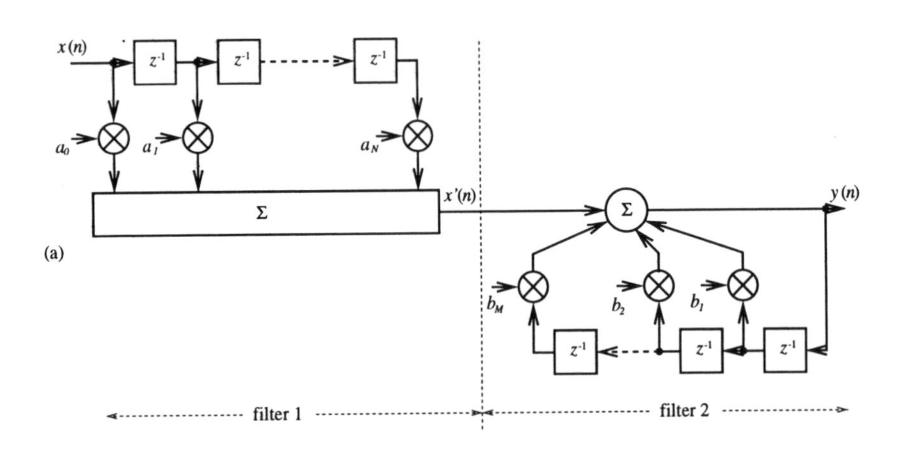

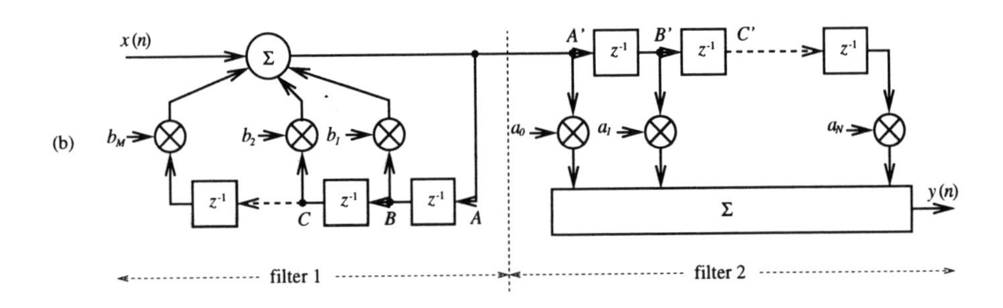

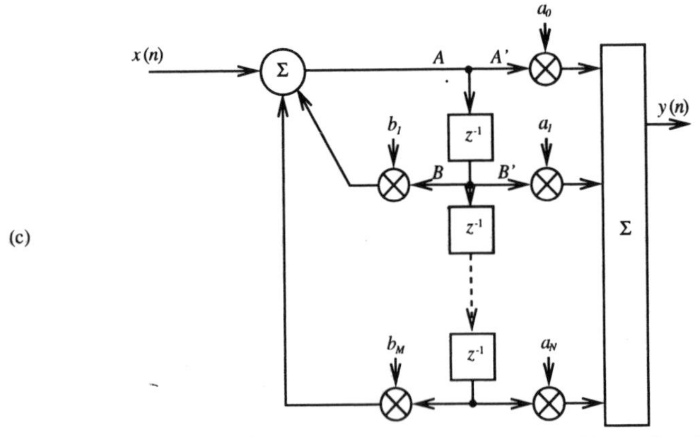

Figure 5.5 *Digital filter realisations: (a) direct form; (b) reordered; (c) canonical form.*

The formal derivation of the canonical form of the Nth order digital filter may be obtained by examining again the transfer function $H(z)$ of equation (4.9):

$$H(z) = \frac{Y(z)}{X(z)} = \frac{\sum_{i=0}^{N} a_i z^{-i}}{\left(1 - \sum_{i=1}^{M} b_i z^{-i}\right)} \tag{5.11}$$

Thus, for input signal $X(z)$, the output is given by:

$$Y(z) = H(z)\, X(z) \tag{5.12}$$

Let:

$$Y'(z) = \frac{X(z)}{\left(1 - \sum_{i=1}^{M} b_i z^{-i}\right)}$$

then:

$$Y(z) = \sum_{i=0}^{N} a_i z^{-i}\, Y'(z)$$

or, in sampled data notation, with some mathematical manipulation:

$$y'(n) = x(n) + \sum_{i=1}^{M} b_i y'(n-i) \tag{5.13}$$

and:

$$y(n) = \sum_{i=0}^{N} a_i y'(n-i) \tag{5.14}$$

Note in Figure 5.5(b) that the sampled data signal A', B', etc., is $y'(n)$ and:

$$y(n) = \sum_{i=0}^{N} a_i x(n-i) + \sum_{i=1}^{M} b_i y(n-i) \tag{5.15}$$

The design of the digital filter involves the calculation of the constants a_i and b_i. The b values control the positions of the filter poles, the a values the zero positions, and the values themselves have to be derived from the prototype analogue filters of section 5.2.

If at least one of the coefficients b_i in equation (5.15) is non-zero, then the filter is classified as a recursive filter. If, however, all the coefficients b_i are identically zero, then the filter is called a non-recursive filter.

Self assessment question 5.7: What is the theoretical duration of the impulse response for a recursive filter?

Note that the recursive filters have an infinite memory since the present output is a function of the past outputs, and they result in IIR filters. Non-recursive filters have no past output memory and they result in finite impulse response (FIR) filters, whose impulse response duration is controlled by the overall filter delay (Chapter 6).

5.3.3 Parallel and cascade realisations

The two significant digital filter realisations are the series cascade and parallel connection. For the cascaded realisation, the transfer function is factorised as a product of second order terms $K_n(z)$. Thus:

$$H(z) = C \prod_{n=1}^{N} K_n(z)$$

where C is a constant. The second order sections are then placed in cascade because the factors are multiplied to form the transfer function (Figure 5.6). For the parallel realisation, the transfer function is expressed as a partial-fraction expansion of second order terms. Thus:

$$H(z) = C + \sum_{n=1}^{N} K_n(z)$$

where C is again a constant. The second order terms are placed in parallel because they are added to form the transfer function of Figure 5.7. The basic filter type used in such a cascade is the so-called biquadratic filter shown in Figure 5.8. This is the canonical form for a second order system.

In principle a high-order digital filter design can thus be synthesised as either a series cascade or parallel connection of second order sections. Although the parallel connection is generally less sensitive to filter coefficient errors, the series cascade is favoured because many second order sections have simple integer coefficients and, more importantly, most commercial software design packages assume such a cascaded design.

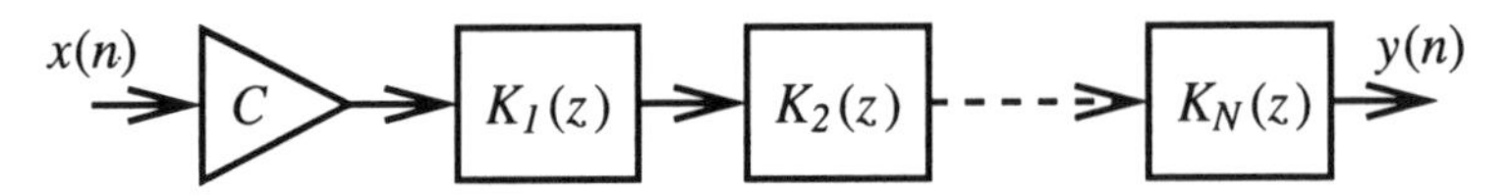

Figure 5.6 *Series cascade of lower-order filters.*

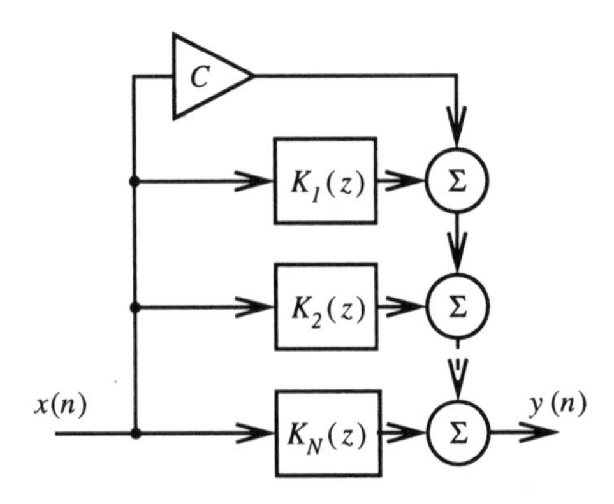

Figure 5.7 *Parallel connection of lower-order filters.*

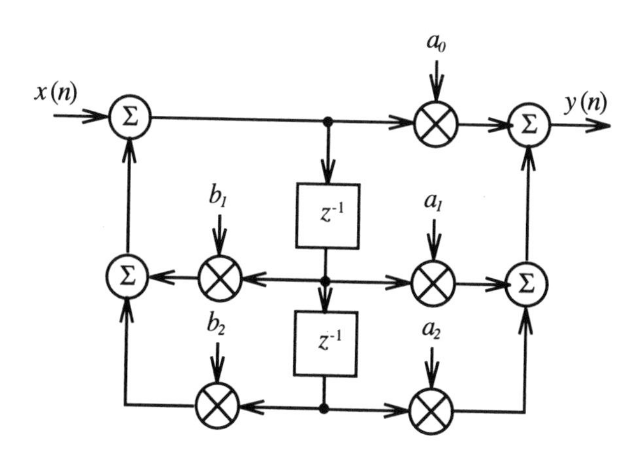

Figure 5.8 *Biquadratic digital filter section.*

There are also subtle differences in performance between the direct and canonical forms of Figure 5.5 and, if low output noise is of paramount importance, then the direct form is usually selected and the extra penalty of more arithmetic operations is tolerated.

5.4 Filter design methods

5.4.1 Introduction

Digital filter designs are predominantly based on analogue prototype frequency response functions, e.g. Butterworth and Chebyshev low-pass filter functions, as described in section 5.2. No single transform exists to perfectly map $H(s)$ to $H(z)$. They all suffer some defect: aliasing, frequency warping, etc. Three principal design approaches are:

(i) Impulse invariant – produces an $H(z)$ whose impulse response is identical to a sampled version of $h(t)$.

(ii) Matched z-transform – poles and zeros of $H(s)$ are directly mapped into poles and zeros of $H(z)$.

(iii) Bilinear z-transform – the whole left half s-plane is mapped into the unit circle in the z-plane in one go, i.e. the whole $|H(\omega)|$ from DC to infinite frequency is telescoped into the range $\omega = 0$ to $\omega_s/2$, where ω_s is the sample rate.

The impulse invariant design is the simplest approach, requiring the prototype (e.g. Butterworth) polynomial to be scaled to the desired filter cut-off frequency and then solved for the a_i and b_i coefficient values. While this gives an accurate filter representation, and is useful for implementing low-pass and bandpass filters, the aliased components due to the sampling at ω_s (Figure 4.11) make this filter impractical for many applications when the $H(\omega)$ response does not approach zero for large ω values.

The matched z-transform is a special case of the impulse invariant technique which maps both zeros and poles directly from the s- into the z-plane. It is usable, but it possesses no particular virtues in terms of its time- or frequency-domain properties and hence is not widely applied.

Superior closed form digital filter designs use the bilinear z-transformation as this avoids any deleterious aliasing effects (Figure 4.11). It is thus the most frequently used design technique and is described further below.

5.4.2 The bilinear z-transform

This transformation makes use of a mapping of the complete imaginary s-plane axis between $\pm\infty$ onto the z-plane unit circle. To facilitate this, the continuous transformation:

$$\omega_a = \frac{2}{\Delta t}\tan\left(\frac{\omega_d \Delta t}{2}\right) \tag{5.16}$$

is used to map the continuous analogue frequency variable ω_a into a modified frequency variable ω_d. This effectively compresses or warps the entire frequency spectrum of the analogue function, described in terms of ω_a, into a much restricted finite interval of $\omega_s/2$ in Figure 5.9. This removes completely the possibility of any aliasing as the digital filter forces the frequency response to be zero at a frequency corresponding to half the sample rate. The warping algorithm is conveniently stated in terms of the variable s by expanding equation (5.16):

$$s = \frac{2}{\Delta t}\frac{j\sin\left(\frac{\omega_d \Delta t}{2}\right)}{\cos\left(\frac{\omega_d \Delta t}{2}\right)} \tag{5.17}$$

which may be rewritten using $\exp(jx) = \cos x + j\sin x$ as:

$$s = \frac{2}{\Delta t}\frac{\frac{1}{2}\left[\exp\left(\frac{j\omega_d \Delta t}{2}\right) - \exp\left(\frac{-j\omega_d \Delta t}{2}\right)\right]}{\frac{1}{2}\left[\exp\left(\frac{j\omega_d \Delta t}{2}\right) + \exp\left(\frac{-j\omega_d \Delta t}{2}\right)\right]}$$

or:

$$s = \frac{2}{\Delta t}\frac{[1 - \exp(-j\omega_d \Delta t)]}{[1 + \exp(-j\omega_d \Delta t)]} \tag{5.18}$$

Applying the standard z-transform, $z = \exp(s_d \Delta t) = \exp(j\omega_d \Delta t)$, yields:

$$s = \frac{2(1 - z^{-1})}{\Delta t(1 + z^{-1})} \tag{5.19}$$

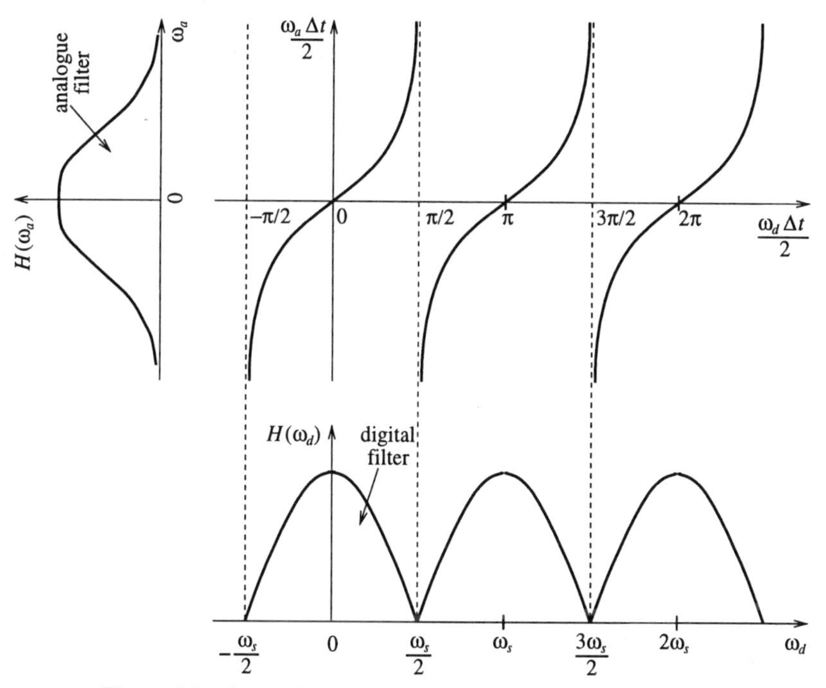

Figure 5.9 *Spectral compression due to the bilinear z-transform.*

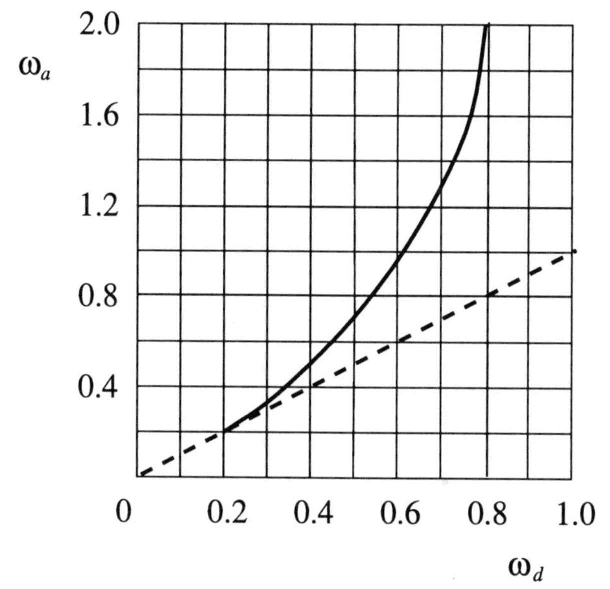

Figure 5.10 *Frequency warping effect of the bilinear z-transform.*

Equation (5.19) provides the bilinear z-transform.

This now permits the design of a digital filter transfer function $H(z)$ from the provided analogue filter function $H(s)$ by direct use of the substitution given in equation (5.19). However, as a consequence of the frequency warping, introduced by equation (5.16), the same transfer function (frequency response) shape will not be obtained. Thus the analogue function must be 'pre-warped' to retain the correct positions for the critical break point (cut-off) frequencies. This is conveniently achieved by the tan function in equation (5.16) and the effect of this is shown in Figure 5.10, and illustrated further in the example 5.2.

Self assessment question 5.8: A 5th order digital filter is to be designed using the bilinear z-transform with a sample rate of 50 kHz and a −3 dB cut-off frequency of 5 kHz. What is the filter attenuation at 25 kHz?

EXAMPLE 5.2

Design the digital equivalent of a 2nd order Butterworth low-pass filter with cut-off frequency ω_d = 628 rad/s and sampling frequency ω_s = 5024 rad/s.

Solution

The analogue prototype 2nd order filter is given from Table 5.1 as:

$$H(s) = \frac{1}{s^2 + \sqrt{2}s + 1}$$

The pre-warping substitution of equation (5.16) is first applied to the critical cut-off frequency ω_d to yield the analogue filter cut-off frequency prior to warping:

$$\omega_a = \left(\frac{2}{\frac{1}{800}} \right) \tan\left(\frac{2\pi\ 100}{2 \times 800} \right) = 663 \quad \text{rad/s}$$

This calculation indicates that to achieve a −3 dB cut-off frequency of 628 rad/s in the digital filter requires an equivalent analogue cut-off frequency, before warping, of 663

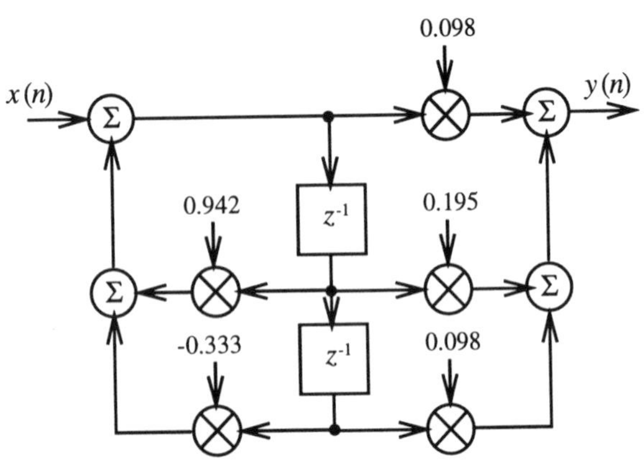

Figure 5.11 *Second order Butterworth coefficients for an $\omega_s/8$ cut-off frequency design.*

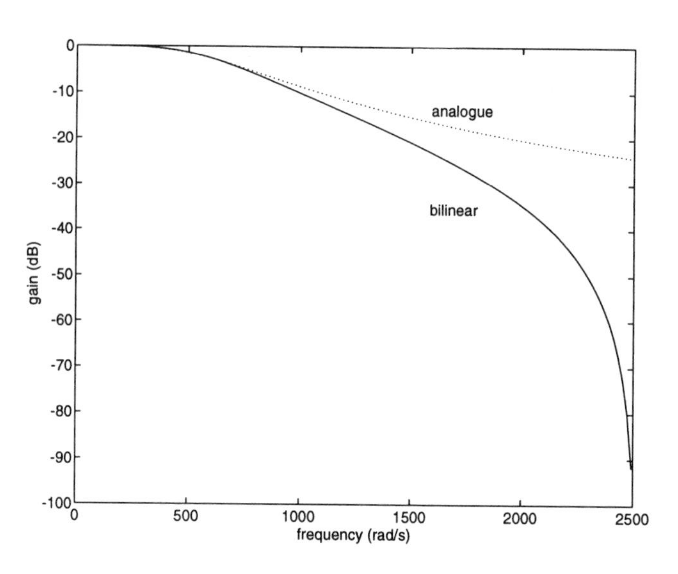

Figure 5.12 *Characteristics for 2nd order analogue and digital Butterworth filters.*

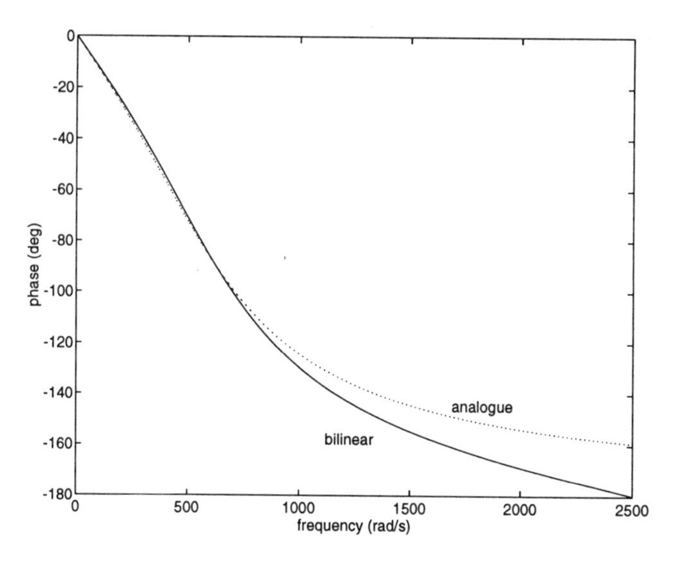

Figure 5.13 *Phase responses for 2nd order analogue and digital Butterworth filters.*

rad/s (Figure 5.10). This resulting 663 rad/s value is thus used to denormalise the prototype filter function of equation (5.20):

$$H(s_d) = \frac{1}{\left(\dfrac{s}{663}\right)^2 + \dfrac{\sqrt{2}s}{663} + 1}$$

The bilinear z-transform, equation (5.19), is now applied to equation (5.21) to yield, after algebraic simplification, the equivalent digital filter transfer function:

$$H(z) = \frac{0.098\, z^2 + 0.195\, z + 0.098}{z^2 - 0.942\, z + 0.333}$$

The physical realisation of this function is shown in Figure 5.11 and the magnitude $|H(\omega)|$ and phase responses for the analogue prototype and resulting digital filters are shown in Figures 5.12 and 5.13. As it is a second order filter design the total phase excursion is 180°. The effect of the frequency warp may be seen here as introducing a sharper roll-off rate after the defined critical cut-off frequency, and providing infinite attenuation at the Nyquist frequency, as shown previously in Figure 5.9. This decrease in gain effectively *increases* the order of the digital filter, compared with the original analogue prototype design.

Self assessment question 5.9: For the above filter design, calculate the pole and zero locations.

EXAMPLE 5.3

Design IIR low-pass filters of 4th and 10th order with a cut-off frequency at 10% of the sampling frequency.

Solution

Figure 5.14, which is derived from MATLAB™ software, shows these low-pass magnitude responses. Clearly the increased complexity of the 10th order design gives much faster roll-off into the stopband.

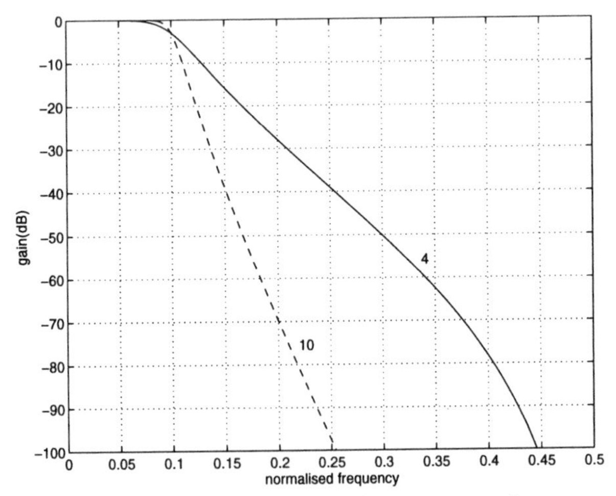

Figure 5.14 *Fourth and 10th order IIR filters based on Butterworth prototype and bilinear z-transform designs.*

5.4.3 Filter transformation

The low-pass sampled-data filter with a cut-off frequency of $\omega_{c_{LP}}$, which is a fraction $1/k$ of the sample rate, ω_s, e.g.:

$$\omega_{c_{LP}} = \frac{\omega_s}{k}$$

can be transformed into a high-pass filter of identical shape with a high frequency cut-off of $\omega_{c_{HP}} = ½ - \omega_{c_{LP}}$ (reflecting the low-pass characteristic about $\omega_s/2$) simply by changing the sign of the zeros and the poles. Thus $z_{HP} = -z_{LP}$ and $p_{HP} = -p_{LP}$, and the corresponding z-transform for the high-pass filter is derived from the low-pass prototype as:

$$H_{HP}(z) = H_{LP}(z)\,|_{z \to -z}$$

Self assessment question 5.10: Convert the filter in Figure 5.11 into an equivalent high-pass design and calculate the new a/b coefficient values.

To achieve a bandpass design with a lower and upper cut-off frequency of ω_{cl} and ω_{cu} and corresponding stopband edge frequencies of ω_{sl} and ω_{su}, all normalised to the sample rate, ω_s, the transformation is effected by replacing z^{-1} in the low-pass design by:

$$z_{LP}^{-1} = \frac{-z^{-1}(z^{-1} - \alpha)}{(1 - \alpha\, z^{-1})} \tag{5.20}$$

where:

$$\alpha = \frac{\cos\left[\dfrac{\omega_{cu}}{2} + \dfrac{\omega_{cl}}{2}\right]}{\cos\left[\dfrac{\omega_{cu}}{2} - \dfrac{\omega_{cl}}{2}\right]} \tag{5.21}$$

The design technique requires the bandpass characteristic to be converted into a low-pass response again using angular frequencies normalised to the sample rate ω_s:

$$\omega_{c_{LP}} = \omega_{cu} - \omega_{cl}$$

$$\omega_{s_{LP}} = \omega_{c_{LP}} + \Delta\omega$$

where the transition bandwidth is $\Delta\omega = \omega_{cl} - \omega_{sl} = \omega_{su} - \omega_{cu}$.

It can be shown that the substitution of equation (5.20) gives:

$$z_{LP} = \frac{-(1 - \alpha\, z^{-1})}{z^{-1}\,(z^{-1} - \alpha)}$$

$$z^{-2}\, z_{LP} - \alpha\, z^{-1}\, z_{LP} = -1 + \alpha\, z^{-1}$$

$$z_{LP} - \alpha\, z\, z_{LP} = -z^2 + \alpha\, z$$

$$z^2 - \alpha\,(z_{LP} + 1)\,z + z_{LP} = 0$$

Solving this quadratic in z yields:

$$z = z_{BP} = \frac{1}{2}\left[\alpha(z_{LP} + 1) \pm \sqrt{\alpha^2\,(z_{LP} + 1) - 4\, z_{LP}}\,\right] \tag{5.22(a)}$$

Using a similar analysis, it can be shown that:

$$p_{BP} = \frac{1}{2}\left[-\alpha(p_{LP}+1) \pm \sqrt{\alpha^2\,(p_{LP}+1) - 4\,p_{LP}}\right] \tag{5.22(b)}$$

The above steps permit a bandpass filter to be synthesised from a low-pass prototype but, as there is an upper and lower roll-off region, the filter order requirement is doubled in the low-pass to bandpass transformation.

To achieve a bandstop design with the same normalised values, ω_{cl}, ω_{cu}, ω_{sl} and ω_{su} as defined before, then the same value of α is calculated as for the bandpass design. This time z^{-1} in the low-pass prototype filter is replaced by:

$$z_{LP}^{-1} = \frac{z^{-1}(z^{-1}-\alpha)}{(1-\alpha\,z^{-1})} \tag{5.23}$$

then:

$$\omega_{c_{LP}} = ½ - (\omega_{cu} - \omega_{cl})$$

$$\omega_{s_{LP}} = \omega_{c_{LP}} + \Delta\omega$$

This gives zeros in the bandstop filter at:

$$z_{BS} = \frac{1}{2}\left[-\alpha(z_{LP}-1) \pm \sqrt{\alpha^2(z_{LP}-1)^2 + 4\,z_{LP}}\right] \tag{5.24(a)}$$

and poles at:

$$p_{BS} = \frac{1}{2}\left[-\alpha(p_{LP}-1) \pm \sqrt{\alpha^2(p_{LP}-1)^2 + 4\,p_{LP}}\right] \tag{5.24(b)}$$

and again the filter order of the bandstop implementation is twice that of the original low-pass prototype because the substitution involves z^{-2} terms.

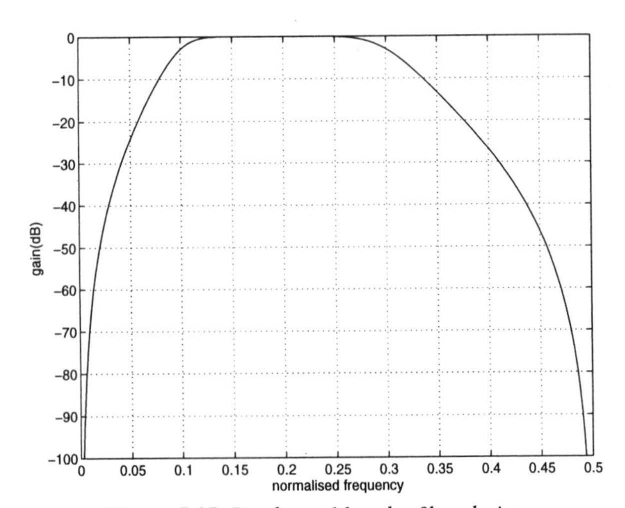

Figure 5.15 *Bandpass 6th order filter design.*

EXAMPLE 5.4
Design a bandpass filter with a lower cut-off frequency of $f_s/10$ and an upper cut-off frequency of $3f_s/10$.

Solution
Figure 5.15 shows a typical bandpass design employing an order 6 design. This again employs the bilinear z-transform design technique and is synthesised using the MATLAB™ commercial design suite [Burrus *et al.*].

5.5 Finite precision effects

5.5.1 Filter coefficient quantisation errors

Since IIR filters are based on a closed loop (recursive) feedback structure they can become unstable, and additional considerations are required. The IIR filter is usually configured as a cascade of second order sections (Figure 5.6). The poles must lie inside the unit circle for a stable filter design to result (Figure 4.22). For the cascade configuration of second order sections, the locations of the poles in each section can be easily made stable by placing the poles at the set of allowable locations, determined by the quantisation accuracy, that lie inside the unit circle. The desired pole must then be transferred to one of the allowable pole locations in the hardware implementation (Figure 5.16(a)), resulting in an alteration in either the resonant frequency or bandwidth.

Figure 5.16 shows, for the IIR filter developed from equation (4.12) and Figure 4.22:

$$H(z) = \frac{1}{(1-0.9\exp(j70°)z^{-1})\,(1-0.9\exp(-j70°)z^{-1})} \tag{5.25}$$

the movement of the pole locations and the corresponding magnitude frequency responses [Kuc]. The responses in Figure 5.16(b) are obtained using the constructions of Figure 4.24(a). Note that the height of the peak depends on the distance from the unit circle, and the frequency of the peak depends on the precise pole location in the z-plane, with positions 2 and 3 providing the lower frequency resonance, and 3 and 4 giving the sharper responses.

For filters which have poles close to the unit circle, e.g.:

$$H(z) = \frac{1}{(1-0.901z^{-1})\,(1-0.943z^{-1})} \tag{5.26}$$

the response can be rounded to the nearest 0.05 to result in stable poles at $z = 0.9$ and 0.95. However, the direct form realisation:

$$H(z) = \frac{1}{1-1.844z^{-1}+0.8496z^{-2}} \tag{5.27}$$

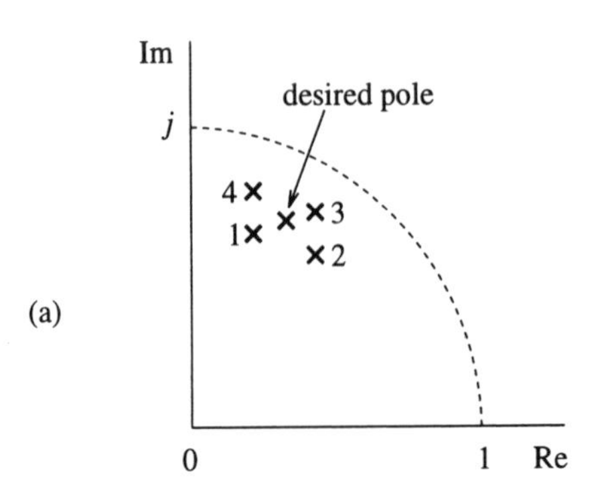

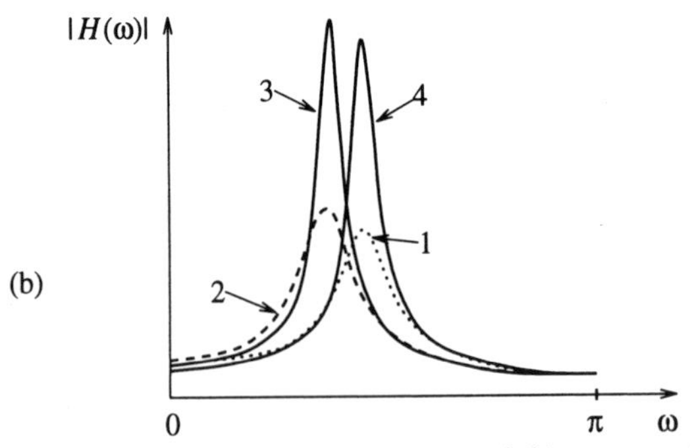

Figure 5.16 *(a) Effect of quantisation on pole placement and (b) corresponding filter magnitude responses for quantising to one of the four distinct pole locations.*

factors into poles at $z = 0.8496$ and unity, causing a pole to lie *on* the unit circle which causes instability (Figure 4.22). This problem increases with the order of the filter.

Several authors have provided estimates for the required precision, b_c, for the internal IIR filter coefficients when using fixed-point arithmetic and here we cite Bellanger's summary as:

$$b_c = \log_2 \left[\frac{1}{(\delta_1 - \delta_{10})} \right] + \log_2 \left[\frac{\omega_s}{\Delta\omega} \frac{1}{2\sin(2\pi\omega_{cl}/\omega_{sl})} \right] \tag{5.28}$$

where δ_1 is the designed passband ripple, δ_{10} is the passband ripple after quantisation, $\Delta\omega$ is again the transition bandwidth $(\omega_{cl} - \omega_{sl})$ and $\omega_s = 2\pi/\Delta t$ is the sample frequency, see later definition in Figure 6.10. Thus in the IIR filter the b_c requirement increases for narrower passband designs. Bellanger also discusses the requirements for the size of the internal memories in IIR and FIR filters. However, these considerations do not apply when the more expensive floating-point arithmetic processors are used, and this is a major attraction for using floating-point arithmetic processors □.

Self assessment question 5.11: What practical effect does quantisation have on the impulse response of an IIR filter?

5.5.2 Limit cycles

Another problem in IIR filters is limit cycles, where the filter can give an oscillatory output when no input is present. This can easily be shown in the first order recursive filter of Figure 4.19 where $a = 0.9$:

$$y(n) = -0.9y(n-1) + x(n) \tag{5.29}$$

With infinite precision this impulse response decays to zero, but finite precision coarse rounding to 0.1 accuracy introduces the impulse response shown in Figure 5.17. Here the output of magnitude 0.45 gets rounded to 0.5 and there is a continuous limit cycle oscillation of the output at ± 0.5 due to the rounding operation.

Limit cycles are avoided in practical IIR filter designs by arranging the filter as a cascade of second order sections (Figure 5.6). If each section is stable (poles inside unit circle) and free from limit cycles then the cascaded higher-order design retains these properties. Many authors [Jackson, Kuc] have shown, for the second order section, the range of permitted values in the recursive coefficients in the b_1, b_2 plane. When:

$$0 < |b_2| < 1$$

and also:

$$b_2 = \frac{b_1^2}{4} \tag{5.30}$$

then the complex conjugate poles will lie within the unit circle. Jackson and Kuc show the range of permitted values for real and complex conjugate poles.

Care must be taken within the arithmetic when the internal values attempt to overflow the available precision in an accumulator. Saturation is the preferred approach as this preserves the magnitude of the value. Forcing the value to zero is usually less attractive.

5.5.3 IIR filter hardware

IIR filter designs are predominantly written in software and implemented in standard DSP microprocessor parts, such as Texas Instruments TMS 320 series, Motorola 56000

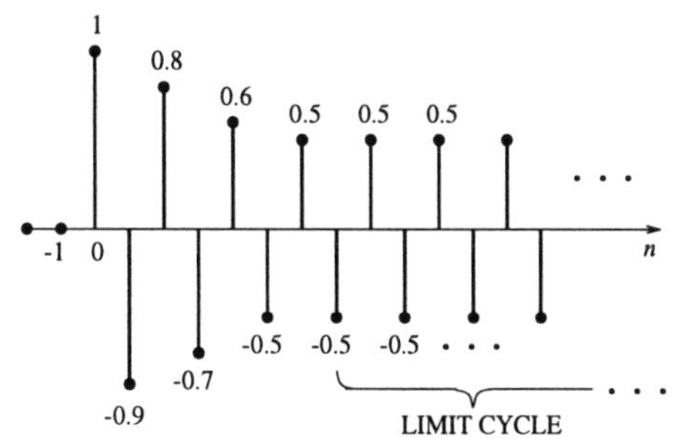

Figure 5.17 *Limit cycle oscillation with rounding in a 1st order recursive filter.*

series or similar products from other vendors. These are fully programmable parts which are available in 16-bit and 24-bit fixed-point and 32-bit floating-point arithmetic, □. The former parts are less expensive and generally can still be used to implement high performance IIR filter designs, with clock rates for a second order section of a few Msample/s. These standard parts are so efficient at implementing IIR filters that, in general, custom parts have not been developed. One exception was the GEC-Plessey MS2014, from the early 1980s, which was a biquadratic IIR filter section with a 64 ksample/s input data rate for PCM telephony applications.

5.6 Summary

This chapter has described the Butterworth and Chebyschev filter prototypes which form the basis for the design of many analogue and digital filters. Following this, the development of the recursive or infinite impulse response digital filter was introduced, from the difference equation which defines the frequency characteristic or response of the filter. The major design methods for arriving at the filter coefficient values were discussed and examples were presented for the bilinear z-transform closed form design technique, which gives the most usable frequency responses. Finally, the chapter concluded with some discussion of finite-precision arithmetic limitations in fixed-point arithmetic and their effects on filter operation and performance were outlined.

5.7 Problems

5.1 The transfer function of a digital filter is $z/(z - ½)$: (a) determine the locations in the z-plane of the filters poles and zeros; (b) state whether or not the filter is stable; (c) sketch the steady-state frequency response of the filter between DC and $\omega = \omega_s/2$, and comment; derive exact values for the magnitude and phase at $\omega = \omega_s/4$; (d) determine the linear difference equation that defines the filter time-domain response; (e) compute the first five samples of the impulse response of the filter; (f) draw a block diagram of a suitable hardware realisation. [(a) 0, 0.5; (b) yes; (c) 0.89, –26.6°; (d) 1, 0.5, 0.25, 0.125 etc.]

5.2 Repeat problem 5.1 for the transfer function $z^2/(z^2 + ½)$.
[0, 0, $\pm j/\sqrt{2}$; yes; 2, ; 0°; 1, 0, –0. 5, 0, 0.25 etc.]

5.3 A second order Butterworth prototype filter having a –3 dB low-pass cut-off at 1 rad/s is defined in Table 5.1. Using the bilinear z-transform, show that an equivalent sampled-data filter sampling at 10^3 rad/s with a cut-off at 100 rad/s has the transfer function:

$$H(z) = \frac{0.067z^{-2} + 0.135z^{-1} + 0.067}{0.413z^{-2} - 1.141z^{-1} + 1}$$

and hence derive a linear difference equation for this sampled-data filter.
[$y(n) = 0.067x(n) + 0.135x(n-1) + 0.067x(n-2) + 1.141y(n-1) - 0.413y(n-2$

5.4 Use the bilinear z-transform to design an infinite impulse response digital low-pass filter based on the third order Butterworth prototype function of Table 5.1. The digital filter is to be designed to operate at a sample rate of 8 kHz with a –3 dB cut-off frequency at 2 kHz. Show that $H(z)$ reduces to:

$$H(z) = \frac{(z+1)^3}{6z\left(z - j\frac{1}{\sqrt{3}}\right)\left(z + j\frac{1}{\sqrt{3}}\right)}$$

Sketch the z-plane pole/zero diagram and the expected shape of the magnitude of the digital filter frequency response from DC to 8 kHz. Calculate from $H(z)$ the individual multiplier weight values, and show clearly on a block diagram where the different weights are employed. What is the roll-off rate in dB/octave in the transition band for this filter design? [1/6, 1/2, 1/2, 1/6, 1/3, 60 dB/decade]

5.5 Use the bilinear transformation method to design a digital filter based on the prototype function $H(s) = s\tau/(1 + s\tau)$. Assume that the desired −3 dB cut-off frequency $(1/\tau)$ is: (i) much lower than the Nyquist frequency, e.g. $\ll \omega_s/100$; and (ii) at $\omega_s/4$.

$$\left[(i)\ \ H(z) = \frac{100(1 - z^{-1})}{(\pi + 100) + z^{-1}\,(\pi - 100)}\ ; \qquad (ii)\ \ H(z) = \frac{z - 1}{2z} \right]$$

CHAPTER 6

Finite impulse response digital filters

6.1 Introduction

The recursive filters of Chapter 5 are very efficient structures for achieving steep-sided filter designs, with low transition bandwidth. However, in many situations, their poor phase response is a severe limitation. Finite impulse response (FIR) filters are non-recursive, making them unconditionally stable, and further they offer the possibility of achieving a linear phase characteristic. However they will require more stages of delay and multiply than an infinite impulse response (IIR) filter with a similar passband magnitude specification. In this chapter both time and frequency descriptions of FIR transversal filters will be examined. Symmetry conditions, to ensure linear phase response, will be outlined and design techniques for linear phase FIR filters will be presented. Finally, there will be a discussion of the effects of finite precision arithmetic on the filter performance and the major applications for FIR filters will be outlined.

6.2 Filter theory and frequency response

6.2.1 Transfer function

The FIR filter is achieved by removing the b coefficients from the general filter difference equation (5.15) to yield the N-coefficient ($N-1$th order) FIR filter:

$$y(n) = \sum_{i=0}^{N-1} a_i \, x(n-i) \tag{6.1}$$

Figure 6.1 shows the physical structure which will realise equation (6.1). Equation (6.1) represents the convolution of the finite impulse response sequence $\{a_n\}$ with input sequence $\{x(n)\}$. In order to obtain the transfer function, $H(z)$, of the FIR filter in Figure 6.1, we must take the z-transform of both sides of equation (6.1):

$$\begin{aligned} Y(z) &= \sum_{i=0}^{N-1} a_i \, X(z) \, z^{-i} \\ &= X(z) \sum_{i=0}^{N-1} a_i \, z^{-i} \end{aligned}$$

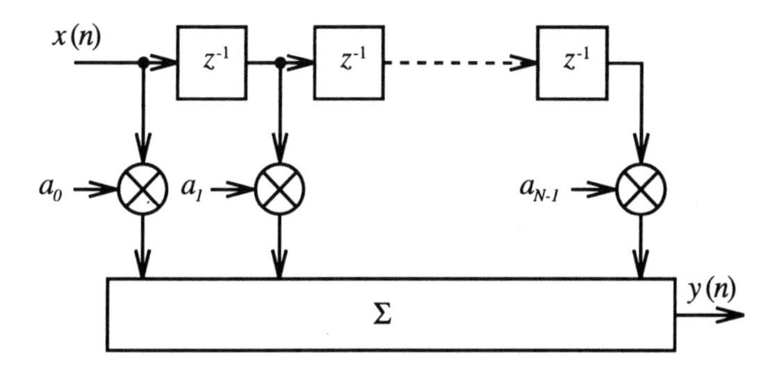

Figure 6.1 *Finite impulse response (non-recursive) filter.*

Hence:

$$H(z) = \frac{Y(z)}{X(z)} = \sum_{i=0}^{N-1} a_i \, z^{-i} \tag{6.2}$$

Now the poles and zeros of this polynomial are identified by expressing equation (6.2) in powers of z which are all greater than or equal to zero:

$$\begin{aligned} H(z) &= a_0 \, z^0 + a_1 \, z^{-1} + \cdots + a_{N-1} \, z^{-(N-1)} \\ &= \frac{z^{N-1}}{z^{N-1}} \left(a_0 \, z^0 + a_1 \, z^{-1} + \cdots + a_{N-1} \, z^{-(N-1)} \right) \\ &= \frac{a_0 \, z^{N-1} + a_1 \, z^{N-2} + \cdots + a_{N-1} \, z^0}{z^{N-1}} \end{aligned}$$

Therefore $H(z)$ has $N-1$ zeros and $N-1$ poles. All the denominator poles are at the origin in the z-plane and hence FIR filters are unconditionally stable as the poles cannot be placed outside the unit circle.

6.2.2 Frequency response

The frequency response, $H(\omega)$, of the FIR filter is obtained directly by replacing z in equation (6.2) with $\exp(j\omega\Delta t)$, see section 4.6:

$$H(\omega) = \sum_{n=0}^{N-1} a_n \exp(-jn\omega\Delta t) \tag{6.3}$$

Thus, given a set of weights, a_n, we can evaluate $H(\omega)$ directly from equation (6.3). Note also that equation (6.3) is a Fourier series expansion of the function $H(\omega)$, (equations (1.6) and (1.12)), which is a periodic function (Figure 6.2) of ω with period $2\pi/\Delta t$ rad/s (i.e. the sampling frequency is $1/\Delta t$ Hz). Hence the Fourier coefficients a_n may be *calculated* by integrating in the *frequency* domain over one period of the function $H(\omega)$, as a direct application of equations (1.7) and (1.8):

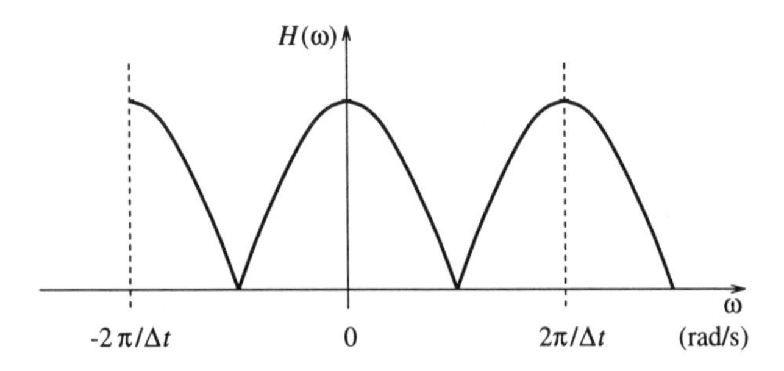

Figure 6.2 *FIR filter frequency response.*

$$a_n = \frac{\Delta t}{2\pi} \int_0^{2\pi/\Delta t} H(\omega) \exp(jn\omega\Delta t)\, d\omega \tag{6.4}$$

Self assessment question 6.1: For the 1st order filter with $a_0 = a_1 = 1$ sketch the frequency response from $\omega = 0$ to ω_s.

6.2.3 Phase response

The phase response of these filters can take many distinct forms. Consider the trivial case with impulse response $h(z) = 1 + ½z^{-1}$. This is equivalent in communications to summing a direct path signal with an attenuated delayed version of the same signal, and it gives rise to a minimum phase channel or filter response. If the multipath delay had preceded the main response, i.e. $h(z) = ½ + z^{-1}$, then the channel of filter would be of a non-minimum phase design. (Later in section 7.5 there is a more comprehensive discussion on filter phase responses.)

There is thus considerable flexibility in the FIR filter to control the phase response and a particularly important type of filter, which will be discussed next, is those which are designed to possess a linear phase response.

Self assessment question 6.2: Plot the phase response for filters with impulse response: (a) $1 + ½z^{-1}$ and (b) $½ + z^{-1}$ over the frequency range 0 to ω_s.

6.3 Linear phase filters

6.3.1 Principles

The achievement of a linear phase characteristic implies a constraint that the FIR coefficients must possess conjugate-even symmetry about their centre point value [Jackson]. Consider a filter comprising a single tap with 10 stages of delay and multiplier weight value of unity. This clearly possesses linear phase as all signal components experience the same 10 samples of delay. The rate of change of the linear phase response is controlled by the 10 delay samples. Now extend this to a three-tap filter with the first tap having eight samples of delay and a multiplier weight magnitude of ½. Clearly, to achieve an overall delay of 10 samples, we require to set the third tap at 10 + 2 samples of delay and the associated multiplier weight value also at ½ to achieve even symmetry.

Now if a filter has linear phase then its frequency response, $H(\omega)$, may be written as the product of a real amplitude response function, $H_A(\omega)$, and a complex exponential term, $\exp(-j\alpha\omega\Delta t)$:

$$H(\omega) = H_A(\omega)\exp(-j\alpha\omega\Delta t)$$

The complex exponential term represents a time delay of $\alpha\Delta t$ seconds. Because $H_A(\omega)$ is real, the phase response of the filter is determined by the exponential term:

$$\angle\{H(\omega)\} = -\alpha\omega\Delta t$$

The phase response is thus a linear function of frequency, i.e. it possesses a linear phase response, but usually the phase is restricted to the range ±180°, as previously shown in Figures 1.6, 5.3(b) and 5.4(b).

We are now interested in knowing what filter impulse response will provide the real amplitude response as required for $H_A(\omega)$. Consider first a digital filter with a real impulse response sequence, $\{c_n\}$. The frequency response, $H_A(\omega)$, is given generally by:

$$H_A(\omega) = \sum_{n=-\infty}^{+\infty} c_n \exp(-jn\omega\Delta t)$$

$$= \sum_{n=-\infty}^{+\infty} \left[c_n \cos(n\omega\Delta t) - j\, c_n \sin(n\omega\Delta t)\right]$$

Now if $\{c_n\}$ has even symmetry:

$$c_n = c_{-n} \tag{6.5}$$

as in Figure 6.3(a), so the imaginary part will disappear and equation (6.3) can be further simplified, i.e.:

$$H_A(\omega) = \sum_{n=-\infty}^{+\infty} c_n \cos(n\omega\Delta t)$$

$H_A(\omega)$ is now a weighted sum of even functions of frequency (as given by the $\cos(n\omega\Delta t)$ terms) and hence it will possess a corresponding even function response in the frequency domain. The condition, summarised in equation (6.5), which ensures that $H_A(\omega)$ is real implies a non-causal, i.e. physically unrealisable, impulse response sequence. This arises from the negative time samples in Figure 6.3(a), which imply that output samples occur *in advance* of the input stimulus to the FIR filter. The non-causal nature of $H_A(\omega)$ is not a problem when dealing with FIR linear phase filters since the delay of $\alpha\Delta t$ seconds can be incorporated readily, to convert the non-causal filter into a causal one.

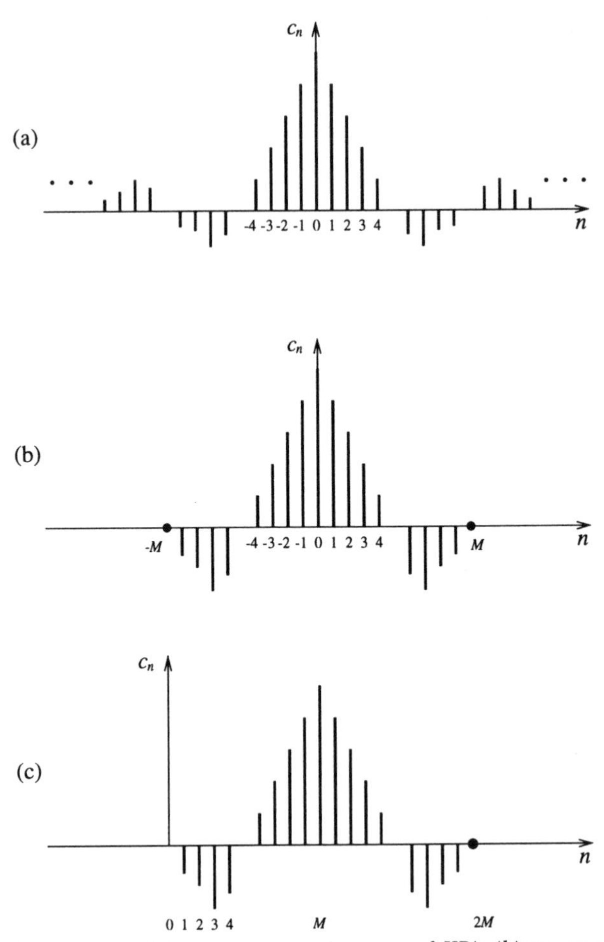

Figure 6.3 *Impulse responses: (a) even symmetry (non-casual IIR); (b) truncated even symmetry (non-causal FIR); (c) left shift M times to achieve the casual FIR design.*

The transfer function, $H_A(z)$, of the non-causal filter illustrated in Figure 6.3(a) is:

$$H_A(z) = \sum_{n=-\infty}^{+\infty} c_n z^{-n}$$

The impulse response series, $\{c_n\}$, can be truncated conveniently to only $2M + 1$ terms (Figure 6.3(b)), where M is a positive integer, without removing the symmetry property from equation (6.5):

$$H_A(z) = \sum_{n=-M}^{M} c_n z^{-n}$$

Following truncation by the finite sequence $-M$ to M, $H_A(z)$ now represents an FIR filter. The filter has $2M + 1$ taps, M of which have indices less than zero (i.e. they refer to negative time – Figure 6.3(b)) implying a non-causal design. The filter can thus be made causal by simply cascading it with a delay of M samples, i.e. $\alpha = M$:

$$H(z) = z^{-M} H_A(z)$$

$$= z^{-M} \sum_{n=-M}^{M} c_n z^{-n}$$

$$= \sum_{n=-M}^{M} c_n z^{-(n+M)}$$

The resultant impulse response is illustrated in Figure 6.3(c). This restriction to having a symmetrical impulse response or weight coefficient set means that the number of multipliers in Figure 6.1 can be halved. The linear phase FIR filter realisation can thus be accomplished with a folded delay line where the delayed signal sample at $-m$ is summed with sample $+m$ before entering the appropriate multiplier element.

There are, in fact, four ways of achieving linear phase FIR filters depending on whether an odd or even number of taps are used or whether the symmetry of the impulse response is odd or even. Only one case has been considered here, i.e. an odd number of taps and even symmetry of the impulse response.

6.3.2 Linear and nonlinear phase filters

The linear phase filter ensures that all input signal components, irrespective of their frequency, experience the same delay when passing through the filter. In the causal filter of Figure 6.3(c) the delay to the centre tap is $M\Delta t$ and the impulse response is symmetrical about this value. Figure 6.4 shows the effect on a rectangular pulse, which has a (sin x)/x spectrum, on filtering this signal in two separate filters with *identical* amplitude responses. It is only the linear phase filter which maintains the delay and phase relationship between the frequency components of the pulse and hence provides the sharp-edged filter output. The nonlinear phase filter completely distorts the pulsed signal.

Self assessment question 6.3: Where is the main application of linear phase filters and what is the significance of the linear phase operation?

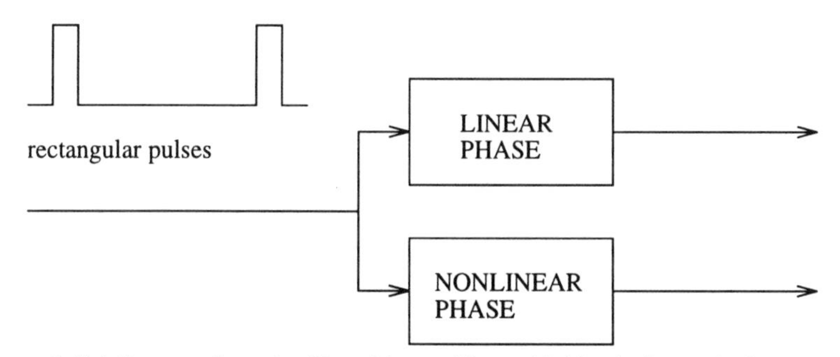

Figure 6.4(a) *Rectangular pulse filtered in two filters with identical magnitude responses.*

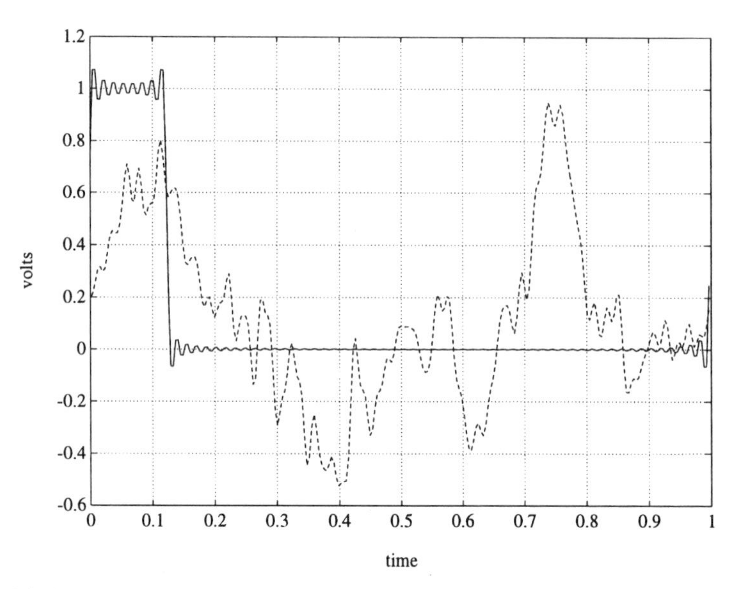

Figure 6.4(b) *Output responses for rectangular pulse filtered in the above linear phase filter (solid plot) and nonlinear phase filter (dashed response).*

6.4 Linear phase filter design

6.4.1 Fourier series method

Linear phase FIR filter design centres on selecting the appropriate set of $2M+1$ coefficients in the filter, c_n. This gives the even amplitude response, $H_D(\omega)$. The set of required coefficients can be calculated conveniently using equation (6.4) which can now be simplified because $H_D(\omega)$ is real and possesses only even symmetry:

$$c_n = \frac{\Delta t}{2\pi} \int_0^{2\pi/\Delta t} H_D(\omega) \exp(jn\omega\Delta t)\, d\omega$$

$$= \frac{\Delta t}{2\pi} \left(\int_0^{2\pi/\Delta t} H_D(\omega) \cos(n\omega\Delta t)\, d\omega + j \int_0^{2\pi/\Delta t} H_D(\omega) \sin(n\omega\Delta t)\, d\omega \right)$$

$$= \frac{\Delta t}{2\pi} \int_0^{2\pi/\Delta t} H_D(\omega) \cos(n\omega\Delta t)\, d\omega$$

Now, reducing the upper integration limit from the sampling to the folding frequency:

$$c_n = \frac{\Delta t}{\pi} \int_0^{\pi/\Delta t} H_D(\omega) \cos(n\omega\Delta t)\, d\omega \tag{6.6}$$

Owing to the restriction of employing only a finite number of coefficients (i.e. $n \leq M$) the achieved amplitude response, $H_A(\omega)$, will be slightly different from the desired amplitude response, $H_D(\omega)$. The achieved amplitude response is provided by equation (6.3) which can be simplified using equation (6.5):

$$\begin{aligned} H_A(\omega) &= \sum_{n=-M}^{M} c_n \cos(n\omega\Delta t) \\ &= c_0 + 2 \sum_{n=1}^{M} c_n \cos(n\omega\Delta t) \end{aligned} \tag{6.7}$$

Self assessment question 6.4: What key feature will equation (6.7) impress on the FIR filter impulse response?

EXAMPLE 6.1

For the 'brick wall' or rectangular low-pass response of Figure 5.1, with a 1 kHz sample rate, a desired magnitude of unity in the passband and a 125 Hz –3 dB cut-off frequency, find the coefficient values appropriate to an order 20 FIR filter design.

Solution

The frequency response falls ideally to zero, when:

$$\frac{\omega_s}{2\pi} = 1000 \text{ Hz} \quad \text{and} \quad 125 \text{ Hz} = \frac{\omega_s}{8 \times 2\pi}$$

Equation (6.6) can be used directly to provide the (sin x)/x weights of Figures 6.3 and 6.5. Thus:

$$\begin{aligned} c_n &= \frac{\Delta t}{2\pi} \int_0^{2\pi/\Delta t} H_D(\omega) \cos \frac{2\pi n\omega}{\omega_s}\, d\omega \\ &= \frac{\Delta t}{\pi} \int_0^{\omega_s/8} 1 \times \cos \frac{2\pi n\omega}{\omega_s}\, d\omega \\ &= \left[\frac{\Delta t}{\pi} \frac{\omega_s}{2\pi n} \sin \frac{2\pi n\omega}{\omega_s} \right]_0^{\omega_s/8} \end{aligned}$$

Now since $\omega_s/(2\pi) = 1/\Delta t$:

$$c_n = \frac{\sin \dfrac{2\pi n\omega_s}{8\omega_s}}{\pi n} = \frac{\dfrac{1}{4} \sin \dfrac{\pi n}{4}}{\dfrac{\pi n}{4}}$$

This solution is a repeat of Figures 4.5 and 4.6, except that here the desired frequency

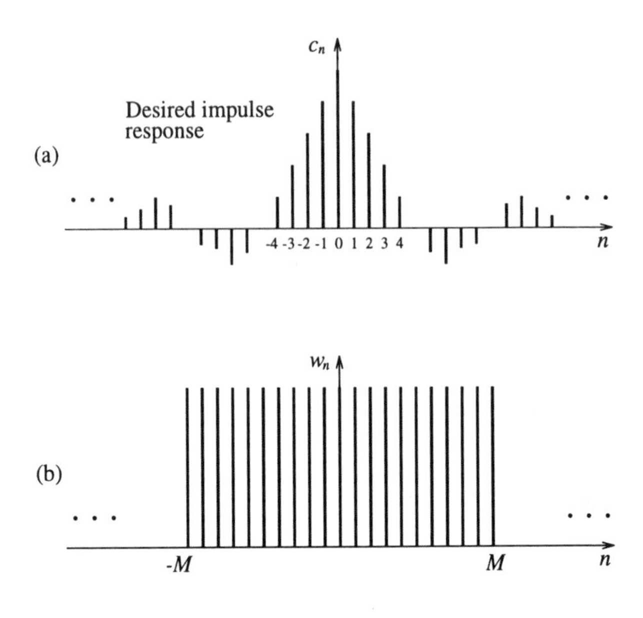

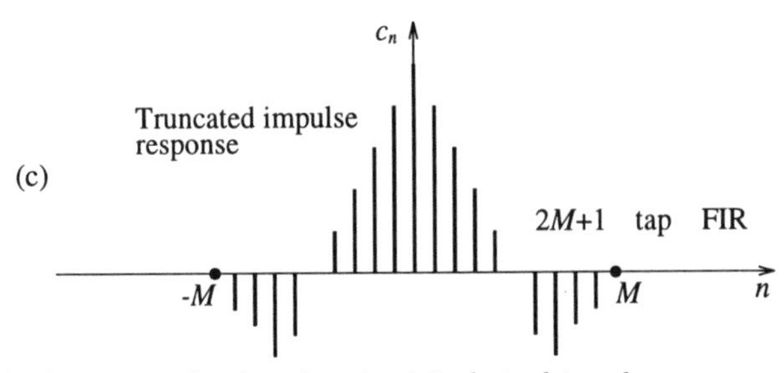

Figure 6.5 *Multiplication in the time domain: (a) desired impulse response; (b) rectangular window; (c) truncated impulse response.*

response possesses the rectangular shape. The number of coefficients in the main lobe is related to the desired bandwidth of the prototype filter. Also the filter weights must be truncated to the M values. Therefore for a 21-stage filter with 11 coefficients $c_0, \cdots, c_{10}$ there are 21 filter weights $a_0, \cdots, a_{20}$. First we substitute in the above equation for $n = 0$ to $n = 10$ to obtain $c_0, \cdots, c_{10}$. The 21 filter weight coefficient values $a_0, \cdots, a_{20}$ are obtained from the $c_0, \cdots, c_{10}$ values as follows:

$$a_0 = a_{20} = c_{10}$$

$$a_1 = a_{19} = c_9$$

$$\cdots\cdots\cdots\cdots$$

$$a_9 = a_{11} = c_1$$

$$a_{10} = c_0$$

As these are related in a (sin x)/x shape, they take the form shown in Figure 6.5(a).

Self assessment question 6.5: If the filter design in Example 6.1 had a gain of 20 dB, how would the coefficient values have differed?

Self assessment question 6.6: The filter weights shown in Figure 6.6 have been drawn with incorrect zero values at c_5, c_{10} etc. instead of multiples of c_4. What should the specified filter −3 dB bandwidth be to match the impulse response in Figure 6.6?

Self assessment question 6.7: If the filter design in Example 6.1 had required a narrower passband width, with a cut-off frequency of 62.5 Hz, how would the impulse response differ from the solution obtained in Example 6.1?

6.4.2 Window effects

The infinite time-domain sequence has been effectively multiplied by a rectangular function of time w_n (Figure 6.5(b)) to obtain the M coefficient truncated response:

$$w_n = 1, \quad |n| \leq M$$

$$w_n = 0, \quad |n| > M$$

Truncation of the time-domain impulse response sequence from c_n into c_n' is achieved

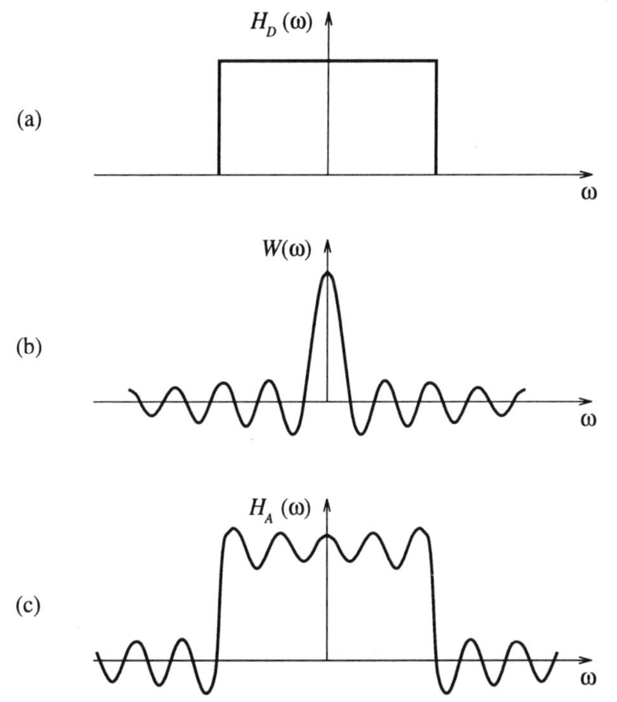

Figure 6.6 *Convolution in the frequency domain: (a) desired amplitude response; (b) frequency response of rectangular window; (c) actual amplitude response.*

by multiplying by the $w_n = 1$ values:

$$c_n' = w_n \, c_n$$
$$= c_n$$

This is equivalent to convolving the desired frequency response, $H_D(\omega)$, with the frequency response of the rectangular truncation window, i.e. $W(\omega)$:

$$H_A(\omega) = H_D(\omega) * W(\omega)$$

This effect is illustrated in Figure 6.6. The frequency response of the wide rectangular truncation window is a narrow $\sin(x)/x$ or sinc function. The sidelobes of this sinc function (Figure 6.6(b)) result in a 'ringing' of the actual frequency response $H_A(\omega)$ (Figure 6.6(c)) at the edges of the desired frequency response $H_D(\omega)$. This result is characteristic of the truncation of a Fourier series. It is also called Gibb's phenomenon.

EXAMPLE 6.2

For a 'brick wall' filter design with a cut-off frequency at 10% of the sample frequency examine the effect, in a rectangular windowed filter, of using different filter lengths or orders.

Solution

We calculate the filter weight coefficient values as in Example 6.1 and then the actual frequency response can be evaluated by inserting the filter weight coefficient values into equations (6.3) or (6.7). Figure 6.7 shows such a frequency response. Note how the

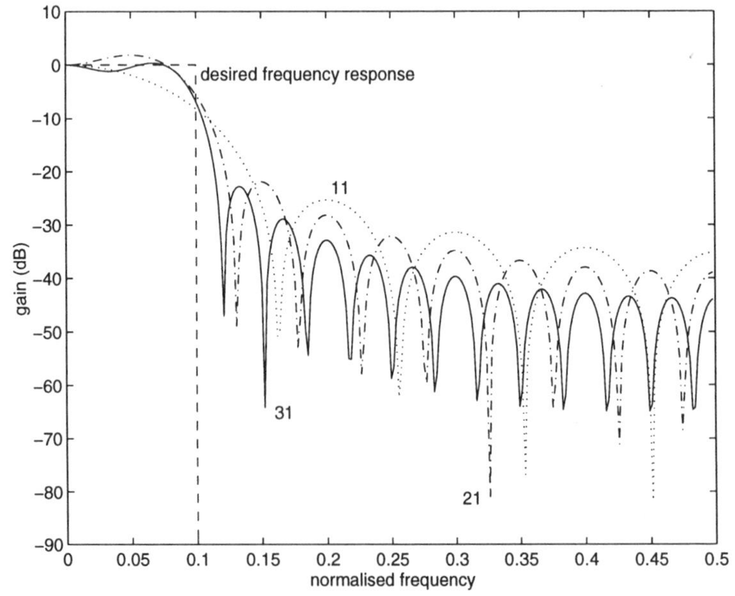

Figure 6.7 *Effect of filter length, N, on frequency response for the low-pass filter design of Figure 5.1 with a rectangular weighting function.*

frequency response of a low-pass filter is more closely approximated as the number of taps, N (and hence the complexity of the filter), is increased. For this rectangular window the first sidelobe is only 21 dB below the passband gain value at 0 Hz.

A vastly improved frequency response is obtained if the impulse response sequence is multiplied by a *tapered* window function, rather than the rectangular truncation function, to smooth out the time-domain discontinuities. This will achieve lower sidelobe levels in the frequency domain, than with the sinc x truncation function. The design of appropriate window functions is a widely researched area.

One simple window is the triangular or Bartlett window. This is of limited practical value as the most important windows belong to the $\cos^{\alpha}$ family (Figure 6.8), which is defined over the variable n where $0 < n \le N-1$:

$$w_n = \sin^{\alpha}\left(\frac{n\pi}{N}\right) \tag{6.8}$$

The member of the family with $\alpha = 2.0$ is known as the sine squared window. It is also commonly referred to as the Hanning window or the raised cosine window since it may be regarded as a cosine function on a pedestal of height 0.5:

$$\sin^2\left(\frac{n\pi}{N}\right) = 0.5\left[1.0 - \cos\left(\frac{2n\pi}{N}\right)\right] \tag{6.9}$$

When this is converted to the revised limits $-M < n < M$, where $M = (N-1)/2$, corresponding to the symmetric linear phase filter responses, then the window becomes:

$$w_n = \frac{1}{2}\left[1 + \cos\left(\frac{n\pi}{M}\right)\right] \tag{6.10}$$

The Hamming window is a slight modification of the raised cosine window and is given by:

$$w_n = 0.54 + 0.46\cos\left(\frac{n\pi}{M}\right) \tag{6.11}$$

The Hamming window differs from the Hanning window in that there is a pedestal

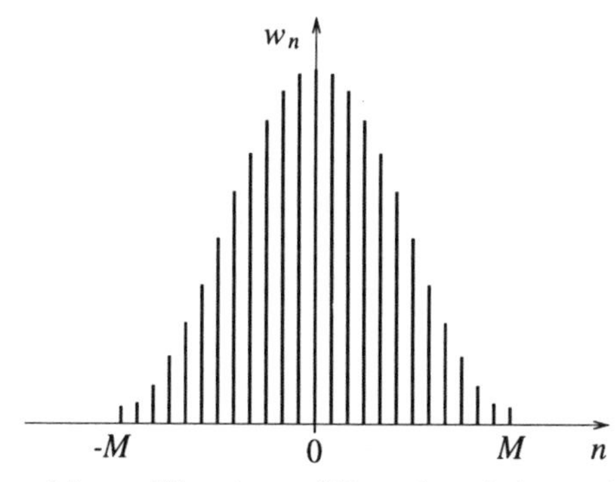

Figure 6.8 *General form of Hanning and Hamming window weighting functions.*

value of 0.08 at the window edge when $n = \pm M$. This sounds like quite a small modification but it has a significant effect on the stopband performance of the window.

EXAMPLE 6.3

For the 21-coefficient 'brick wall' filter design of Figure 6.7, examine the effect of Hamming weighting this FIR filter.

Solution

The Hamming weighting coefficient values are obtained by multiplying the unweighted (rectangular) values by the Hamming window. Figure 6.9 shows the resulting frequency response. The Hamming window design has a maximum sidelobe level of –53 dB, which is 11 dB lower than the corresponding maximum of the Hanning window, but the trade-off here is that far away from the main lobe the stopband does not decrease as rapidly as with the Hanning window. The Hamming window 'bottoms out' at about 60 dB stopband rejection (Figure 6.9). This is quite acceptable for an FIR filter design but, later in Chapter 10, we will prefer the Hanning window for DFT design.

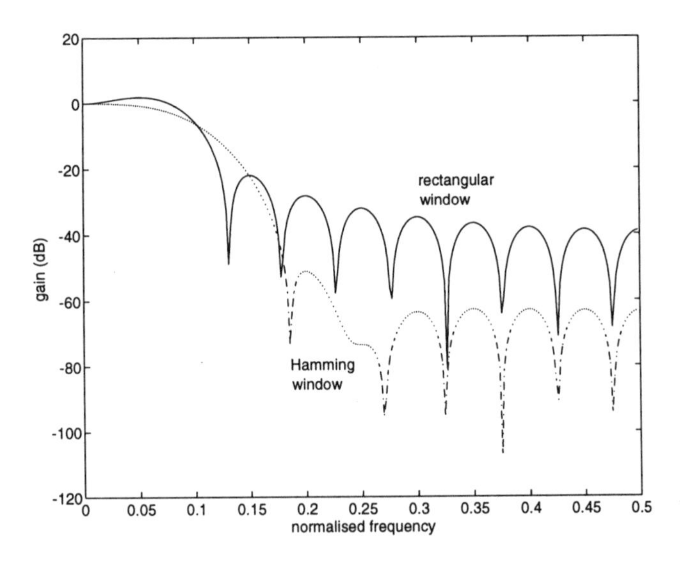

Figure 6.9 *Effect of window on frequency response for a 21-coefficient filter design.*

Another popular window function is the Blackman design:

$$w_n = 0.42 + 0.5 \cos\left(\frac{n\pi}{M}\right) + 0.08 \cos\left(\frac{2n\pi}{M}\right) \tag{6.12}$$

A final window is the Kaiser design:

$$w_n = \frac{I_0[\beta\sqrt{1-(2n/M)^2}]}{I_0[\beta]} \tag{6.13}$$

Here β is a constant controlling the trade-off between the transition bandwidth and the worst case sidelobe level, at a given value of filter length N. $I_0[\ ]$ is a modified zeroth-order Bessel function of the first kind.

Although all these windows reduce the stopband rejection, this is not obtained without a penalty. If you examine closely the width of the responses in Figure 6.9 in comparison with Figure 6.7 then you will observe that the width of the transition band is widened in the windowed filter responses. This is obvious as the window function has reduced the duration of the impulse response, which also affects the frequency response. The transition bandwidth, Δf, and other filter specifications are summarised in Figure 6.10 and the characteristics of these windows are summarised in Table 6.1. In Figure 6.10, f_1 represents the passband cut-off frequency and f_2 corresponds to the edge of the stopband.

6.4.3 Design summary

The window design of linear phase FIR filters involves six steps:

(i) First select a window to give the required stopband rejection using Table 6.1. If a stopband rejection of 40 dB was required then any of the Hanning, Hamming, Kaiser or Blackman, etc. windows could be employed.

(ii) The overall length or complexity N is also obtained from Table 6.1 to achieve the required transition band, Δf, as defined in Figure 6.10.
Continuing with the above example, if a Hanning window had been chosen then:

$$\Delta f = \frac{3.1}{N\,\Delta t}$$

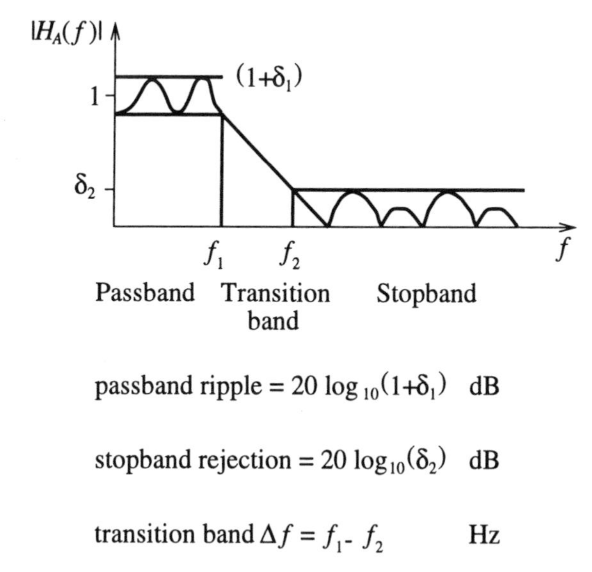

Figure 6.10 *Filter passband ripple and stopband rejection specifications.*

Table 6.1 *Window characteristics.*

Window	*Transition band* (Hz)	*Stopband rejection* (dB)
rectangular	$\frac{1}{N\Delta t}$	21
Hanning	$\frac{3.1}{N\Delta t}$	44
Hamming	$\frac{3.3}{N\Delta t}$	53
Kaiser, $\beta = 6$	$\frac{4}{N\Delta t}$	63
Blackman	$\frac{5.5}{N\Delta t}$	74
Kaiser, $\beta = 9$	$\frac{5.7}{N\Delta t}$	90

and hence:

$$N = \frac{3.1}{\Delta f \; \Delta t}$$

The transition band represents the width of the mainlobe of the appropriate window, transformed to the frequency domain.

(iii) Now calculate the window coefficient values for your N-tap filter using the selected window and the appropriate earlier equations (6.10) to (6.13). When calculating the set of $N = 2M + 1$ weight coefficients $\{w_n\}$ note that, because of the symmetry in the impulse response, as evidenced by equation (6.5), only $M + 1$ calculations are required to obtain the $2M + 1$ coefficients.

(iv) To obtain a required frequency response, $H_D(\omega)$, we calculate a set of $N = 2M + 1$ coefficients, $\{c_n\}$. Again, because of the symmetry in the impulse response, only $M + 1$ calculations are required to obtain the $2M + 1$ coefficients:

$$c_n = \frac{\Delta t}{\pi} \int_0^{\pi/\Delta t} H_D(\omega) \cos(n\omega\Delta t) \, d\omega \tag{6.14}$$

(v) Now each point in the impulse response sequence, $\{c_n\}$, must be multiplied by the corresponding point in the window sequence, $\{w_n\}$, to produce the windowed impulse response sequence, $\{c_n{}'\}$:

$$c_n{}' = w_n \, c_n \tag{6.15}$$

(vi) Finally, to verify the accuracy of the filter design, the actual achieved frequency response, $H_A(\omega)$, of the designed filter must be calculated and compared with the original desired frequency response, $H_D(\omega)$:

$$H_A(\omega) = c_0' + 2 \sum_{n=1}^{M} c_n' \cos(n\omega\Delta t) \tag{6.16}$$

The integration in equation (6.14) is usually straightforward as the desired frequency response is often a simple rectangular shape.

The modification of the filter performance by the window function is illustrated in Figure 6.9, in comparison with Figure 6.7. For a particular filter length, reducing the stopband rejection tends to increase the width of the transition band. The above choice of window function and the selection of the number of taps required to meet a given specification are not an accurate design technique, as the process really only provides a good starting point for an iterative design procedure.

Self assessment question 6.8: You are to design a 21-tap Hamming weighted filter with a −3 dB cut-off frequency of 100 Hz at a 1 kHz sample rate. What is the transition bandwidth in Hz?

Self assessment question 6.9: A Hanning windowed filter, with the same design parameters as in SAQ 6.7, must have a transition bandwidth which is only 50% of the 100 Hz cut-off frequency. What order of filter is now required?

6.4.4 Design optimisation techniques

Optimised design can be achieved by iterative application of DFT/IDFT processing on a filter design specification, such as that shown in Figure 6.10. In this method one starts with a set of coefficients, c_n or $h(n)$, and performs a DFT (Chapter 9) to get the corresponding $H(\omega)$ values. The filter performance limits from Figure 6.10 are then imposed on parts of the $H(\omega)$ which deviate from specification and then the updated $H(\omega)$ vector is given an inverse DFT to yield an updated set of $h(n)$ values. This process will result in the new $h(n)$ having significant component values beyond the restricted length, N, permitted for the FIR filter design. These additional values are simply truncated and then the iterative process is repeated to get a new $H(\omega)$ vector, which is again compared with the specification. This optimisation technique is guaranteed to converge, subject to certain constraints [Parks and Burrus].

Another optimisation technique is the Parks–McClellan algorithm. This technique, which reduces significantly the filter complexity N compared with window designs, for a given required bandshape, uses a minimax design approach. This uses a minimum weighted Chebyshev error to approximate the desired frequency response by iteration, i.e.:

$$\underset{\{c_n\}}{\text{minimum}} \left\{ \underset{\omega}{\text{maximum}} \;|\; L(\omega)\,[\, H_D(\omega) - H_A(\omega) \,] \;|\right\}$$

The set of weight values, $\{c_n\}$, which *min*imises the *max*imum error between the desired frequency response and actual frequency response, provides directly the optimal filter design. The positive weighting function, $L(\omega)$, allows the designer to emphasise some areas of the frequency response more than others. The minimisation is performed iteratively using a computer program based on the Remez exchange algorithm. This method is now used as the basis of most commercial filter design software packages. As we are minimising the maximum error at each iteration, by adjusting the filter coefficients, this is termed a minimax design procedure. The technique reduces significantly the order of the final filter design compared with window design approaches.

EXAMPLE 6.4

Design a low-pass filter example with a stopband above 0.18 of the sampling frequency with 50 dB attenuation, and a passband for frequencies below 0.1 of the sampling frequency. Compare a window filter design with a Remez exchange algorithm with 1 dB of permitted passband ripple.

Solution

The Hamming weighted frequency domain window design requires a 31-tap filter and achieves a ripple-free passband (Figure 6.11(b)). In contrast, the minimax Remez MATLAB™ design requires only a 20-tap filter design to achieve the same stopband performance goal (Figure 6.11(a), below), but it introduces small ripples in the passband. However, it is a very significant filter design technique.

In some applications the stopband in Figure 6.10 is not continuous. In other words, there are frequencies at which we do not need to specify a high stopband rejection and we do not care what response is provided by the filter at these frequencies as there are not expected to be any signal components there. This allows the use of multiband FIR designs with further savings on the filter complexity or order.

The number of coefficients, N, or order, $N-1$, of the FIR filter is an important parameter as it controls directly the computational complexity needed to implement the filter which satisfies the required specification. For a low-pass filter the specifications

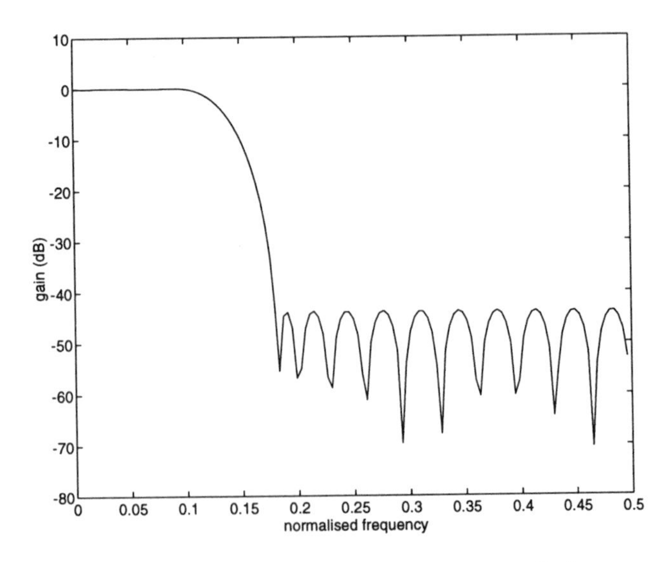

Figure 6.11(a) *20-coefficient minimax design FIR filter.*

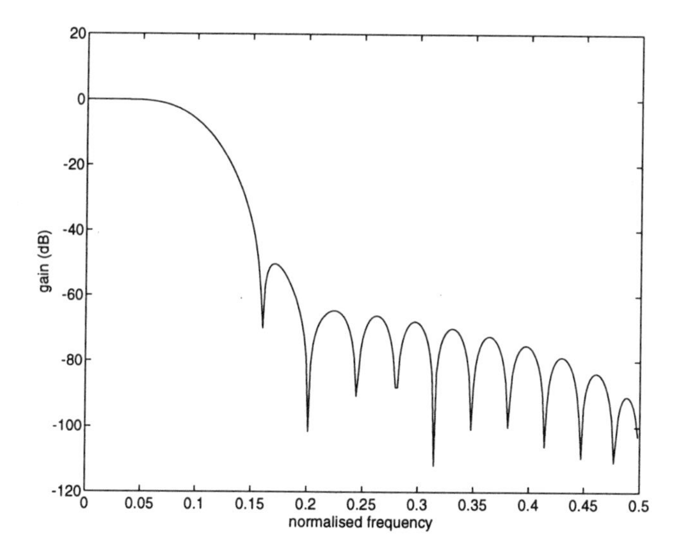

Figure 6.11(b) *31-coefficient Hanning window design FIR filter.*

are summarised in Figure 6.10. Thus to minimise the filter complexity, or maximise the throughput rate, the order should be minimised. It has been show [Bellanger] that an estimate $\hat{N}$ of the complexity or order required for a minimax design is given by:

$$\hat{N} = \frac{2}{3} \frac{1}{\Delta t \, \Delta f} \log_{10}\left(\frac{1}{10\delta_1 \delta_2} \right) \tag{6.17}$$

The transition band, Δf, is the most sensitive parameter, with the permitted passband ripple, δ_1, and required stopband rejection, δ_2, (Figure 6.10) having a less significant impact. However it is worth emphasising that, according to this estimate, the filter complexity is independent of the passband width. Formulae also exist for estimating the required order for bandpass filter designs.

Self assessment question 6.10: For a filter design with a sampling frequency of 1 kHz, transition bandwidth of 50 Hz with a passband ripple of ½ dB and stopband rejection of 40 dB, estimate the required filter order.

EXAMPLE 6.5

Compare the filter frequency responses for Hamming window designs with 31 and 256 coefficient values.

Solution

Figure 6.12 shows examples of low-pass filters designed with MATLAB™ for N values of 31 and 256. Clearly there is a much faster roll-off and reduced transition bandwidth, as predicted by Table 6.1, with the increased filter complexity.

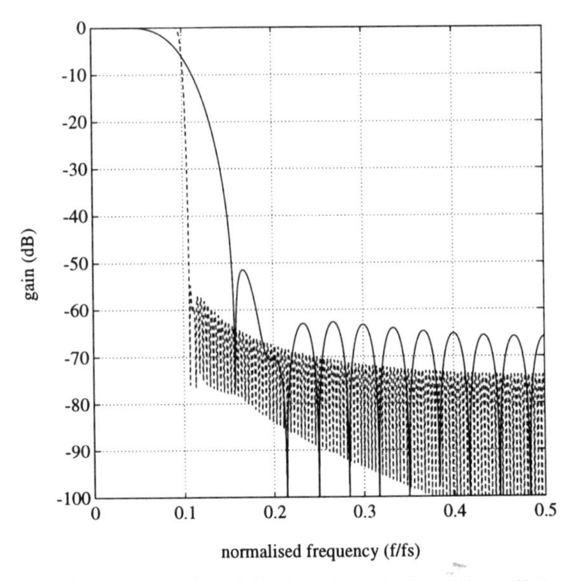

Figure 6.12 *FIR filter design examples with Hamming window. 31-coefficient design in the solid curve and the 256-coefficient design shown dashed.*

6.5 Finite precision effects

6.5.1 Noise reduction through the filter

Later, in section 7.6.1, the effects of quantisation noise are introduced. The noise or errors present at the input of the FIR filter due to quantisation in the input analogue-to-digital converter (ADC), $e_q(n)$, are usually reduced on the output signal. In the simple FIR filter of equation (6.1) which performs the convolution operation (equation (2.10)):

$$\begin{aligned} y_q(n) &= x_q(n) * h(n) \\ &= x(n) * h(n) + e_q(n) * h(n) \end{aligned} \quad (6.18)$$

The output noise component is thus given by the convolution:

$$e_q(n) * h(n)$$

and hence for an input variance, σ_e^2, the output variance, σ_v^2, is given by:

$$\sigma_v^2 = \sigma_e^2 \sum_{i=0}^{N-1} h^2(k) \quad (6.19)$$

Self assessment question 6.11: If the input variance is 1/12, what will the output variance be in a unity gain three-stage FIR filter averager where: $y(n) = 1/3?\ [x(n) + x(n+1) + x(n+2)]$

6.5.2 Filter coefficient quantisation errors

First consider a second order FIR filter response, factorised to show the zero locations in the z plane:

$$H(z) = (1 - r\exp(j\theta)z^{-1})(1 - r\exp(-j\theta)z^{-1}) \tag{6.20}$$
$$= (1 - 2r\cos(\theta)z^{-1} + r^2 z^{-2})$$

When represented in a finite-precision number system, the coefficients $2r\cos(\theta)$ and r^2 can take on only certain values, determined by the precision of the number system. To illustrate the effects of this quantisation, consider the zero locations in the right half of the z-plane, when very coarsely quantised to only 4-bit precision.

Quantising r^2 to have only a prescribed set of values defines a set of circles in the z-plane on which the zeros must lie. The quantisation coefficient $2r\cos(\theta)$ in equation (6.20) further defines a set of vertical lines in the z-plane passing through $z = 2r\cos(\theta)$, which indicate the pairs of allowable r and θ values that can occur. The allowable zero locations for the second order section can then fall only at the points of intersection between these circles and vertical lines, which for the 4-bit quantisation are shown in Figure 6.13. This produces a non-uniform distribution of the locations that tend to cluster around $z = \pm j$.

Because of the coefficient quantisation, the required position of the zero must be shifted from that required by theory to one of the adjacent allowable positions. The filter designer must choose one of these positions or increase the number of bits in the finite-precision representation. This latter choice comes with an increase in hardware cost and possibly slower operation. Branch-and-bound techniques are generally required to achieve an optimum solution [Kuc].

Figure 6.14 shows how moving to this quantisation grid affects the response of a filter. This models the response of equation (6.20) where $\theta = 70°$. Figure 6.14(b) shows how moving the zero to one of the four nearest 4-bit quantised locations affects

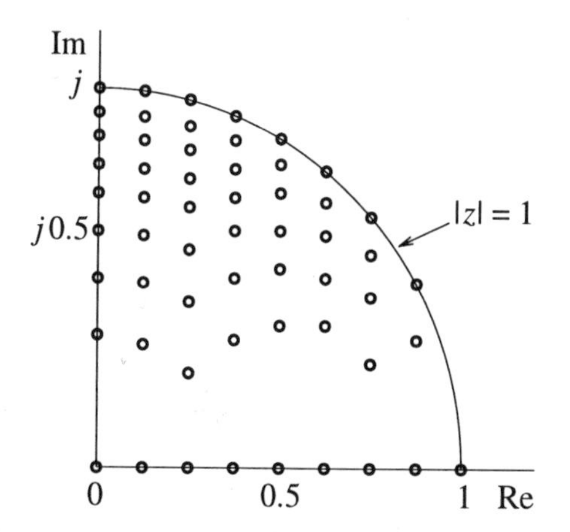

Figure 6.13 *FIR filter coefficient quantisation for 4-bit arithmetic.*

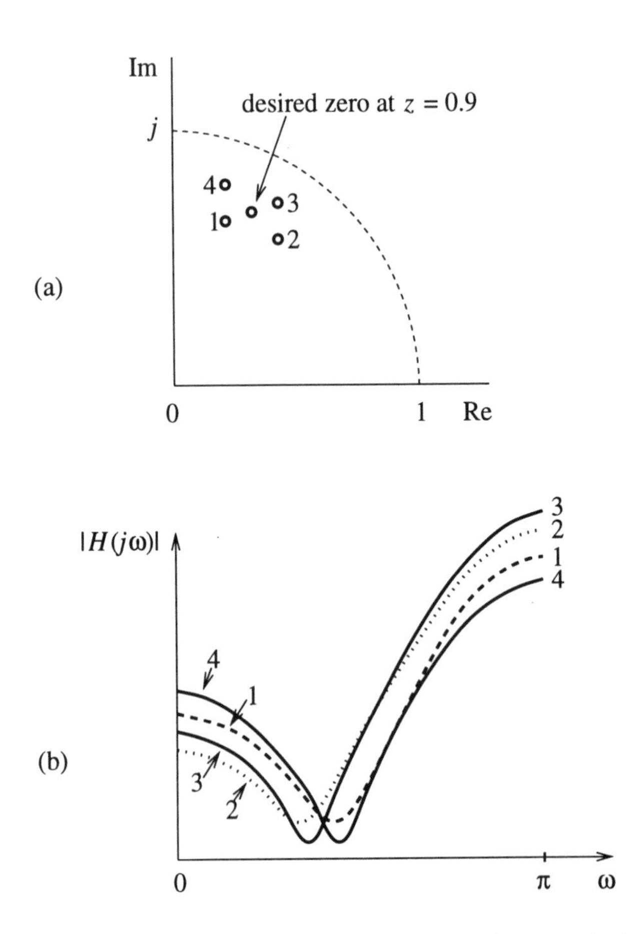

Figure 6.14 *Effect of quantisation on zero placement and resulting magnitude responses.*

the frequency response both in the precise depth and exact frequency at which the null occurs, in a manner similar to the precision effects in IIR filters (Chapter 5).

EXAMPLE 6.6

Investigate the effect of limited precision on a low-pass FIR filter design.

Solution

Figure 6.15 shows, on a 63-tap low-pass FIR filter, the effect of reducing the coefficient accuracy between the more practical levels of 16-bit, 12-bit and 8-bit fixed-point precision. These show that for FIR filters, the noise averaging effects of equation (6.19) do permit us to work comfortably with 16-bit arithmetic as supplied in many fixed-point precision DSP microprocessor parts ◻, and it gives an indication of the degradation which would be experienced by the lower precision, e.g. 8-bit, high sample rate integrated FIR filters when these are used for frequency filtering.

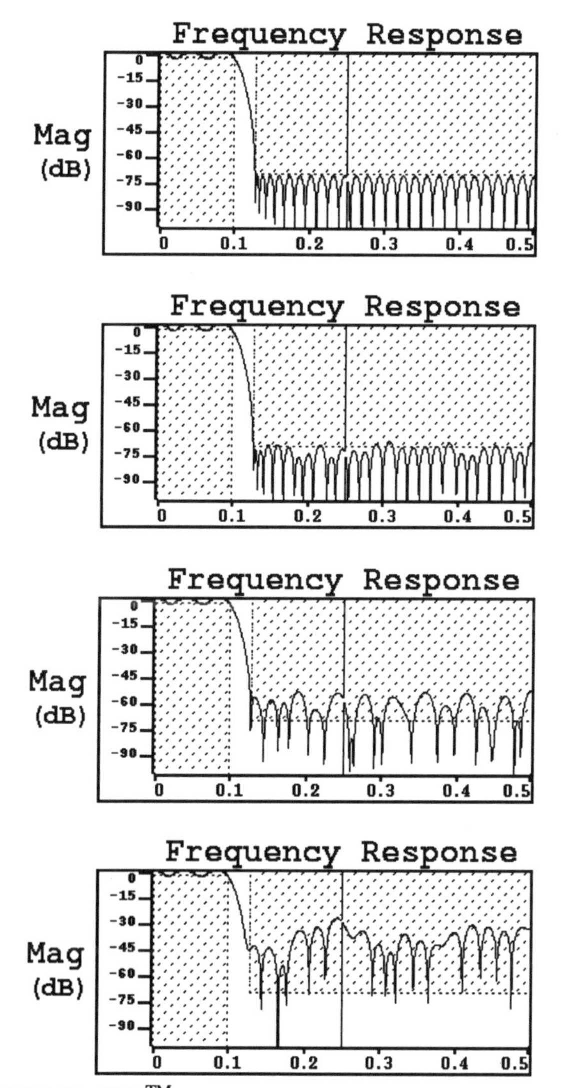

Figure 6.15 COMDISCO SPW™ *modelling of floating-point arithmetic compared with 16-, 12- and 8-bit finite precision filter coefficient quantisation in a 63-tap FIR filter.*

When this example is analysed in more detail there are found to be small errors (1/20 dB) in the passband due to filter quantisation effects. The stopband performance can be assessed either as a deviation from the theoretical response or, more significantly, by examining the minimum achievable stopband rejection. Performance with 8-bit quantisation produces most deviation and, while this may be acceptable in short (31-tap) low-pass filters, the minimum stopband rejection is typically degraded from –40 to –32 dB in longer (255-tap) low-pass designs.

For a more accurate estimation of the number of bits required for representing the FIR filter coefficients, the following expression [Bellanger] can be used:

$$b_c = 1 + \log_2 \left[\frac{\Delta t(f_1 + f_2)}{(\delta_m - \delta_0)} \sqrt{\left(\frac{N}{3}\right)} \right] \tag{6.21}$$

where:

b_c is the number of bits used to represent the coefficients (including the sign),
f_1 is the passband edge,
f_2 is the stopband edge,
δ_m is the limit imposed on the amplitude of the ripple,
δ_0 is the amplitude of the ripple of the filter before the number of bits is limited.

This equation also applies to high-pass and low-pass filters.

6.5.3 FIR filter hardware

Just as with IIR filters, FIR designs can again use the standard DSP microprocessor parts for real-time hardware implementations □. Motorola have customised their DSP56000 series into a DSP56200 specifically for FIR filtering applications. The inherent stability of the feedforward FIR design also allows the fixed-point arithmetic precision to be degraded below 16- or 24-bit for some applications, with a consequent simplification in complexity and increase in speed or sample rate. The Inmos A100 is a 32-tap filter with 2.5 Msample/s rate at 16-bit precision, increasing progressively to 10 Msample/s at 4-bit precision. More impressive in the mid 1990s was the Harris HSP43168 16-tap FIR filter which processed at 45 Msample/s! The other approach to implementing *long* FIR filter or convolution operations is via FFT processing (section 10.7), which greatly simplifies the arithmetic complexity and hence again increases the speed or real-time filter bandwidth.

6.6 FIR filter applications

6.6.1 Matched filter detector

Consider a signal input of $f(t)$ corrupted by input additive white Gaussian noise of $n_I(t)$ which is processed in a receiver filter. The filter output is $g(t)$ with corresponding output noise $n_O(t)$. The matched filter frequency response, which *maximises* the output signal to noise ratio, is defined as:

$$H(\omega) = k\, F^*(\omega) \exp(-j\omega t_m) \tag{6.22(a)}$$

and its impulse response $h(t)$ can also be defined as:

$$h(t) = k\, f^*(t_m - t) \tag{6.22(b)}$$

Again the * superscript defines the complex conjugate operation (section 3.4). This arrangement implies that the impulse response of the matched filter is time-reversed or comprises a mirror image of the expected signal $f(t)$ delayed by a term t_m, as shown in Figure 4.21. Note that this condition ensures that $|H(\omega)| = |F(\omega)|$, i.e. the frequency response or –3 dB bandwidth of the filter is carefully *matched* to the expected signal. Also the complex conjugate operation ensures that all the different signal components add coherently, i.e. in phase, as shown for the pulsed signal in Figure 6.4. This filter impulse response is thus the time reverse of signal, for matched filter operation.

Self assessment question 6.12: How does the frequency response of the matched filter relate to that of the signal to which it is matched?

Self assessment question 6.13: What is the optimum filter to detect a signal buried in white Gaussian noise?

6.6.2 Matched filter applications

One example of matched filter detection occurs in the receivers used in coded radar and communications systems. In spread spectrum communications, slow speed narrow-band data are multiplied by a high speed spreading code before transmission. The following trivial example shows how a short 3-bit code is added modulo-2 to the slow speed data, where each data bit has a duration equal to the entire 3-bit spreading code, and is synchronised to it, to obtain the transmitted encoded product sequence:

Data	+1	+1	+1	–1	–1	–1	+1	+1	+1
Spreading code	+1	–1	+1	+1	–1	+1	+1	–1	+1
Product sequence	+1	–1	+1	–1	+1	–1	+1	–1	+1

This modulates the data with the spreading code. One receiver to decode this transmitted data is a 3-stage FIR filter whose coefficients are coded with the time reverse of the expected sequence, but note here that the code is symmetric. Thus in Figure 6.1 $a_0 = +1$ and $a_1 = -1$, etc. This receiver filter is then matched to the expected transmissions and it is the optimum receiver from a signal-to-noise ratio standpoint. The output is given by the convolution of the received signal with the stored weight values, as shown previously in the problems in Chapter 2.

Self assessment question 6.14: Plot, either manually or with MATLAB™, the convolution of the above product sequence with a receiver FIR filter whose impulse response is: $1 - z^{-1} + z^{-2}$.

Note that with *binary* stored weight values, the FIR design is simplified and suppliers such as TRW market these single chip 'correlators' with up to 64 binary weighted taps at 30 Msample/s input sample rates.

In radar systems the coding is more sophisticated but the same principle is still widely applied. Here the waveforms are often modulated with a linear increase in frequency with time and the receiver is coded with the time-reversed signal, i.e. it has decreasing frequency with time. These signals [Cook and Bernfeld] are often called chirp waveforms. With multiple samples within this waveform then complex $\{h_n\}$

weight coefficient values are needed but the processing operation is still as described previously. Thus the radar digital matched filter requires a full FIR multibit filter capability, extended with parallel channels to accommodate complex processing.

6.6.3 Other receiver designs

Following the more detailed treatment of Chapters 7 and 8 on random signals and Wiener filters, the reader will realise that the FIR matched filter is a specific case of the more general receiver, $H(z)$, or Wiener filter for the signal $F(z)$:

$$H(z) = \frac{F^*(z)}{F(z)F^*(z) + n_O^2} \tag{6.23}$$

When the filter output noise $n_O{}^2$ term predominates in the denominator then:

$$H(z) = \frac{F^*(z)}{n_O^2} \tag{6.24}$$

to give the matched filter condition. In systems where noise is low and interference or radar clutter is the major degradation then:

$$H(z) = \frac{1}{F(z)} \tag{6.25}$$

to give the inverse filter, where the required receiver filter impulse response can be obtained either by long division or by the adaptive filter techniques, described later in Chapter 8. The inverse filter fundamentally uses all the signal bandwidth (well beyond the signal –3 dB point) and thus obtains a much sharper output pulse than the matched filter – but it *cannot* operate in high noise. It is applicable in communications for equalisation and in radar for resolving overlapping returns or for reducing clutter echoes when interference is the degradation mechanism rather than noise.

6.7 Summary

The major advantages of FIR digital filters are that they are unconditionally stable and they offer the possibility of designing a linear phase filter response to minimise signal distortions. Finite precision effects are simpler to analyse than in IIR designs and, in general, the FIR filter can be designed with lower precision arithmetic than for an IIR design. Although closed form design techniques are not possible for FIR filters, the window design method can produce reasonable results when computational efficiency is not of prime importance. More efficient designs can be obtained using the minimax technique and this is now widely available in software packages for DSP applications. Minimax also allows the synthesis of arbitrary frequency response characteristics, which is not easy to achieve with IIR filters. The only disadvantage of FIR designs is that relatively long filters will be required to provide sharp cut-off in the frequency domain, but with multirate techniques (Chapter 11) the complexity can be reduced. The need for linear phase response in communication and radar systems, to detect coded waveforms, makes the FIR filter a very popular choice for these receivers.

6.8 Problems

6.1 An FIR digital low-pass filter is to be designed with a 10 kHz sample rate and an ideal (brick wall) frequency response with a gain of 4 from DC to 1.25 kHz. Calculate the coefficient values for a 21-tap unwindowed filter and comment on the shape of the impulse response envelope for this filter. [$a_0, a_1, a_2, \cdots, a_{20}$ = 0.13, 0.1, 0, –0. 13, –0. 21, –0. 18, 0, 0.3, 0.64, 0.9, 1, 0.9, 0.64, 0.3, 0, –0. 18, etc.]

6.2 In problem 6.1, if the filter design had required a 40 dB stopband rejection, pick an appropriate window function and calculate the range of frequencies over which the stopband achieves this rejection value. [2.73 to 5 kHz]

6.3 Design an FIR digital low-pass filter for an 8 kHz sampling rate whose frequency response has a cosinusoidal shape below 1 kHz:

$$H_D(\omega) = \begin{cases} \cos\left[\dfrac{\omega}{4\times 10^3}\right], & 0 \le \omega < 2\pi \times 10^3 \\ 0, & \text{elsewhere} \end{cases}$$

Calculate the values of the filter coefficients for a 15-tap FIR filter. Sketch the actual frequency response from DC to 8 kHz. Describe with reference to a block diagram, how such filters can be implemented by multiplexing a single high speed hardware multiplier. [$c_0{}'$ to $c_7{}'$ = 0.16, 0.15, 0.12, 0.09, 0.05, 0.021, 0.0, –0. 01]

6.4 Design a 17-tap linear phase filter to approximate a unity gain bandpass response with –3 dB cut-off frequencies at $0.1/\Delta t$ and $0.2/\Delta t$ incorporating a Hanning window function. [a_0 to a_{16} = 0.0, 0.005, 0.016, 0.0, –0. 067, –0. 118, –0. 05, 0.112, 0.2, 0.112, –0. 05, –0. 118, etc.]

6.5 You are required to design a FIR bandpass filter with a passband gain of 10 dB from $0.1f_s$ to $0.2f_s$, where f_s is the sampling frequency.
It is suggested to you that this can be achieved simply as two low-pass filters and a subtractor, operating as in the figure below, where $H_1(\omega)$ is a filter with cut-off at $0.2f_s$ and $H_2(\omega)$ is a filter with cut-off at $0.1f_s$. Calculate the rectangular windowed coefficient values for the center 12 taps of these filters. If the appropriate amount of stopband rejection could be achieved with 31 coefficients in the $H_1(\omega)$ design, how many coefficients should be used in the second low-pass filter, $H_2(\omega)$?
Using the properties of linear systems, the two filters can be combined into a single FIR bandpass filter. Produce, from your earlier calculations, the weights for the centre 12 taps of the single bandpass filter design.
[c_0 to c_6 = 0.632, 0.36, –0. 18, – 0. 517, – 0. 389, 0, 0.26]

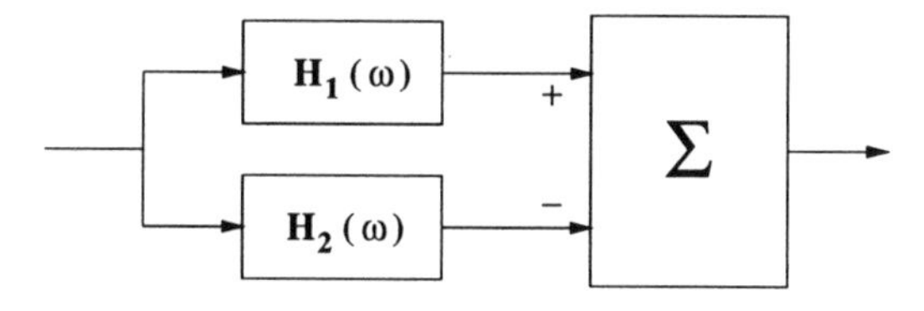

CHAPTER 7

Random signal analysis

7.1 Introduction

Much of the discussion in Chapters 1 to 4 has centred on what can be classified as deterministic signals (both continuous and discrete). Such signals can be described as 'weighted sums of complex exponentials' and are thus highly predictable in the following sense: given the Fourier transform of a signal we can work out exactly what the value of that signal would be at any time t. In practical applications other signals are encountered which are not amenable to such a description and are not exactly predictable. Such signals are often termed as 'noise'. However, using concepts from probability theory, the Fourier/Laplace descriptions can be modified to accommodate these 'noise' signals.

In electronic circuits, thermal noise stands out as a naturally occurring example of what probability theory would describe as a purely random signal. In simple terms, thermal noise is the voltages and currents generated by the motion of electrons within a resistor. If we observe the voltage across a resistor it appears totally irregular and unpredictable. Many important practical signals such as speech and music can only be characterised using concepts from probability and statistics.

In section 7.2 some concepts from probability theory are reviewed and their use in the characterisation of signals is considered. In particular, ergodicity and stationarity are considered and the probability density function and joint density function are defined. Statistical averages such as the mean and variance are presented in section 7.3. These lead to the time-domain concepts of correlation and covariance as well as the frequency-domain representation in the form of the power spectral density. In section 7.4 the relationship between random signals and discrete-time linear systems is examined. This leads to the topics of spectral factorisation, inverse filtering and noise whitening filters in section 7.5. Methods for calculating the noise power at the output of a digital filter are presented in section 7.6. Finally, section 7.7 provides a summary.

7.2 Random processes

The result of tossing a coin is a reasonable example of a random event. Before the coin is tossed there is no way of telling or predicting what the outcome will be. If the coin is a fair one, all we can say is that given a large number of coin tossing experiments the average number of heads will equal the average number of tails. More formally, the outcome 'a head is tossed' can be labelled as event A and the outcome 'a tail is tossed' is labelled as event B. The number of times A occurs is N_A and the number of times B occurs is N_B. The total number of trials or experiments is $N_T = N_A + N_B$. For any sequence of N_T trials the relative occurrence frequency of A, defined as N_A/N_T, converges to a limit as N_T becomes large. The value in the limit is the probability of the event A occurring:

$$prob[A] = \lim_{N_T \to \infty} \frac{N_A}{N_T}$$

This relative frequency† approach to probability has great intuitive appeal, however it presents difficulties in deducing appropriate mathematical structure for more complicated situations.

The axiomatic approach to probability provides a sounder theoretical foundation where a probability system consists of the three following axioms:

- a sample space S of elementary events or possible outcomes;
- a class E of events that are a subset of S;
- a probability measure, $prob[A]$, assigned to each event A in the class E and having the following properties: $prob[S] = 1$; $0 \leq prob[A] \leq 1$; if $A + B$ is the union of two mutually exclusive events in the class E, then

$$prob[A + B] = prob[A] + prob[B]$$

The relative frequency interpretation and axiomatic definition of probability tend to co-exist: the former providing the intuitive insight; the latter providing the appropriate rigour.

The coin tossing experiment is a convenient mechanism for introducing the concept of a random process and some of the features associated with it. Consider an individual who tosses the same fair coin once every minute and records a value of +1 for a head and −1 for a tail. The individual has generated a discrete random signal or sequence. A typical example is illustrated in Figure 7.1(a). This is not the only sequence that could have been generated with the coin. Two other possible sequences are illustrated in Figures 7.1(b) and 7.1(c). The collection of all possible sequences is known as an ensemble or a random process. One particular sequence is a single realisation or sample†† function of the random process.

The concept of an ensemble is a challenging one. Insight can be gained by altering the coin tossing example a little. Consider a large number of individuals in a room, each tossing identical coins once per minute and recording +1 or −1 as before. The sequence generated by each individual is a sample function from the ensemble of all possible sequences present in the room. If it was necessary to measure some simple statistic of the process, such as the mean or average value, two choices would be available. Either consider one individual or realisation alone and measure the average value of all the experiments carried out by that individual, or measure the average of all the individuals at one particular point in time. The former is a time average and the latter is an ensemble average – see Figure 7.1. In this example the time average and the

† Not to be confused with the concept of frequency in Fourier analysis.

†† Not to be confused with sampling a continuous-time signal to produce a discrete-time one.

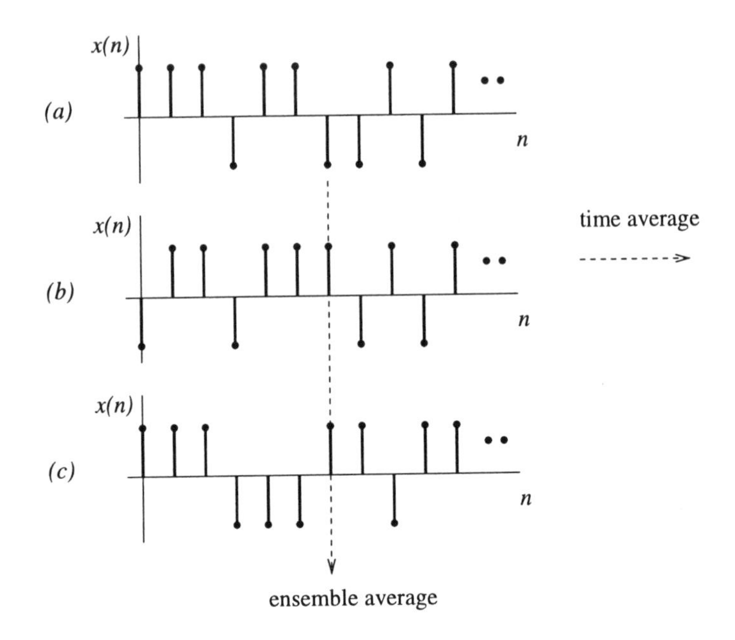

Figure 7.1 *Typical sequences from the ensemble of sequences associated with a Bernoulli random process.*

ensemble average are the same. When a random process is such that ensemble averages and time averages are the same, the process is said to be ergodic. In the example only the simplest average, the mean value, was considered. The definition of ergodic covers all possible averages, some of which will be considered in section 7.3.

In general, ensemble averages and time averages are not the same. In the coin tossing experiment it is possible to envisage a situation where the individuals grow more fatigued with time in such a way that the probability of a head occurring becomes greater than the probability of a tail occurring. The ensemble averages would reflect this change (being different at different times) but the time averages would not highlight the underlying change. If the statistical characteristics of the sample functions do not change with time, the process is said to be stationary. This does not imply however that the process is ergodic. If, for example, a proportion of individuals in the room has coins that are badly balanced such that the probability of a head is greater than the probability of a tail, the presence of these individuals ensures that time and ensemble averages are not the same, despite the sample functions being stationary.

An example of a processes which is not ergodic is illustrated in Figure 7.2. A production line produces signal generators which provide square waves $x(t-\alpha)$ of period 4. The delay α is a random variable such that:

$$prob[\alpha = 0] = 0.75$$

and

$$prob[\alpha = 1] = 0.25$$

Four possible realisations are shown. The time average of each realisation is obviously zero whereas the ensemble average is itself a waveform with period 4.

The random processes of Figures 7.1 and 7.2 are binary in nature. For example, the Bernouilli process of Figure 7.1 has two values, +1 or −1, each with a probability of ½. In general, continuous or discrete random processes can take an infinite number of values. A realisation of a continuous-time random process is illustrated in Figure 7.3. The probability that one realisation $X(t)$ of the signal assumes a particular value such as x is vanishingly small. However it is reasonable to consider the probability that the random signal X is less than a particular value x, i.e. $prob[X < x]$ which is the probability distribution function or cumulative distribution function normally written as $P(x)$. Having defined the probability distribution function it is natural to consider the probability that the signal occupies a small range of values such as that between x and $x + \Delta x$, i.e. $prob[x \le X < x + \Delta x]$. In the limit, as this range of values tends to zero, the ratio of this probability over the size of the range Δx tends to a constant – the probability density function:

$$
\begin{aligned}
p(x) &= \lim_{\Delta x \to 0} \frac{prob[\ x \le X < x + \Delta x\ ]}{\Delta x} \\
&= \lim_{\Delta x \to 0} \frac{P(x + \Delta x) - P(x)}{\Delta x} \\
&= \frac{dP(x)}{dx}
\end{aligned}
$$

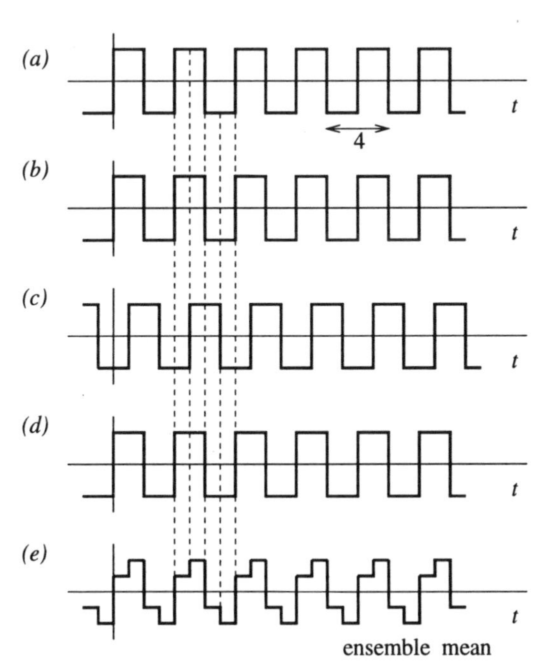

Figure 7.2 *A random process which is not ergodic: four realisations, (a) to (d), of the square wave with ensemble mean (e).*

which is the differential of the probability distribution function. Figure 7.3 illustrates how the probability density function (pdf) could be estimated for a continuous time ergodic random signal. The probability that the signal lies within the range $[x \le X < x + \Delta x]$ is the fraction of the total time spent by the signal within this range, and can be estimated by summing all such time intervals to give $\sum_j \Delta t_j$ and dividing by the total time T for which the signal was observed. Dividing this probability estimate by Δx provides an estimate of the pdf, i.e. $\hat{p}(x)$:

$$\hat{p}(x) = \frac{1}{\Delta x} \frac{1}{T} \sum_j \Delta t_j$$

A probability function which is frequently encountered is the Gaussian or normal distribution/density. The density of a Gaussian random variable is defined as:

$$p(x) = \frac{1}{\sqrt{2\pi}\sigma} \exp\left(- \frac{(x - \mu)^2}{2\sigma^2} \right)$$

where μ and σ are the parameters which define the distribution. The interpretation of these parameters will be considered in section 7.3. An example of a Gaussian probability distribution function and associated probability density function are illustrated in Figure 7.4.

The probability density function plays an important role in defining non-stationary random processes. At each point in time t there is a separate probability density function, $p(x(t))$, which defines the distribution of signal values over all realisations of the process at that particular point in time. Two realisations of a continuous-time non-stationary random process are illustrated in Figure 7.5(a). The process has a uniform distribution (Figure 7.5(b)) whose maximum X_{max} increases linearly with time.

In dealing with signals and sequences, a primary interest is how they evolve with time. Given a signal sample $x(n)$ it is reasonable to ask how it is related to a signal sample earlier or later in time, i.e. $x(m)$. The interdependence of the two random variables $x(n)$ and $x(m)$ is described by the joint probability functions:

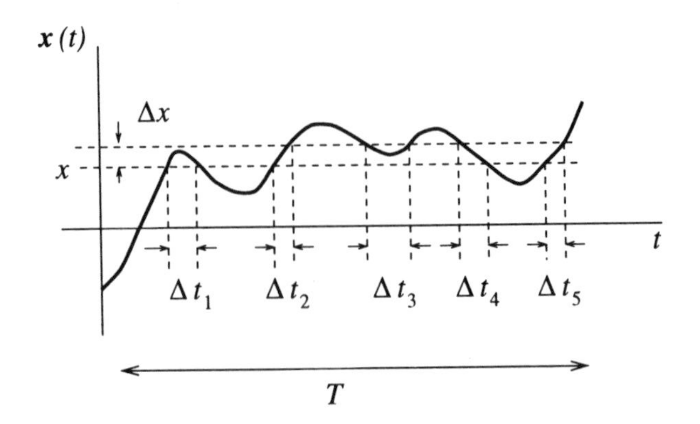

Figure 7.3 *Measuring the probability density function from one realisation of a continuous-time ergodic process.*

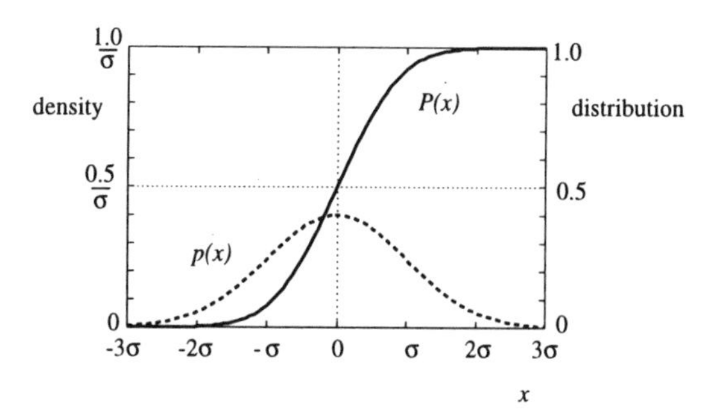

Figure 7.4 *Probability distribution function $P(x)$ and probability density function $p(x)$ of a Gaussian ergodic random process for $\mu = 0$.*

$$P(x(n), x(m)) = prob\ [X(n) \leq x(n) \text{ and } X(m) \leq x(m)]$$

$X(n)$ is the random variable, $x(n)$ is a particular value of $X(n)$, $X(m)$ is the random variable and $x(m)$ is a particular value of $X(m)$. In words, it is the probability that the signal at sample n is less than one particular value and simultaneously the signal at sample m is less than another particular value. The concept can be extended to characterise the interdependence between two different signals at difference points in time, e.g. $P(x(n), y(m))$. The joint probability density function is developed in a similar manner to the probability density function in that the starting point is to consider the probability that $X(n)$ lies in the range $[x(n) \leq X(n) < x(n) + \Delta x]$ and simultaneously $X(m)$ lies in the range $[x(m) \leq X(m) < x(m) + \Delta x]$. The probability density can be estimated by dividing this probability by the area Δx^2. In the limit, as Δx tends to zero, the density estimate becomes the density itself:

$$p(x(n), x(m)) =$$

$$\lim_{\Delta x \to 0} \frac{prob\ [x(n) \leq X(n) < x(n) + \Delta x \text{ and } x(m) \leq X(m) < x(m) + \Delta x]}{\Delta x^2}$$

$$= \frac{\partial^2 P(x(n), x(m))}{\partial x(n)\, \partial x(m)}$$

7.3 Averages and spectral representations

The probability density and joint probability density of a random process provide the basis for defining a variety of averages associated with the process. For example, the mean $m_x(n)$ or expected value ($E[.]$) of $x(n)$ is:

$$m_x(n) = E[x(n)] = \int_{-\infty}^{\infty} x\ p_{x(n)}(x)\ dx \tag{7.1}$$

The subscript $x(n)$ indicates that we are considering the probability density function of

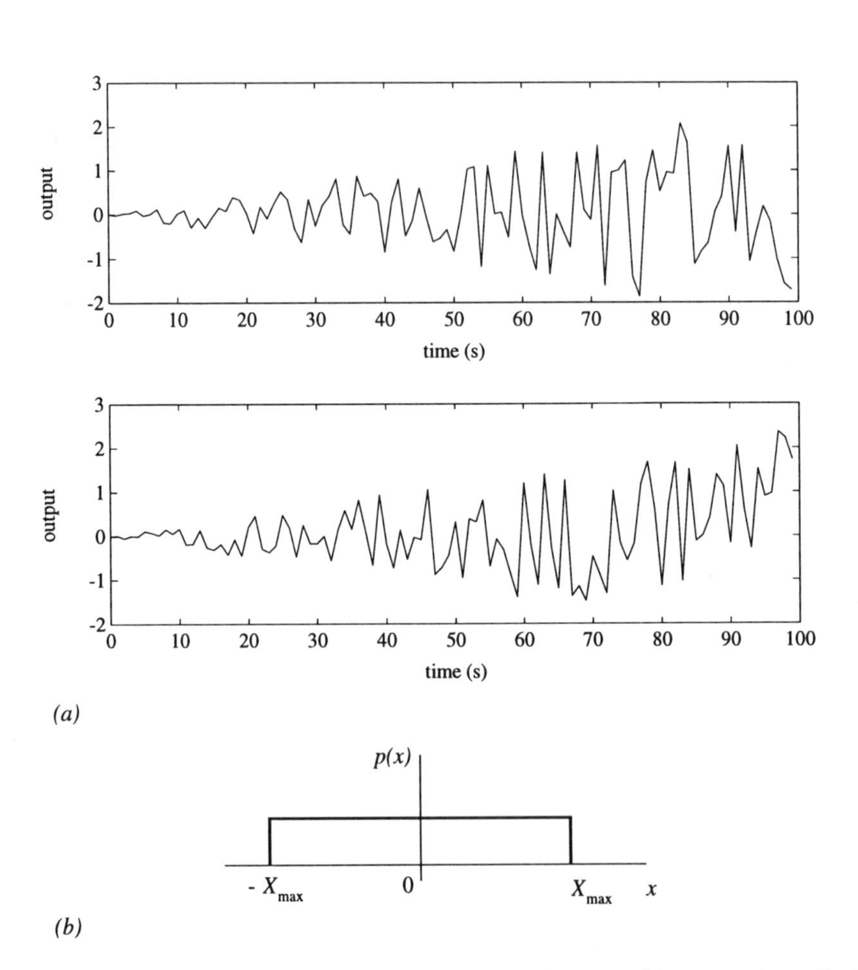

Figure 7.5 *Non-stationary random process: (a) two realisations; (b) time-varying pdf where* $X_{max} = 0.025t$.

the process x at time n. Thus we may be considering a non-stationary processes where the ensemble mean is varying with time. The integration inherent in equation (7.1) is effectively an averaging process over all the realisations of the ensemble at a particular time n. If the process is stationary and ergodic, all reference to the time n can be removed from equation (7.1) to give:

$$m_x = E[x] = \int_{-\infty}^{\infty} x \; p(x) \; dx$$

Under these conditions the ensemble average can be replaced with a time average, i.e.:

$$m_x = E[x] = \lim_{M \to \infty} \frac{1}{2M+1} \sum_{n=-M}^{M} x(n) \tag{7.2}$$

In most practical situations only one realisation of the random process is available and it is only possible to use a finite rather than an infinite segment of data to form an estimate of the average.

The mean-square of $x(n)$ is the average of $x^2(n)$, i.e.:

$$E[x^2(n)] = \int_{-\infty}^{\infty} x^2 \; p_{x(n)} \; dx \tag{7.3}$$

The mean-square value is also the average power in the signal. This can be seen more clearly if the signal is assumed to be ergodic and the ensemble average is replaced with a time average:

$$E[x^2] = \lim_{M \to \infty} \frac{1}{2M+1} \sum_{n=-M}^{M} x^2(n) \tag{7.4}$$

The variance $\sigma_x^2(n)$ of $x(n)$ is the mean-square variation about the mean:

$$\begin{aligned} \sigma_x^2(n) &= E[(x(n) - m_x(n))^2] \\ &= E[x^2(n)] - m_x^2(n) \end{aligned}$$

Self assessment question 7.1: Show that $\sigma_x^2(n) = E[x^2(n)] - m_x^2(n)$.

7.3.1 Autocorrelation and autocovariance

The mean, variance and mean-square are simple averages which give no indication of the dependence of a signal at one point in time with the same signal at a different point in time. The autocorrelation $\phi_{xx}(n, k)$ is a measure of the correlation between values of the random process at different times. It is the average of the product of a sample at time n and a sample at time k and is defined using the joint probability density function:

$$\begin{aligned} \phi_{xx}(n, k) &= E[x(n) \; x(k)] \\ &= \int_{-\infty}^{\infty} \int_{-\infty}^{\infty} x(n) \; x(k) \; p(x(n), x(k)) \; dx(n) \; dx(k) \end{aligned}$$

The autocovariance $\gamma_{xx}(n, k)$ bears a similar relationship to the autocorrelation as the variance has to the mean-square value. The autocovariance is the autocorrelation about the means, i.e.:

$$\begin{aligned} \gamma_{xx}(n, k) &= E[(x(n) - m_x(n))(x(k) - m_x(k))] \\ &= \phi_{xx}(n, k) - m_x(n) m_x(k) \end{aligned}$$

Thus *for zero-mean processes, autocorrelation and autocovariance are the same.*

If a process is ergodic, the mean is in fact the DC level of the signal. For non-stationary signals the time varying mean can be viewed as a deterministic component, e.g. it might be a ramp. When dealing with linear systems, the deterministic and random components can be dealt with separately because superposition applies. Thus *it is common practice when dealing with random processes to make the assumption that the mean is zero and hence the terms autocorrelation and autocovariance are used interchangeably.*

In general for a non-stationary random process, the autocorrelation and autocovariance of a discrete-time random process are two dimensional sequences – dependent upon both n and k. If however the random process is stationary, the autocorrelation or autocovariance depends only on the time difference $k' = k - n$, e.g.:

$$\begin{aligned}\phi_{xx}(\, n, n+k'\,) &= \phi_{xx}(k') \\ &= E[x(n)\; x(n+k')]\end{aligned}$$

As usual, if the process is ergodic the ensemble average can be replaced by a time average:

$$\begin{aligned}\phi_{xx}(k) &= E[x(n)\; x(n+k)] \\ &= \lim_{M \to \infty} \frac{1}{2M+1} \sum_{n=-M}^{M} x(n)\; x(n+k)\end{aligned} \tag{7.5}$$

A typical autocorrelation sequence associated with a zero-mean ergodic process is illustrated in Figure 7.6. There are several properties of autocorrelation and autocovariance sequences that are worth noting:

- they are symmetrical about the origin, e.g.:

$$\phi_{xx}(k) = \phi_{xx}(-k)$$

- they contain the variance:

$$\sigma_x^2 = \gamma_{xx}(0) = \phi_{xx}(0) - m_x^2 \tag{7.6}$$

- (iii) for many random sequences, individual samples become less correlated as they become more separated in time:

$$\lim_{m \to \infty} \gamma_{xx}(m) = 0$$

In a similar manner, the cross-correlation provides a measure of the dependence between two different random sequences. The cross-correlation of two ergodic signals is:

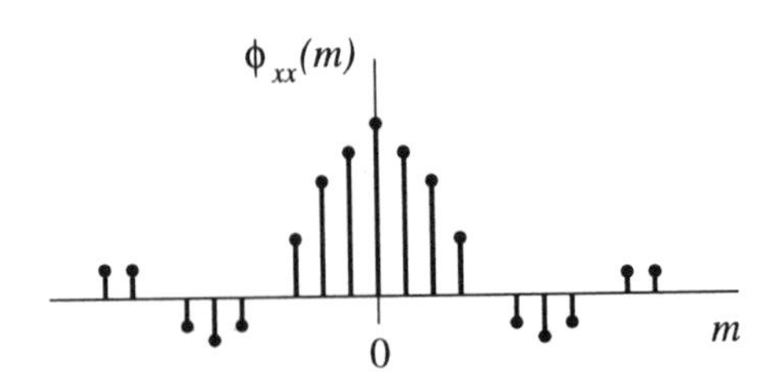

Figure 7.6 *Typical autocorrelation sequence.*

$$\phi_{xy}(k) = E[x(n)\ y(n+k)] \tag{7.7}$$

with associated time average:

$$\begin{aligned}\phi_{xy}(k) &= E[x(n)\ y(n+k)] \\ &= \lim_{M \to \infty} \frac{1}{2M+1} \sum_{n=-M}^{M} x(n)\ y(n+k)\end{aligned}$$

It is worth noting that the cross-correlation is not symmetrical but the following property does apply:

$$\phi_{xy}(m) = \phi_{yx}(-m)$$

Self assessment question 7.2: Estimate the mean and variance of the following stationary sequence: $\{1.3946,\ 1.6394,\ 1.8742,\ 2.7524,\ 0.6799\}$.

7.3.2 Correlation and dependence

Two terms are commonly used to indicated the dependency (or lack of it) of a signal at one point in time with the same signal at a different point in time, or more generally for the dependency of one signal upon another. These terms are *independent* and *uncorrelated.* They are often confused but have strict definitions. Two random processes are linearly independent or uncorrelated if:

$$E[x(n)\ y(k)] = E[x(n)]\ E[y(k)] \tag{7.8}$$

Two random processes are statistically independent if:

$$p(x(n)\ y(m)) = p(x(n))\ p(y(m)) \tag{7.9}$$

Statistically independent random processes are uncorrelated but uncorrelated random processes may be statistically dependent. It is only if the two process are jointly Gaussian that the terms uncorrelated and independent are equivalent.

7.3.3 Power spectral density

Random signals are by definition non-periodic. They also tend to have infinite energy and finite power. Hence the z-transform and Fourier transform of a random sequence do not exist. Autocorrelation sequences on the other hand have finite energy as is indicated by the tendency to decay to zero as the lag m increases. Thus, as in Chapter 4, the z-transform $S_{xx}(z)$ of the autocovariance sequence $\gamma_{xx}(m)$ can be defined as:

$$S_{xx}(z) = \sum_{m=-\infty}^{\infty} \gamma_{xx}(m)\ z^{-m}$$

If we make the assumption that the sequence is zero-mean then $S_{xx}(z)$ is also the z-transform of the autocorrelation sequence $\phi_{xx}(m)$:

$$S_{xx}(z) = \sum_{m=-\infty}^{\infty} \phi_{xx}(m)\, z^{-m} \tag{7.10}$$

The only difference between this and the definition of the z-transform for deterministic signals is the summation which includes negative values of the lag m. This is known as a two-sided z-transform. The inversion of this transform can be accomplished by a partial fraction expansion. However the two-sided nature makes it more complicated than the one-sided transform and it will not be considered further in this text. The interested reader should consult [Oppenheim and Schafer 1975]. The Fourier transform of the autocovariance sequence can be obtained in a similar manner to the Fourier transform of deterministic sequences by simply replacing z by $\exp(j\omega\Delta t)$. Thus for a zero-mean sequence:

$$S_{xx}(\omega) = \sum_{m=-\infty}^{\infty} \phi_{xx}(m) \exp(-j\omega m\Delta t) \tag{7.11}$$

In common with the Fourier transform of any discrete sequence, $S_{xx}(\omega)$ is periodic in ω with period $2\pi/\Delta t$. Equation (7.11) can be interpreted as a weighted sum of harmonically related complex phasors – the autocorrelation coefficients $\phi_{xx}(m)$ being the weights. Thus the weights can be calculated using the Fourier series of equation (1.11) in the usual way:

$$\phi_{xx}(m) = \frac{\Delta t}{2\pi} \int_{0}^{2\pi/\Delta t} S_{xx}(\omega) \exp(j\omega m\Delta t)\, d\omega \tag{7.12}$$

Equations (7.11) and (7.12) define forward and reverse Fourier transforms which relate autocorrelation to $S_{xx}(\omega)$. Often the sampling period Δt is normalised to unity.

Having defined the Fourier transform of the autocorrelation it is natural to seek some physical interpretation of it. A useful insight comes from the relationship between the variance and the autocorrelation. For a zero-mean stationary random process, the variance is the average power:

$$\begin{aligned}\sigma_x^2 &= E[x^2(n)] \\ &= \lim_{n \to \infty} \frac{1}{2N+1} \sum_{n=-N}^{N} x^2(n)\end{aligned}$$

The variance can also be obtained from the autocorrelation which is related to $S_{xx}(\omega)$:

$$\begin{aligned}\sigma_x^2 &= \phi_{xx}(0) \\ &= \frac{\Delta t}{2\pi} \int_{0}^{2\pi/\Delta t} S_{xx}(\omega)\, d\omega\end{aligned}$$

The average power in the signal is the integral of $S_{xx}(\omega)$ over the whole frequency range. Thus $S_{xx}(\omega)$ is the distribution of average power with respect to frequency, i.e. the power spectral density (PSD). A simple system for estimating the PSD of a stationary random process $x(n)$ at a particular frequency ω_0 is illustrated in Figure 7.7. The signal is applied to a narrow-band filter centred at ω_0 with a bandwidth of $\Delta\omega$. The power at the output of this filter is estimated by first squaring the output and then performing a time average. The output of the time average operation is a scaled estimate of the required spectral density – scaled by the bandwidth $\Delta\omega$. Techniques for estimating PSD will be considered further in Chapter 9.

Self assessment question 7.3: A random signal has a PSD which is constant between 0 and 10 kHz and is almost zero elsewhere. The signal is applied to a filter which has a similar frequency response to that illustrated in Figure 7.7. The centre frequency is 7.5 kHz, the bandwidth is 1 kHz and the passband gain is 0 dB. Is the power in the output signal more or less than the power in the input signal? By what factor?

7.3.4 Alternative representations of a random process

Several representations of discrete random processes have been considered. These include: realisations of a random process; probability density; autocorrelation; and power spectral density. Figure 7.8 illustrates these representations for a white random process taken from a uniform distribution. One possible realisation is shown in Figure 7.8(a). Figure 7.8(b) shows the uniform probability density function – the signal is equally likely to have any value between -0.5 and 0.5. Closer examination of Figure 7.8(a) confirms that the signal never attains a value greater than 0.5 or less than -0.5. Since the signal is white, the PSD has a constant value between plus and minus half the sampling frequency – the power in the signal is distributed uniformly in the frequency domain. Since the autocorrelation is the inverse Fourier transform of the PSD, the autocorrelation sequence must contain a single non-zero value at lag $m = 0$. The value of the single value is of course the variance of the signal. It is left as an exercise for the reader to confirm that the probability density function shown implies a variance of 1/12 as indicated by the autocorrelation sequence.

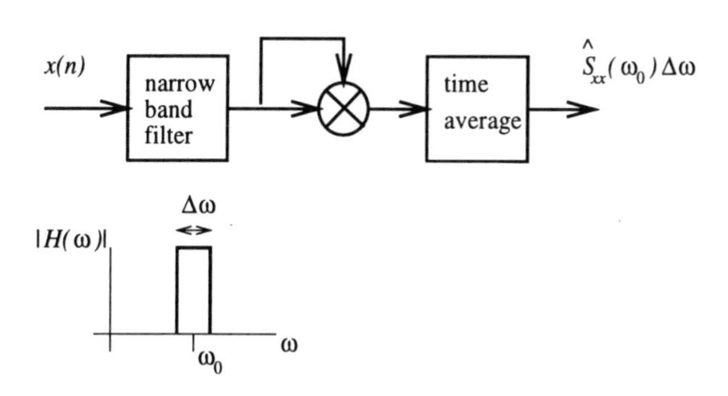

Figure 7.7 *Estimate of the PSD at frequency ω_0.*

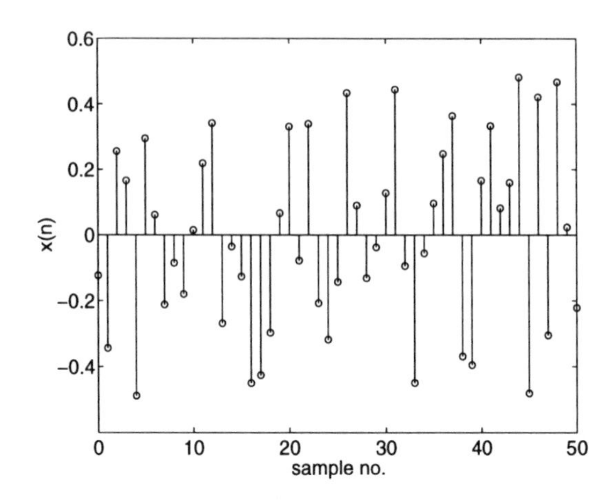

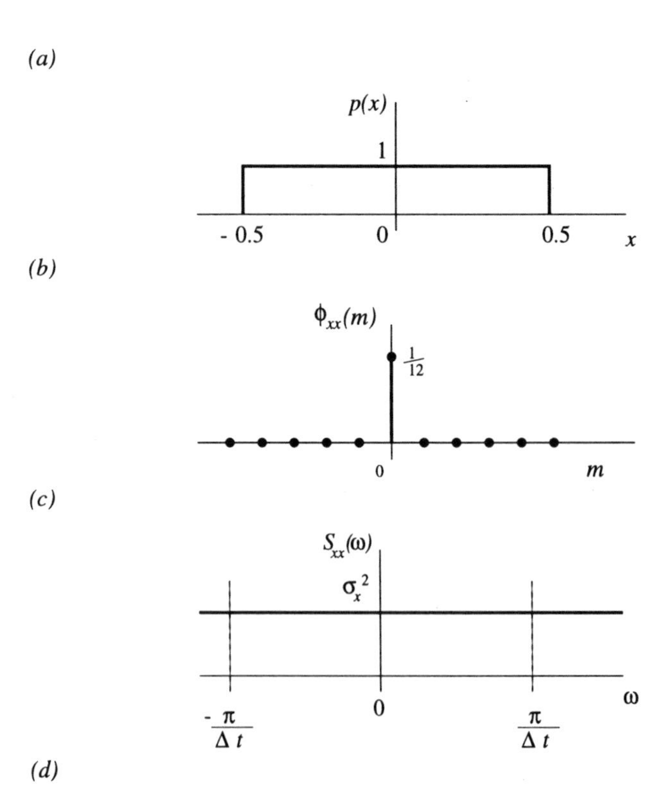

Figure 7.8 *Several representations of a uniform white random sequence: (a) one realisation; (b) probability density function; (c) autocorrelation sequence; (d) power spectral density.*

7.4 Random signal and discrete linear systems

A discrete-time linear system with ergodic input $x(n)$ and output $y(n)$ is illustrated in Figure 7.9. The system is characterised by its impulse response sequence $\{h_n\}$ and hence the input and output are related by the convolution operation of equation (6.1):

$$y(n) = \sum_{k=0}^{\infty} h_k \, x(n-k)$$

If the mean value of the input is m_x the mean of the output is straightforward to evaluate, i.e.:

$$\begin{aligned} m_y &= E[\sum_{k=0}^{\infty} h_k \, x(n-k)] \\ &= \sum_{k=0}^{\infty} h_k \, E[x(n-k)] \\ &= m_x \sum_{k=0}^{\infty} h_k \end{aligned}$$

It is worth noting that the same result would be obtained if we calculated the steady-state response of the filter to a constant DC level of m_x. Using the ideas from Chapter 3, the frequency response of the filter is $\sum_k h_k \exp(-jk\omega\Delta t)$. The gain at DC is obtained by setting $\omega = 0$, i.e. $\sum_k h_k$ as expected. Because the system is linear and superposition applies, the response to a random process can be evaluated in two parts and combined to give the overall response. Thus for convenience we will assume that the means of all random processes are zero, e.g. $E[x(n)] = 0$. Hence autocorrelation and autocovariance sequences are identical. For a stationary input signal, the z-transform of the input and output autocorrelation sequences are related as follows:

$$S_{yy}(z) = H(z) \, H(z^{-1}) \, S_{xx}(z) \tag{7.13}$$

It is appropriate to contrast this result with the equivalent one for deterministic signals, i.e. $Y(z) = H(z)X(z)$. For a deterministic signal, $H(z)$ relates the transform of the output signal directly to the transform of the input signal. For a stationary random signal, $H(z)H(z^{-1})$ relates the transform of the output autocorrelation to the transform of the input autocorrelation. To relate the PSD at the output to the input PSD it is

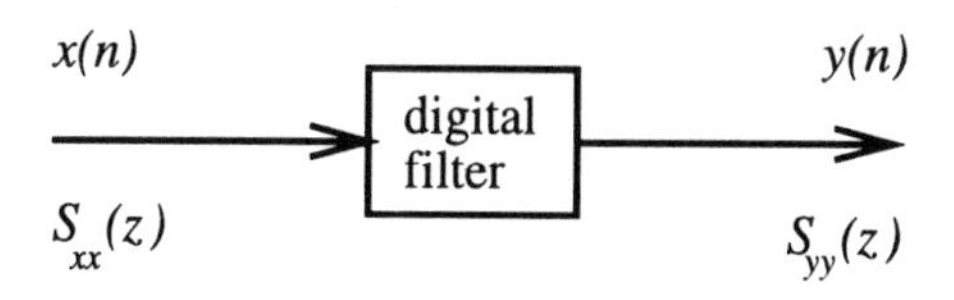

Figure 7.9 *A digital filter with random input and output signals.*

simply a matter of replacing z with $\exp(-j\omega\Delta t)$. Thus:

$$S_{yy}(\omega) = H(\omega)\ H^*(\omega)\ S_{xx}(\omega)$$
$$= |\ H(\omega)\ |^2\ S_{xx}(\omega) \tag{7.14}$$

To illustrate the power of these relationships, consider the following simple example.

EXAMPLE 7.1

A zero-mean stationary white noise signal $x(n)$ is applied to a FIR filter with impulse response sequence $\{1, 0.5\}$ as shown in Figure 7.10. Since the input is white, the autocorrelation at the input is by definition:

$$\phi_{xx}(m) = \sigma_x^2\ \delta(m)$$

with z-transform:

$$S_{xx}(z) = \sigma_x^2$$

Using equation (7.13) the z-transform of the output autocorrelation is:

$$S_{yy}(z) = H(z)\ H(z^{-1})\ S_{xx}(z)$$
$$= (1 + 0.5\ z^{-1})\ (1 + 0.5\ z)\ \sigma_x^2$$
$$= (0.5\ z + 1.25 + 0.5\ z^{-1})\ \sigma_x^2$$

The autocorrelation at the output is the inverse z-transform of this:

$$\phi_{yy}(-1) = 0.5\sigma_x^2;\quad \phi_{yy}(0) = 1.25\sigma_x^2;\quad \phi_{yy}(1) = 0.5\sigma_x^2.$$

From equation (7.5), the variance of the output is identifiable from the zeroth autocorrelation term as $1.25\sigma_x^2$ with a corresponding RMS value of $\sqrt{1.25}\sigma_x$. The PSD at the output can be found by either setting z to $\exp(-j\omega\Delta t)$ or by direct application of equation (7.11). Using the former:

$$S_{yy}(\omega) = (0.5\ e^{-j\omega\Delta t} + 1.25 + 0.5\ e^{j\omega\Delta t})\ \sigma_x^2$$

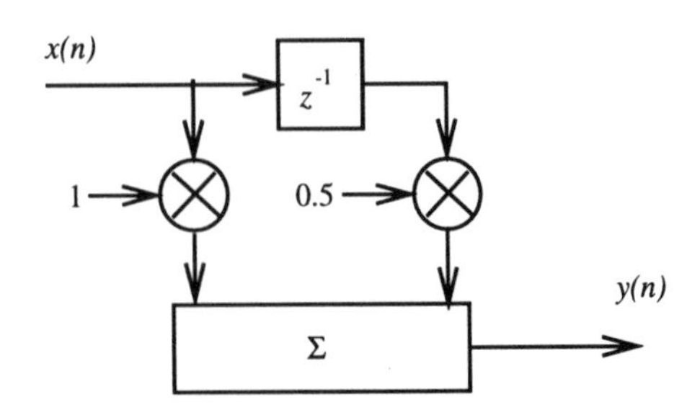

Figure 7.10 *A 2-tap FIR filter with random input.*

$$= (1.25 + \cos(\omega \Delta t))\ \sigma_x^2$$

7.4.1 Cross-correlation between the input and output of a filter

Another result which is particularly useful is that which relates the cross-correlation between the input and the output with the autocorrelation of the input. The cross-correlation between the input and output can be defined as:

$$\phi_{xy}(m) = E[x(n)\ y(n+m)]$$

The desired relationship is:

$$S_{xy}(z) = H(z)\ S_{xx}(z) \tag{7.15}$$

where $S_{xy}(z)$ is the z-transform of the cross-correlation sequence, i.e.:

$$S_{xy}(z) = \sum_{m=-\infty}^{\infty} \phi_{xy}(m)\ z^{-m}$$

If the input is white we obtain a simple relationship between the z-transform of cross-correlation and the transfer function:

$$S_{xy}(z) = H(z)\ \sigma_x^2$$

Taking inverse transforms gives:

$$\phi_{xy}(m) = h_m\ \sigma_x^2$$

Thus the cross-correlation sequence between the input and output is simply a scaled version of the impulse response sequence. This result can be utilised in the important practical problem of system identification. Given a unknown linear system with white stationary input $x(n)$ and output $y(n)$, the impulse response can be measured by estimating the cross-correlation directly from the data, e.g.:

$$\hat{\phi}_{xy}(m) = \frac{1}{M} \sum_{n=0}^{M-1} x(n)\ y(n+m)$$

and then scaling with an appropriate estimate of the variance σ_x^2, e.g.:

$$\hat{\sigma}_x^2 = \frac{1}{M} \sum_{n=0}^{M-1} x^2(n)$$

The estimate of the impulse response is thus:

$$\hat{h}_m = \frac{\hat{\phi}_{xy}(m)}{\hat{\sigma}_x^2}$$

Self assessment question 7.4: What is the difference between the PSD and the pdf of a signal?

Self assessment question 7.5: Repeat example 7.1 for a FIR filter with impulse response sequence $\{0.5, 0.75\}$.

7.5 Spectral factorisation, inverse and whitening filters

It is clear from equation (7.14) that it is the amplitude response of the digital filter, i.e. $|H(\omega)|$, rather than the phase response $\angle H(\omega)$ that controls the relationship between the input PSD and the output PSD. In fact, the output PSD is totally independent of the phase response of the filter. Thus if a stationary random signal is applied to two filters with the same amplitude response but different phase responses, the PSDs at their outputs would be identical. Further, there could be a family of such digital filters with different phase responses and identical amplitude responses. A simple example illustrates this point.

EXAMPLE 7.2

A zero-mean stationary white noise $x(n)$ is applied to a filter with transfer function $H_1(z)$:

$$H_1(z) = \frac{(z - ½)\,(z - 3)}{z^2}$$

The filter has zeros at 0.5 and 3. The z-transform of the output autocorrelation sequence is obtained by application of equation (7.13):

$$\begin{aligned} S_{yy}(z) &= H_1(z)\, H_1(z^{-1})\, \sigma_x^2 \\ &= z^{-2}\,(z - ½)\,(z - 3)\, z^2\,(z^{-1} - ½)\,(z^{-1} - 3)\,\sigma_x^2 \end{aligned}$$

The transform $S_{yy}(z)$ has four zeros at 1/2, 1/3, 3 and 2, whereas the filter has two at 1/2 and 3. (N.B. The terms z^{-2} and z^2 cancel but are left in for convenience.) Re-ordering the factors to emphasis the zeros at 1/2 and 1/3 gives:

$$S_{yy}(z) = \left(z^{-2}\,(z - ½)\,(z^{-1} - 3) \right)\left(z^2\,(z^{-1} - ½)\,(z - 3) \right)\sigma_x^2$$

Thus an alternative filter $H_0(z)$ could be used to generate a sequence with the same PSD using white noise as a source:

$$S_{yy}(z) = H_0(z)\, H_0(z^{-1})\, \sigma_x^2$$

where:

$$H_0(z) = \frac{(z - ½)\,(z^{-1} - 3)}{z^2}$$

For this example there are four possible filters which could produce the PSD $S_{yy}(z)$ from the white noise sequence $x(n)$. The filters are summarised in Table 7.1. The filters are classified as minimum, mixed and maximum phase. In general, a minimum phase filter has all its zeros inside the unit circle in the z-plane. A mixed phase filter has some of its zeros outside the unit circle and a maximum phase filter has all of its zeros outside the unit circle.

Table 7.1 *Family of 2-tap FIR filters with identical amplitude responses.*

filter	*zeros*		*classification*
$H_0(z)$	$\frac{1}{2}$	$\frac{1}{3}$	minimum phase
$H_1(z)$	$\frac{1}{2}$	3	mixed phase
$H_2(z)$	2	$\frac{1}{3}$	mixed phase
$H_3(z)$	2	3	maximum phase

As illustrated in Figure 7.11, all the filters in Table 7.1 have the same gain or amplitude response. The minimum phase filter $H_0(z)$ has the smallest phase shift of all the filters at any frequency – hence the term minimum phase. In a similar manner, the maximum phase filter $H_3(z)$ has the largest phase shift of all the filters at any frequency – hence the term maximum phase. The other two mixed phase filters lie between these two extremes.

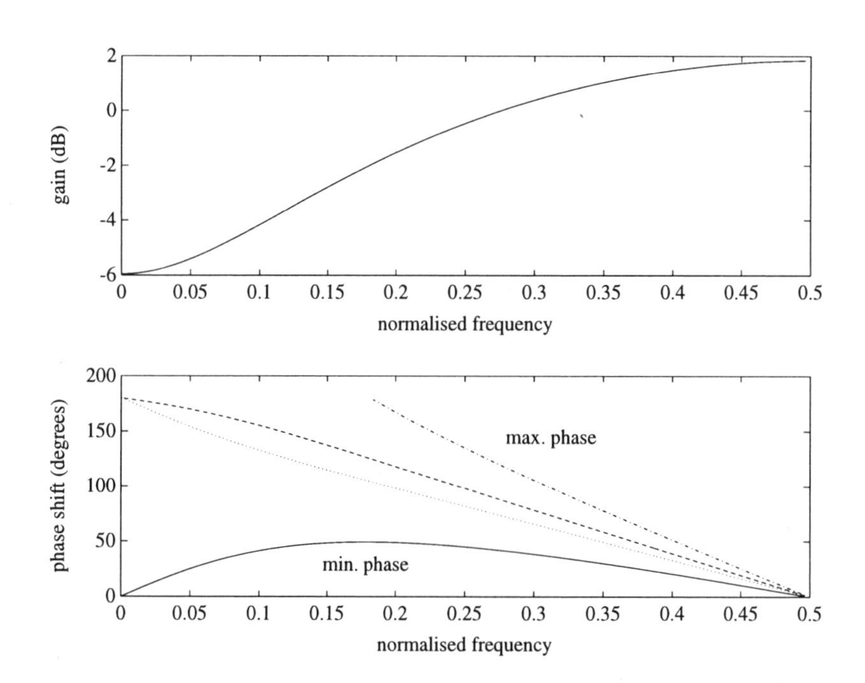

Figure 7.11 *Amplitude and phase responses of minimum, mixed and maximum phase filters.*

In the more general case where we have a set of poles and zeros which exist in complex conjugate pairs as in Figure 7.12, choosing all poles and zeros inside the unit circle provides the minimum phase spectral factorisation of $S_{yy}(z)$.

7.5.1 Inverse filters

Consider the problem illustrated in Figure 7.13. A stationary sequence $x(n)$ is applied to a filter with transfer function $H(z)$ to produce an output $y(n)$. The requirement is to design a filter which will reconstruct or reconstitute the original input from the observed output. The reconstruction may have an arbitrary gain A and a delay of d. This problem is found in many application areas under a variety of names. The simplest definition of the required inverse filter $H^{-1}(z)$ is:

$$H(z)\,H^{-1}(z) = Az^{-d} \tag{7.16}$$

The inverse filter $H^{-1}(z)$ is said to 'equalise' the amplitude and phase response of $H(z)$ since the inverse filter has equal and opposite amplitude and phase response to those of $H(z)$. The filter $H^{-1}(z)$ is also said to perform 'deconvolution' on the output $y(n)$ to reconstruct $x(n)$ at its own output. The phase characteristics of $H(z)$ are particularly important in this context. If it is minimum phase and stable, it has all its poles and zeros inside the unit circle. Hence the inverse filter will also be minimum phase and *stable* because the poles of $H(z)$ become the zeros of $H^{-1}(z)$, and the zeros of $H(z)$ become the poles of $H^{-1}(z)$. If, however, the filter $H(z)$ is mixed or maximum (non-minimum) phase, the zeros outside the unit circle become poles outside the unit circle and the inverse filter is unstable.

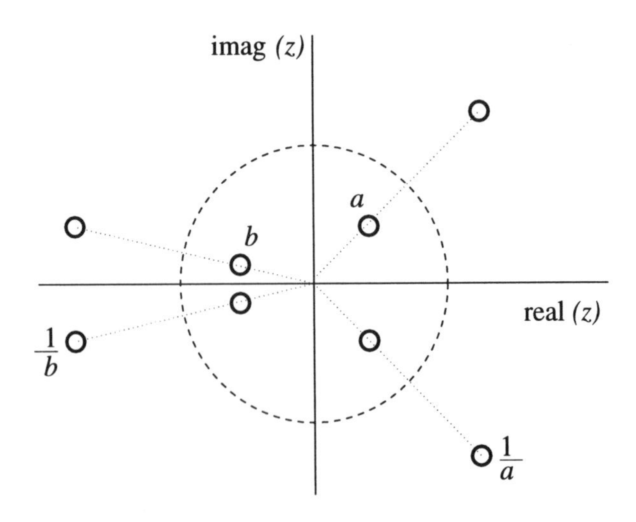

Figure 7.12 *Family of possible zeros and their locations in the z-plane.*

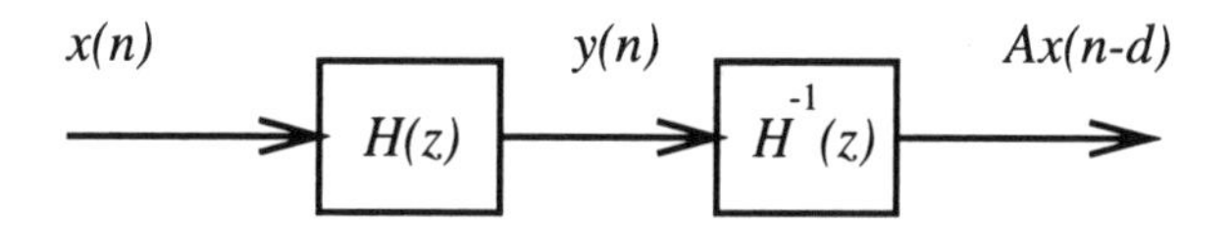

Figure 7.13 *An inverse filter operation.*

7.5.2 Noise whitening

A related problem to inverse filtering is noise whitening. Given a correlated sequence $y(n)$ with PSD $S_{yy}(\omega)$, how do we design a filter whose output $u(n)$ will be white? Applications include: estimator design; spectral analysis; linear predictive coding of speech; and matched filtering in correlated noise. By way of an example, consider the signal $y(n)$ generated by applying the white stationary sequence to any of the filters of Table 7.1. As before, the z-transform of the autocorrelation is:

$$S_{yy}(z) = z^{-2}(z - ½)\,(z - 3)z^2(z^{-1} - ½)\,(z^{-1} - 3)\,\sigma_x^2$$

The inverse of any of the filters $H_0(z)$, $H_1(z)$, $H_2(z)$ or $H_3(z)$ will whiten the signal, i.e.:

$$S_{uu}(z) = \sigma_u^2 = H_i^{-1}(z)\,H_i^{-1}(z^{-1})\,S_{yy}(z)$$

but only the inverse $H_0^{-1}(z)$ will be stable. The filter $H_0^{-1}(z)$ is said to be the minimum phase whitening filter for the signal $y(n)$.

It is worth emphasising that deconvolution and whitening are not the same thing. Figure 7.14 illustrates a simple deconvolution problem with a stable inverse filter. The only condition placed on $x(n)$ is that it should be stationary.

Figure 7.15 illustrates the use of a whitening filter on the output of a non-minimum phase system. A white random sequence is applied to a filter $H_1(z)$. The filter $H_0^{-1}(z)$ will whiten the sequence $y(n)$ to produce a white sequence $u(n)$. $x(n)$ and $u(n)$ are not the same.

Self assessment question 7.6: A unit variance white noise source is applied to a digital filter with transfer function $H(z) = 0.1 - 0.8z^{-1}$. Design a second filter to whiten the output of the first. Is this whitening filter an inverse filter for $H(z)$?

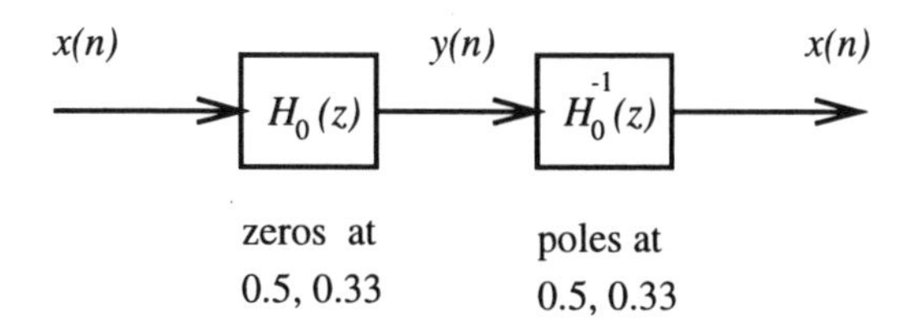

Figure 7.14 *Stable inverse filter for a minimum phase FIR system.*

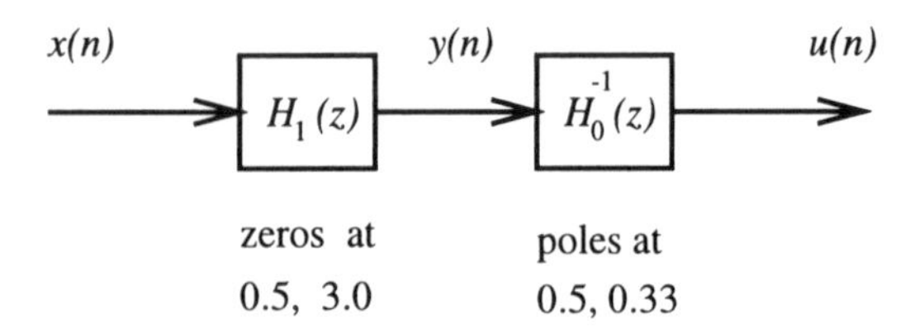

Figure 7.15 *Whitening filter for a non-minimum phase FIR signal generation model.*

7.5.3 Cross-correlation between two filter outputs

Another useful result relates to Figure 7.16(a). A stationary random signal is applied to two filters with transfer functions $H_1(z)$ and $H_2(z)$ and respective outputs $y_1(n)$ and $y_2(n)$. The two outputs are obviously related since they originate from the one source. One measure of the relationship between $y_1(n)$ and $y_2(n)$ is the cross-correlation $\phi_{y_2y_1}(m) = E[y_2(n)\, y_1(n+m)]$. The first step in obtaining this cross-correlation is to re-draw Figure 7.16(a) as the more familiar problem of Figure 7.16(b). Here the inverse filter $H_2^{-1}(z)$ has been used to obtain the input $x(n)$ from the output $y_2(n)$. The cross-correlation between the new input $y_2(n)$ and the new output $y_1(n)$ is obtained by direct application of equation (7.15) to Figure 7.16(b):

$$\begin{aligned} S_{y_2y_1}(z) &= H_2^{-1}(z)\, H_1(z)\, S_{y_2y_2}(z) \\ &= H_2^{-1}(z)\, H_1(z) \left(H_2(z)\, H_2(z^{-1}) S_{xx}(z) \right) \\ &= H_1(z)\, H_2(z^{-1}) S_{xx}(z) \end{aligned} \tag{7.17}$$

Note that, because $H_2^{-1}(z)$ cancels with $H_2(z)$, 'the non-minimum phase or otherwise'

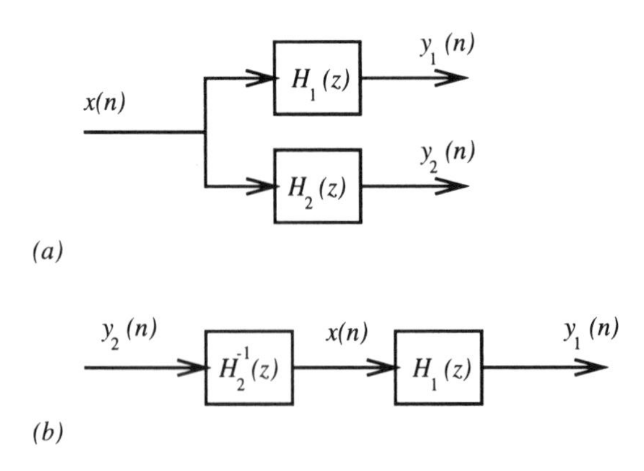

Figure 7.16 *Random signal applied to filter bank: (a) original problem; (b) re-structured problem using inverse filter.*

issues are of no significance here. Finally, the cross-correlation is obtained by taking the inverse transform:

$$\phi_{y_2 y_1}(m) = Z^{-1}[S_{y_2 y_1}(z)]$$

7.6 Filter noise calculations

In several areas, such as: digital phase lock loop design; assessment of the affects of quantisation noise; performance of sigma-delta (ΣΔ) A/D converters; and normalisation of fading generators for mobile radio simulation – it is important to be able to calculate the noise power at the filter output given a white input. For non-white inputs, spectral factorisation can be used to derive an equivalent 'white noise plus filter' model. Thus we have a situation where a white stationary sequence of variance σ_x^2 is applied to a filter with transfer function $H(z)$ and we wish to calculate the variance σ_y^2. Using equation (7.13) as before, the z-transform of the output sequence is:

$$S_{yy}(z) = H(z)\, H(z^{-1})\, \sigma_x^2$$

Thus the autocorrelation sequence is:

$$\phi_{yy}(m) = Z^{-1}[S_{yy}(z)]$$

and the variance is the zero lag term in the autocorrelation:

$$\begin{aligned}\sigma_y^2 &= \phi_{yy}(0) \\ &= \int_C S_{yy}(z)\, z^{-1} dz \end{aligned} \qquad (7.18)$$

The most obvious approach to solving equation (7.18) is to use calculus of residues [Oppenheim and Schafer 1975]. However this involves root finding and may be numerically unstable in the case of multiple poles. For a FIR filter with transfer function:

$$H(z) = \sum_{i=0}^{N} h_i\, z^{-i}$$

Equation (7.18) reduces, as in equation (6.19), to:

$$\sigma_y^2 = \sigma_x^2 \sum_{i=0}^{N-1} h_i^2 \qquad (7.19)$$

In other words, the noise gain is simply the sum of the squares of the coefficient values.

EXAMPLE 7.3

For the specific example where $y(n) = 1/3\, [x(n+2) + x(n+1) + x(n)]$, what is the FIR filter output noise variance with respect to the input variance?

Solution

Using equation (7.19):

$$\sigma_y^2 = \sigma_x^2 \sum_{k=0}^{2} 1/9 = \sigma_x^2 \times 1/3 = \Rightarrow \sigma_y^2/\sigma_x^2 = 1/3$$

If the filter is IIR with transfer function as in equation (4.9):

$$H(z) = \frac{\sum_{i=0}^{N} a_i z^{-i}}{1 - \sum_{i=1}^{N} b_i z^{-1}}$$

a better alternative to calculus of residues is to embed the problem in a system of linear equations. The noise variance at the output is given by the solution to a set of $(N+1)$ linear equations:

$$\mathbf{B\,c} = \mathbf{a} \tag{7.20}$$

where:

$$\mathbf{B} = \left\{ \begin{bmatrix} 1 & -b_1 & -b_2 & \cdots & -b_N \\ 0 & 1 & -b_1 & \cdots & -b_{N-1} \\ 0 & 0 & 1 & \cdots & -b_{N-2} \\ \cdot & \cdot & \cdot & & \cdot \\ \cdot & \cdot & \cdot & \cdots & \cdot \\ 0 & 0 & 0 & 0 & 1 \end{bmatrix} + \begin{bmatrix} 1 & \cdots & -b_{N-2} & -b_{N-1} & -b_N \\ -b_1 & \cdots & -b_{N-1} & -b_N & 0 \\ -b_2 & \cdots & -b_N & 0 & 0 \\ \cdot & & \cdot & \cdot & \cdot \\ \cdot & \cdots & \cdot & \cdot & \cdot \\ -b_N & \cdots & 0 & 0 & 0 \end{bmatrix} \right\};$$

$$\mathbf{a} = \begin{bmatrix} a_0 & a_1 & a_2 & \cdots & a_N \\ 0 & a_0 & a_1 & \cdots & a_{N-1} \\ 0 & 0 & a_0 & \cdots & a_{N-2} \\ \cdot & \cdot & \cdot & & \cdot \\ \cdot & \cdot & \cdot & \cdots & \cdot \\ 0 & 0 & 0 & 0 & a_0 \end{bmatrix} \begin{bmatrix} a_0 \\ a_1 \\ a_2 \\ \cdot \\ \cdot \\ a_N \end{bmatrix}$$

and:

$$\mathbf{c} = [\, c_0 \; c_1 \; c_2 \cdots c_N \,]^T$$

Solving a set of linear equations to yield **c** is inherently simpler that finding roots of polynomials in the calculus of residues method. The noise variance is directly related to the first element of the vector **c**:

$$\sigma_y^2 = \sigma_x^2 \, 2c_0$$

The method is also capable of handling multiple poles.

7.6.1 Quantisation noise

In section 4.1, the operation of an A/D converter was described as a combination of two processes: sampling and quantisation. The former was considered in detail in section

4.2. Here the emphasis will be on the effect of quantisation on the signal and subsequent signal processing.

Figure 7.17(a) illustrates the sampling and quantisation elements of many A/D conversion systems. The operation of a simple 4-bit quantiser is illustrated in Figure 4.3. The function of the quantiser is to represent analogue samples from the S/H as a binary number – in the example of Figure 4.3, a 4-bit binary number. Since it is a 4-bit quantiser there are $2^4 = 16$ separate voltage levels. In general, for an M-bit quantiser there are 2^M separate voltage levels. If an analogue voltage is presented to the quantiser by the S/H, the quantiser represents it with the nearest of these 2^M voltage levels. Obviously as the number of bits and hence the number of levels increases, the quality of the approximation improves. A perfect quantiser would have an infinite number of levels and produce the ideal discrete-time sequence $\{x(n)\}$ from the analogue signal $x(t)$ as considered in Chapter 4. However a practical quantiser produces the quantised sequence $\{x_q(n)\}$ from the same analogue signal as indicated in Figure 7.18. The quantised sequence is a nonlinear function $q(\ )$ of the ideal sequence:

$$x_q(n) = q(x(n))$$

The error introduced by the quantisation process is thus:

$$e(n) = x_q(n) - x(n)$$

The presence of the nonlinear function $q(\ )$ makes the analysis of the quantiser difficult. It is common practice to model the quantisation error $e(n)$ as a stationary additive noise process which is uncorrelated with the ideal signal $x(n)$. Thus:

$$x_q(n) = x(n) + e(n)$$

and

$$E[x(n)\ e(n)] = 0$$

This stochastic model is illustrated in Figure 7.18(b). The size of the error $e(n)$ is limited by:

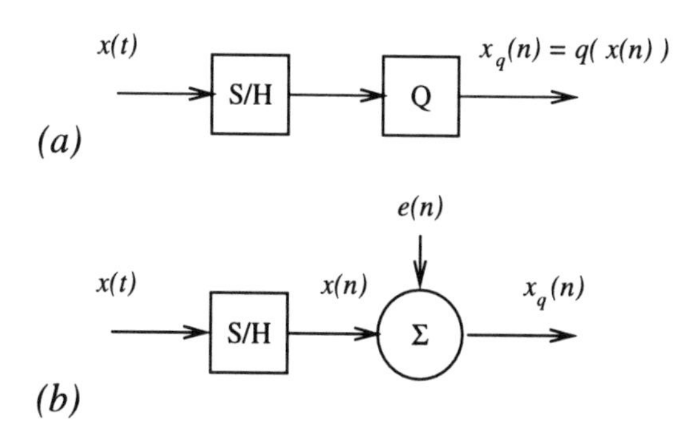

Figure 7.17 *Analogue-to-digital conversion: (a) block diagram; (b) stochastic quantisation model.*

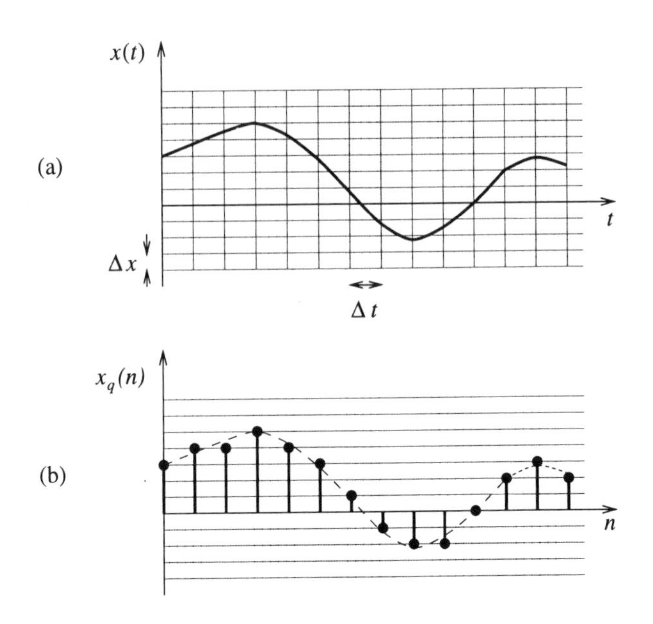

Figure 7.18 *Sampling and quantisation: (a) analogue input; (b) discrete output sequence.*

$$|e(n)| \le \frac{\Delta x}{2}$$

where Δx, the quantisation step size, is the difference between adjacent voltage levels of the quantiser (Figure 7.18(b)). Further, the error process is assumed to be white and to have a probability density function which is uniform over the range of the quantisation error, i.e. $-\Delta x/2 \le e \le \Delta x/2$. While these assumptions are not always valid, they are reasonable provided the number of levels is large and the degree of correlation between successive samples of the ideal sequence is not excessive. The value of the assumptions is that they make analysis tractable. Such analysis can provide a starting point for the design of a sampled data system.

7.6.2 Dynamic range

Dynamic range is an important concept that can be applied to both analogue and sampled data systems. Basically it is the ratio of the power in the biggest signal that can be represented to the power in the smallest signal that can be represented. This ratio is usually expressed in deciBels (dB). In an analogue system, the biggest signal is usually determined by the power supplies. Thus in a typical operational amplifier circuit with a +15 V and a −15 V supply, the biggest sine wave that we can use before clipping occurs has an amplitude of 15 V. The smallest signal that we can use or detect is determined by how much noise or unwanted interference is present. If the signal is smaller than the noise then we cannot easily detect it. Thus in an analogue system, the dynamic range is determined by the power supply and the amount of interference or noise.

In a digital discrete-time system, the largest signal that can be represented is usually determined by the power supply to the A/D converter. Thus if a 4-bit two's complement A/D operates off a +15 V and a −15 V supply, then an analogue voltage of +15 V will be represented in binary as 0111 and a −15 V analogue voltage will be represented by 1000. If the analogue voltage becomes bigger than +15 V it still gets represented as 0111, i.e. clipping occurs. Thus the largest signal that we can represent is a sine wave of amplitude 15 V. The smallest signal that can be represented has the same power as the quantisation error. In the following paragraphs we shall develop a simple relationship between the number of bits and the dynamic range of the digital data system.

Consider an A/D converter running off a power supply of $\pm A$ V. The largest sine wave that can be represented has amplitude A. The power in this sine wave is $A^2/2$. Thus we have an expression for the power in the maximum signal that we can represent. If we are using M-bit binary numbers then we can represent 2^M voltage levels. The voltage step size, Δx, between each of these levels, which is the voltage equivalent of 1 LSB, is given by the following relationship:

$$\Delta x = \frac{2A}{2^M}$$

The smallest signal that can be represented has the same power as the quantisation noise. Assuming this noise has a uniform distribution, its variance is:

$$\sigma_e^2 = \frac{(\Delta x)^2}{12}$$

The dynamic range R_D is a signal-to-noise ratio of the maximum representable signal with respect to the minimum representable signal. In this case:

$$\begin{aligned} R_D &= 10\log_{10}\left(\frac{A^2/2}{\sigma_e^2}\right) \\ &= 1.76 + 6M \quad \text{dB} \end{aligned}$$

There are many alternative forms of the above analysis. Each variation has a different set of assumption with respect to the signal and the quantisation noise. For example, it might be more realistic to model the signal as a zero-mean Gaussian process with a variance of σ_x^2. For a Gaussian process there will always be a finite probability that the signal $x(t)$ will be greater than $+A$ or less than $-A$, which will result in the quantiser clipping. However the probability that the signal will exceed three times the RMS value σ_x is very low and hence we might suggest that $\sigma_x = A/3$ is characteristic of the largest signal to be converted. With regard to the quantiser, it is more precise to say that the number of quantisation steps is $2^M - 1$ rather than 2^M, although the difference becomes less significant as M increases. While these analyses yield slightly different results with regard to how the absolute dynamic range is obtained for a particular number of bits, they all show the same trend, i.e. there is a 6 dB improvement in dynamic range for every bit that is added to the quantiser.

Self assessment question 7.7: A 1 kHz sine wave of amplitude 5 V is applied to a 14-bit A/D converter operating from a ±10 V supply at a sampling frequency of 10 kHz. What is the signal-to-quantisation noise ratio at the output of the A/D in dB?

EXAMPLE 7.4

A sine wave of exactly one eighth of the sampling frequency is applied to an A/D converter, the output of which is applied to a digital filter with difference equation:

$$y(n) = x(n) - x(n-2) + 1.2728y(n-1) - 0.81y(n-2)$$

How many extra bits are required to represent the output of the digital filter compared with its input?

Solution

The transfer function of the digital filter is written in the same form as equation (4.9):

$$H(z) = \frac{1 - z^{-2}}{1 - 1.2728z^{-1} + 0.81z^{-2}}$$

Assuming that the quantisation noise at the output of the A/D is white, equation (7.20) can be used to calculate the quantisation noise gain through the digital filter. Thus:

$$\left\{\begin{bmatrix} 1 & -1.2728 & 0.81 \\ 0 & 1 & -1.2728 \\ 0 & 0 & 1 \end{bmatrix} + \begin{bmatrix} 1 & -1.2728 & 0.81 \\ -1.2728 & 0.81 & 0 \\ 0.81 & 0 & 0 \end{bmatrix}\right\} \begin{bmatrix} c_0 \\ c_1 \\ c_2 \end{bmatrix} = \begin{bmatrix} 1 & 0 & -1 \\ 0 & 1 & 0 \\ 0 & 0 & 1 \end{bmatrix} \begin{bmatrix} 1 \\ 0 \\ -1 \end{bmatrix}$$

$$\begin{bmatrix} 2.0000 & -2.5456 & 1.6200 \\ 1.2728 & 1.8100 & -1.2728 \\ 1.6200 & -1.2728 & 1.0000 \end{bmatrix} \begin{bmatrix} c_0 \\ c_1 \\ c_2 \end{bmatrix} = \begin{bmatrix} 2 \\ 0 \\ -1 \end{bmatrix}$$

Thus:

$$\mathbf{c} = [5.2632 \;\; 0.0 \;\; -5.2632]^T$$

and hence the noise gain through the filter in dB is given by:

$$10 \log_{10}(2\,(5.2632)) = 10.22 \text{ dB}$$

The gain of the sine wave through the digital filter can be obtained using the technique described in Chapter 4. Thus:

$$|H(\omega)| = \left| \frac{1 - \exp(-j2\omega\Delta t)}{1 - 1.2728\exp(-j\omega\Delta t) + 0.81\exp(-j2\omega\Delta t)} \right|$$

The sine wave is at one eighth of the sampling frequency. Hence $\omega\Delta t = \pi/4$. The amplitude gain at this frequency is:

$$\left| \frac{1 - \exp(-j2\pi/4)}{1 - 1.2728\exp(-j\pi/4) + 0.81\exp(-j2\pi/4)} \right| = 10.51$$

The power gain in dB experienced by the sine wave is:

$$20 \log_{10}(10.51) = 20.43 \text{ dB}$$

The improvement in signal power with respect to quantisation noise power between the input and output of the filter is thus:

$$20.43 - 10.22 = 10.21 \text{ dB}$$

At 6 dB/bit, an additional 2 bits are required to represent the output of the digital filter.

7.7 Summary

A random signal is stationary if its ensemble averages can be replaced with time averages. This property is important in practical applications since only one realisation of a signal is usually available. When a signal is non-stationary and its statistics are slowly varying with time, the ensemble average can be estimated by taking a time average over a finite period of time (a window).

The covariance and cross-correlation are averages of the product of a signal sample at one point in time with a signal sample at another point in time. For stationary signals, both the covariance and cross-correlation are dependent on the time difference between the samples alone. They are both measures of the linear dependence between the two samples.

The Fourier transform or z-transform of a finite power discrete-time random signal does not exist. The covariance however is a finite-energy sequence and hence both its transforms exist. The Fourier transform of the autocovariance is the power spectral density which is a measure of how the average power in a signal is distributed with frequency.

The power spectral density at the output of a digital filter can be related to the power spectral density at the input through the amplitude response of the filter. If white noise is applied at the input to a filter, then the cross-correlation between the input and the output is proportional to the impulse response.

The power spectral density at the output of a filter driven by white noise at the input can be factorised into a minimum phase and a maximum phase part. The reciprocal of the minimum phase transfer function is the whitening filter. If the original filter is also minimum phase, then the whitening filter is also the inverse filter which can be used to recover the original input.

If a filter is driven by white noise, the PSD at its output is dependent on the filter transfer function and the variance or power of the white noise. Hence the power at the output of the filter can be calculated from the filter transfer function and the white noise variance. Quantisation noise associated with A/D conversion is often approximated by a white noise process. This implies that an extra 6 dB of signal-to-quantisation noise is gained every time an extra bit is added to the converter.

7.8 Problems

7.1 Estimate: the mean; the variance; the autocorrelation up to and including lag 2; the autocovariance up to and including lag 2 – of the stationary sequence $\{x(n)\}$ given the following samples from the sequence:

$$\{-1.6129,\ -1.2091,\ -0.4379,\ -2.0639,\ -0.6484\}$$

Form an estimate of the power spectral density at a quarter of the sampling frequency. [Mean = -1.1944; variance = 0.36; $\hat{\phi}_{xx}(0) = 1.787$; $\hat{\phi}_{xx}(1) = 1.180$; $\hat{\phi}_{xx}(2) = 1.162$; $\hat{\gamma}_{xx}(0) = 0.360$; $\hat{\gamma}_{xx}(1) = -0.284$; $\hat{\gamma}_{xx}(2) = 0.036$; PSD at quarter sampling frequency = -5.4134 dB]

7.2 Zero-mean white Gaussian noise with variance 2 is applied to two filters simultaneously. Filter 1 has transfer function:

$$H_1(z) = 1 - 2.75z^{-1} - 0.75z^{-2}$$

Filter 2 has transfer function:

$$H_2(z) = 1 - 1.1314z^{-1} + 0.64z^{-2}$$

What is the autocorrelation sequence of the output of each filter? Sketch the sequences. Calculate the cross-correlation sequence $\phi_{y_1y_2}(m)$ and $\phi_{y_2y_1}(m)$. Sketch the sequences. Design a whitening filter for the output of filter 1. Design a whitening filter for the output of filter 2. Draw block diagrams of the two filters. Which of the two whitening filters is also an inverse filter? [Autocorrelation sequence, filter 1: $\{-1.5, -1.375,\ 18.25,\ -1.375, -1.5\}$; filter 2; $\{1.28, -3.711,\ 5.379,\ -3.711,\ 1.28\}$. Cross-correlation sequences: $\{-1.5, -3.803,\ 7.263,\ -5.783,\ 1.28\}$; and $\{1.28,\ -5.783,\ 7.263,\ -3.803, -1.5\}$. $W_1(z) = 1/(1 - 0.0833z^{-1} - 0.0833z^{-2})$, $W_2(z) = 1/(1 - 1.1314z^{-1} + 0.64z^{-2})$]

7.3 (a) For a 16-bit A/D converter, what is the dynamic range for a sine wave input signal? What is the dynamic range for a Gaussian input signal?
(b) A signal processing system is illustrated below. The input to the S/H is a continuous-time zero-mean Gaussian random process. The signal has already been filtered by what can be assumed to be an ideal LPF to ensure that there is no significant power in the signal above B Hz. The amplifier gain A is adjusted to ensure that the probability of clipping for the A/D is 0.01. The sampling rate is $100B$ Hz.

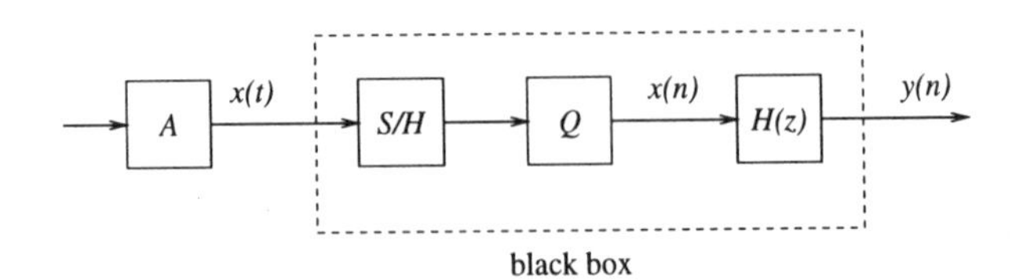

Select a minimum number of bits for the quantiser to ensure that the dynamic range of $x(n)$ is at least 40 dB. The discrete time filter $H(z)$ can be assumed to be a perfect LPF with a cut-off frequency at 1/100 of the sampling rate. Calculate the dynamic range at the output of this filter and the minimum number of bit required to represent $y(n)$. Explain the change in the dynamic range and the number of bits required.
If the A/D converter and the digital filter were placed inside a black box so that the user had access only to the input $x(t)$ and the output $y(n)$, what would be a suitable name for or description of the black box?
It is suggested that only every 50th sample of $y(n)$ is required. The remaining samples are redundant and could be ignored. Is this correct? [(a) 97.8 dB; 91.3 dB. (b) Quantiser requires 8 bits; dynamic range at output of filter is 57 dB; 11 bits are required; an 11-bit A/D

converter; yes]

7.4 A 3 kHz sine wave of amplitude 5 V is applied to an 8-bit A/D convertor operating from a ±15 V supply at a sampling frequency of 8 kHz. The output of the A/D is connected to two digital filters in cascade. The first digital filter is defined by the following difference equation:

$$y_1(n) = x(n) - 1.3435y(n-1) - 0.9025y(n-2)$$

What is the signal-to-quantisation noise ratio in dB at the output of this first filter? The output of the first filter is connected to the input of a second filter with transfer function:

$$H_2(z) = 2 - 2z^{-2}$$

What is the signal-to-quantisation noise ratio in dB at the output of the second filter? [Output of filter 1: 53.3 dB; output of filter 2: 53.5 dB]

CHAPTER 8

Adaptive filters

8.1 Introduction

Filters are commonly used to extract a desired signal from a background of random noise or deterministic interference. The design techniques discussed in previous chapters are based firmly on frequency domain concepts. For example, in a particular application it might be known *a priori* that the desired signal exists within a narrow bandwidth of B Hz centred at a frequency f_0 Hz and is subjected to additive noise or interference whose power is spread over a much wider band of frequencies. The quality of the signal, or more precisely the signal-to-noise ratio can be improved by application of a filter of bandwidth B Hz centred at f_0 Hz.

By contrast, Wiener filters are developed using time-domain concepts. They are designed to minimise the mean-square error between their output and a desired or required output. Thus they are said to be optimum in a mean-square error (MSE) sense. This particular definition of optimality is convenient because it leads to closed form solutions for the filter coefficients in terms of the autocorrelation of the signal at the input to the filter and the cross-correlation between the input and the desired output. Other definitions of optimality are possible but are generally much less tractable.

An adaptive filter is a mechanism for realising the optimal Wiener estimator when explicit knowledge of the auto- and cross-correlation functions is not available. The explicit knowledge required for the optimal estimator is replaced with a requirement for a second input sequence, known as a training or desired input (Figure 8.1(b)). The training signal is in some sense close to, or it approximates, the expected output of an optimal filter. Such an input is available in many adaptive filter applications, see later (section 8.4). The impulse response of the adaptive filter is then progressively altered as more of the observed and training sequence become available, so that the output $\hat{x}(n)$ more closely approximates (in a MSE sense) the training sequence $x(n)$ and hence the output of the optimal filter is approximated.

The optimal minimum mean squared error Wiener FIR filter is derived in section 8.2. An example of its application to the equalisation of a digital communications channel is presented. Algorithms for adaptive filtering are considered in section 8.2.1. The recursive least squares (RLS) and least mean squares (LMS) are compared and contrasted in terms of performance and computational complexity. Applications of adaptive filter techniques are presented in section 8.4. These cover the major modes of operation of adaptive filtering, from prediction to system modelling to inverse system

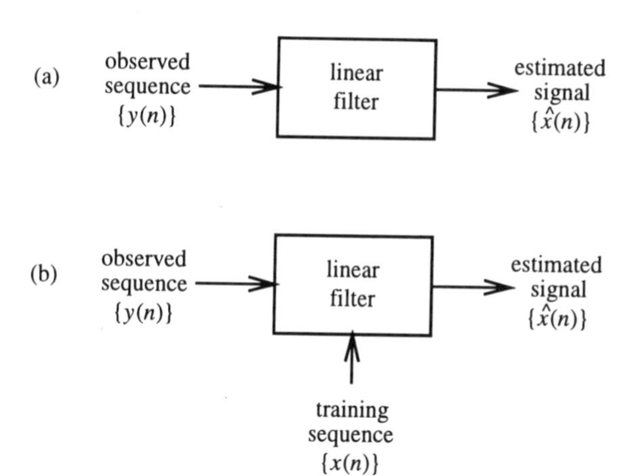

Figure 8.1 *The linear estimator: (a) optimal filter; (b) adaptive filter.*

modelling. Finally a summary of the key points is provided in section 8.5.

8.2 Wiener filters

Figure 8.2 illustrates a common problem in linear estimation. The signal we are interested in is the sequence $\{x(n)\}$. We cannot observe it directly. Rather we can observe the sequence $\{y(n)\}$ which is formed by applying $\{x(n)\}$ to a linear, signal-distorting system, and adding noise or interference to the output. To reconstruct $\{x(n)\}$ from $\{y(n)\}$ we wish to design a linear filter as in Figure 8.3. The output of this new filter is the sequence $\{\hat{x}(n)\}$ which is an estimate of the signal of interest $\{x(n)\}$. The quality of the estimate is a function of the error $\{e(n)\}$, which is the difference between the information-bearing sequence and the estimated sequence:

$$e(n) = x(n) - \hat{x}(n) \tag{8.1}$$

Obviously a small error is an indicator of a good quality estimate and a large error is an indicator of a poor estimate. We wish to choose an impulse response or equivalently a

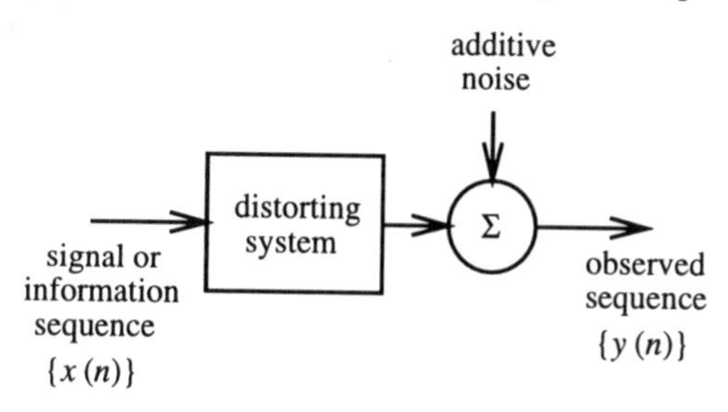

Figure 8.2 *The estimation problem.*

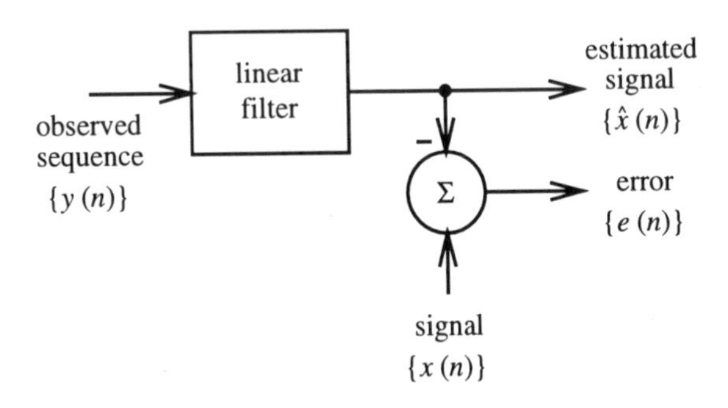

Figure 8.3 *Linear estimation.*

frequency response for the filter in Figure 8.3 which minimises the size of the error as much as possible. This is what is known as an optimisation problem. We wish to minimise some function, $f(e)$, of the error with respect to the impulse response. A convenient way of doing this is to differentiate the function with respect to each point in the impulse response and then attempt to find a value for the impulse response which simultaneously sets all the differentials to zero. Thus an extremely convenient choice is $f(e) = e^2$ as illustrated in Figure 8.4. Such a simple quadratic function is easy to differentiate and makes the task of finding an appropriate impulse response tractable.

One further point worth considering is that frequently we must deal with signals which are random or have a random component. Thus the most common choice for what is called the cost function is the mean-square error or average of the squared error:

$$\xi(n) = E[e^2(n)] \tag{8.2}$$

Thus the optimal filter is defined as that filter of the set of all possible linear filters which minimises the MSE.

8.2.1 Wiener FIR filter

If we initially restrict the linear filter of Figure 8.3 to be FIR, then, as in Chapter 6, its output $\hat{x}(n)$ is formed from a finite summation of N products:

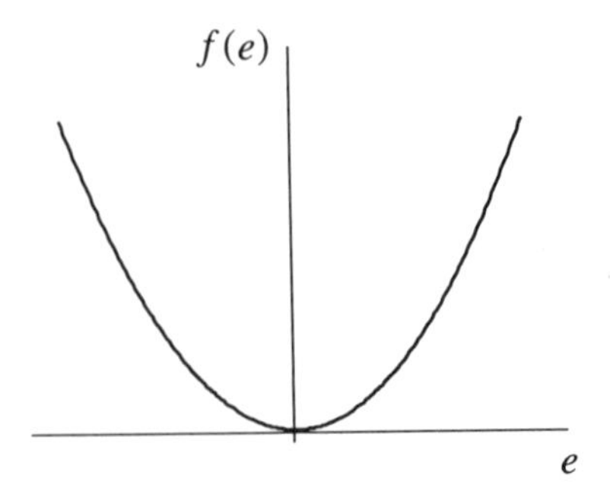

Figure 8.4 *Squared error function.*

$$\hat{x}(n) = \sum_{i=0}^{N-1} h_i \, y(n-i)$$

This sum of products may be written more compactly as a vector inner product:

$$\hat{x}(n) = \mathbf{h}^T \, \mathbf{y}(n) \tag{8.3}$$

or equivalently:

$$\hat{x}(n) = \mathbf{y}^T(n) \, \mathbf{h}$$

where $\mathbf{h}$ is the impulse response vector which is an $(N \times 1)$ matrix with N the first N elements of the impulse response sequence $\{h_n\}$:

$$\mathbf{h} = [\; h_0 \; h_1 \; \cdots \; h_{N-1} \;]^T$$

and $\mathbf{y}(n)$ is the signal vector which contains the last N elements of the input sequence $\{y(n)\}$:

$$\mathbf{y}(n) = [\; y(n) \; y(n-1) \cdots y(n-N+1) \;]^T$$

The superscript T denotes vector or matrix transposition, see Appendix A. The structure of an FIR filter is illustrated in Figure 6.1 and again in Figure 8.5.

If both the sequence of interest $\{x(n)\}$ and the observed sequence $\{y(n)\}$ are stationary, then substitution of equations (8.1) and (8.3) into equation (8.2) yields an expression for the MSE cost function as follows:

$$\begin{aligned} \xi &= E[(x(n) - \mathbf{h}^T \, \mathbf{y}(n))^2] \\ &= E[x^2(n) - 2 \, \mathbf{h}^T \, \mathbf{y}(n) \, x(n) + \mathbf{h}^T \, \mathbf{y}(n) \, \mathbf{y}^T(n) \, \mathbf{h}] \\ &= E[x^2(n)] - 2 \, \mathbf{h}^T \, \Phi_{yx} + \mathbf{h}^T \, \Phi_{yy} \, \mathbf{h} \end{aligned} \tag{8.4}$$

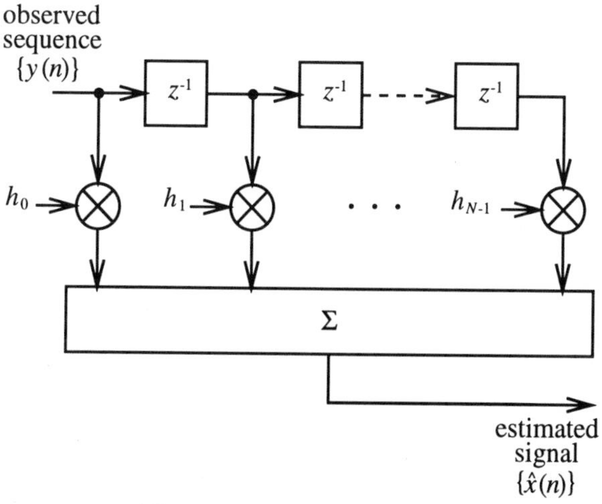

Figure 8.5 *An FIR estimator.*

where $\mathbf{\Phi}_{yy}$ is an $(N \times N)$ autocorrelation matrix:

$$\mathbf{\Phi}_{yy} = E[\mathbf{y}(n)\,\mathbf{y}^T(n)] \tag{8.5}$$

and $\mathbf{\Phi}_{yx}$ is an N-element cross-correlation vector:

$$\mathbf{\Phi}_{yx} = E[\mathbf{y}(n)\,x(n)] \tag{8.6}$$

Thus for an FIR filter, the MSE cost function is a quadratic function of the elements of the impulse response vector **h**. To obtain the minimum MSE, we differentiate the MSE with respect to each element of the impulse response, h_j, in turn and then set all the differentials equal to zero simultaneously:

$$\begin{aligned}\frac{\partial \xi}{\partial h_j} &= E[\,\frac{\partial}{\partial h_j}\{e^2(n)\}\,] \\ &= E[\,2\,e(n)\,\frac{\partial e(n)}{\partial h_j}\,]\end{aligned}$$

Substitute for $e(n)$ in the partial derivative using equations (8.1) and (8.3):

$$\frac{\partial \xi}{\partial h_j} = E[2\,e(n)\,\frac{\partial}{\partial h_j}\{x(n) - \mathbf{h}^T\,\mathbf{y}(n)\}]$$

And, using the fact that the filter output $\mathbf{h}^T\,\mathbf{y}(n)$ can be written as the sum of N products of which only one contains h_j:

$$\begin{aligned}\frac{\partial \xi}{\partial h_j} &= E[2\,e(n)\,\frac{\partial}{\partial h_j}\{-h_j\,y(n-j)\}] \\ &= E[-2\,e(n)\,y(n-j)] \\ &= 0, \qquad \text{for } j = 0, 1, \cdots, N-1\end{aligned}$$

Repeating the above differentiation for all values of j, i.e.: $j = 0, 1, \cdots, N-1$, provides a set of N scalar equations. This set of equations summarises the principle of statistical orthogonality for an FIR estimator. In words, the output error, $e(n)$, associated with an optimal filter is uncorrelated with any of the observations, $y(n), y(n-1), \cdots, y(n-N+1)$, which are currently in the filter.

The set of equations can also be written more compactly in vector form by collecting all the differential terms together into a vector which is known as the gradient vector ∇:

$$\nabla = \begin{bmatrix} \partial\xi/\partial h_0 \\ \partial\xi/\partial h_1 \\ \cdot \\ \partial\xi/\partial h_j \\ \cdot \\ \partial\xi/\partial h_{N-1} \end{bmatrix} = -2\,E\begin{bmatrix} y(n)\,e(n) \\ y(n-1)\,e(n) \\ \cdot \\ y(n-j)\,e(n) \\ \cdot \\ y(n-N+1)\,e(n) \end{bmatrix}$$

Taking out the common term yields:

$$\nabla = -2\,E\left\{\begin{bmatrix} y(n) \\ y(n-1) \\ \cdot \\ y(n-j) \\ \cdot \\ y(n-N+1) \end{bmatrix} e(n)\right\}$$

$$= -2\,E[\mathbf{y}(n)\,e(n)]$$

Substituting for $e(n)$ using equations (8.1) and (8.3):

$$\nabla = -2\,E[\mathbf{y}(n)\,(x(n) - \mathbf{y}^T(n)\,\mathbf{h})]$$

$$= -2\,E[\mathbf{y}(n)\,x(n)] + 2\,E[\mathbf{y}(n)\mathbf{y}^T(n)]\,\mathbf{h}$$

Using the definition of the autocorrelation matrix and cross-correlation vectors from equations (8.5) and (8.6) respectively gives:

$$\nabla = -2\,\mathbf{\Phi}_{yx} + 2\,\mathbf{\Phi}_{yy}\,\mathbf{h} \qquad (8.7)$$

The optimum impulse response $\mathbf{h}_{opt}$ which minimises the MSE is thus the solution to a set of N simultaneous linear equations, i.e.:

$$\mathbf{\Phi}_{yy}\,\mathbf{h}_{opt} = \mathbf{\Phi}_{yx} \qquad (8.8)$$

The filter defined by equation (8.8) is the Wiener FIR filter or Levinson filter. The minimum MSE, ξ_{opt}, is obtained by substitution of equation (8.8) into (8.4):

$$\xi_{opt} = E[\,x^2(n)\,] - \mathbf{h}_{opt}^T\,\mathbf{\Phi}_{yx} \qquad (8.9)$$

Equation (8.8) provides a means for designing optimum linear FIR filters. However, in order to calculate the impulse response of the optimum filter, precise knowledge of the autocorrelation matrix and the cross-correlation vector is required.

Self assessment question 8.1: Verify equation (8.9).

8.2.2 Application to channel equalisation

A classic problem in signal processing is illustrated in Figure 8.6(a). A white random sequence $x(n)$ is applied to a filter or channel with transfer function $C(z) = \sum_i c_i z^{-i}$. White noise $\eta(n)$ is added to the output of the filter to form the observed sequence $y(n)$. The objective is to design a filter with transfer function $H(z)$ whose output is a good estimate of $x(n)$. It is usually acceptable to produce an estimate with a delay or lag d, i.e. the output of the filter is an estimate of $x(n-d)$ rather than $x(n)$. This problem is known as equalisation in digital communications systems and deconvolution in seismic signal processing. The Wiener FIR filter provides a solution. Figure 8.6(a) is re-drawn in Figure 8.6(b) to emphasise the relationship to Figure 8.3. For notational convenience define:

$$x'(n) = x(n-d)$$

$$e'(n) = x(n-d) - \hat{x}(n-d) \tag{8.10}$$

and label the output of the channel filter as $y'(n)$ where:

$$y(n) = y'(n) + \eta(n)$$

The additive noise $\eta(n)$ and the signal $x(n)$ are assumed to be uncorrelated. This is a reasonable assumption in many practical applications. The Wiener filter which minimises the MSE is defined as in section 8.2.1, i.e.:

$$\mathbf{\Phi}_{yy}\,\mathbf{h}_{opt} = \mathbf{\Phi}_{yx'}$$

where:

$$\mathbf{\Phi}_{yy} = E[\mathbf{y}(n)\,\mathbf{y}^T(n)]$$

and:

$$\mathbf{\Phi}_{yx'} = E[\mathbf{y}(n)\,x'(n)]$$

Since the processes are stationary and ergodic the autocorrelation matrix $\mathbf{\Phi}_{yy}$ can be

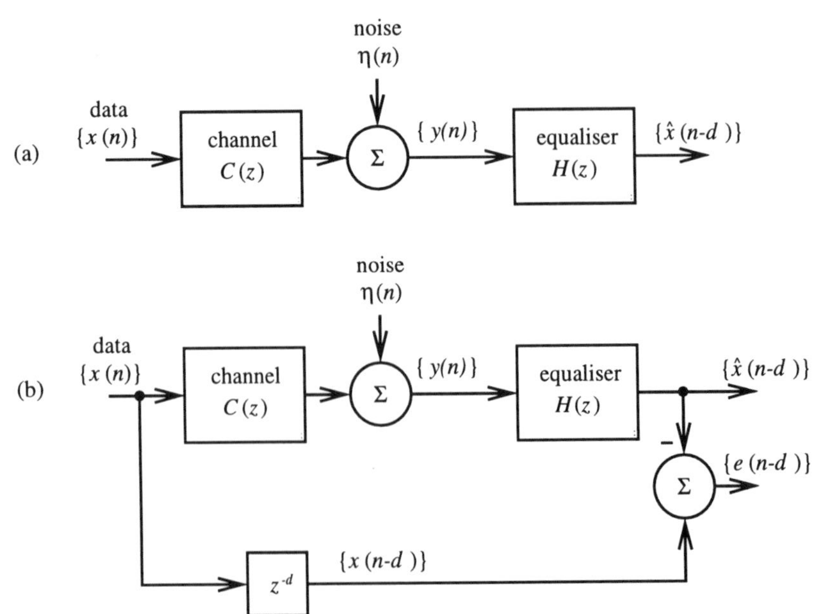

Figure 8.6 *The Wiener equaliser: (a) an estimation problem; (b) alternative formulation.*

deduced from the autocorrelation sequence $\phi_{yy}(m)$ where:

$$\begin{aligned}\phi_{yy}(m) &= E[y(n)\ y(n+m)] \\ &= E[\{y'(n)+\eta(n)\}\{y'(n+m)+\eta(n+m)\}] \\ &= \phi_{y'y'}(m)+\phi_{\eta y'}(m)+\phi_{y'\eta}(m)+\phi_{\eta\eta}(m)\end{aligned} \quad (8.11)$$

Since $y'(n)$ is a linear combination of the signal samples from $x(n)$, i.e.:

$$y'(n) = \sum_{i=0}^{\infty} c_i\ x(n-i)$$

and $x(n)$ and $\eta(n)$ are uncorrelated, it follows that $y'(n)$ and $\eta(n)$ are also uncorrelated thus:

$$\phi_{\eta y'}(m) = E[\eta(n)]\ E[y'(n+m)] = 0$$

for zero-mean processes. Further, since $\eta(n)$ is a white noise process, $\phi_{\eta\eta}(m) = \sigma_\eta^2 \delta(m)$. Hence equation (8.11) can be rewritten as:

$$\phi_{yy}(m) = \phi_{y'y'}(m) + \sigma_\eta^2 \delta(m)$$

Since $y'(n)$ is the output of a linear filter with a random input, the z-transform techniques of section 7.4 can be used to calculate its autocorrelation sequence. The z-transform of the autocorrelation is given in equation (7.13) as:

$$S_{y'y'}(z) = C(z)\ C(z^{-1})\ \sigma_x^2$$

so an inverse transform yields the required autocorrelation:

$$\phi_{y'y'}(m) = Z^{-1}[\ S_{y'y'}(z)\ ]$$

The calculation of the cross-correlation vector requires more care as cross-correlation sequences are not symmetrical. If the Wiener filter has N coefficients, the vector $\Phi_{yx'}$ will have N terms of the form $E[y(n-m)x'(n)]$, where $0 \le m < N$. The first thing to note is that:

$$E[y(n-m)x'(n)] = E[y(n)x'(n+m)] = \phi_{yx'}(m)$$

because the processes are stationary. It is the time difference between the samples which defines the correlation sequence. Further:

$$E[y(n)x'(n+m)] = E[(y'(n)+\eta(n))\ x'(n+m)] = \phi_{y'x'}(m)$$

A convenient way to evaluate this cross-correlation sequence is to note the similarity between Figure 8.6(b) and Figure 7.16(a). If $C(z)$ corresponds to $H_2(z)$ and z^{-d} to $H_1(z)$, the z-transform of the required cross-correlation sequence is:

$$S_{y'x'}(z) = z^{-d}\ C(z^{-1})\ \sigma_x^2$$

Note that the cross-correlation sequence is two sided but only non-negative values are required for the cross-correlation vector. It is also worth noting that $C(z^{-1})$ represents the time reversal of $C(z)$.

EXAMPLE 8.1

A communications channel has transfer function $0.5 + z^{-1}$. The signal $x(n)$ is zero-mean and white with variance 1.0. The additive noise is zero-mean and white with variance 0.1 and is uncorrelated with the signal. Design a 3-tap Wiener equaliser with a lag of 1 and calculate the minimum MSE that it achieves.

Solution

The z-transform of the autocorrelation sequence associated with observed sequence $y(n)$ is:

$$S_{yy}(z) = (0.5 + z^{-1})\,(0.5 + z) + 0.1$$
$$= 0.5z + 1.35 + 0.5z^{-1}$$

Hence $\phi_{yy}(0) = 1.35$, $\phi_{yy}(1) = 0.5$ and $\phi_{yy}(2) = 0$. The first step in calculating the coefficients of the Wiener FIR filter is to generate the autocorrelation matrix. By definition:

$$\mathbf{\Phi}_{yy} = E[\,\mathbf{y}(n)\mathbf{y}^T(n)\,]$$

$$= E\left\{\begin{bmatrix} y(n) \\ y(n-1) \\ y(n-2) \end{bmatrix} [y(n)\ y(n-1)\ y(n-2)]\right\}$$

$$= E\begin{bmatrix} y^2(n) & y(n)\,y(n-1) & y(n)\,y(n-2) \\ y(n-1)\,y(n) & y^2(n-1) & y(n-1)\,y(n-2) \\ y(n-2)\,y(n) & y(n-2)\,y(n-1) & y^2(n-2) \end{bmatrix}$$

If the processes involved are stationary, all the terms on the right-hand side of the above equation may be generated from the first row of the matrix:

$$\phi_{yy}(0) = E[y^2(n)] = E[y^2(n-1)] = E[y^2(n-2)]$$
$$\phi_{yy}(1) = E[y(n)\,y(n-1)] = E[y(n-1)\,y(n-2)]$$

Thus $\mathbf{\Phi}_{yy}$ simplifies to:

$$\mathbf{\Phi}_{yy} = E\begin{bmatrix} y^2(n) & y(n)\,y(n-1) & y(n)\,y(n-2) \\ y(n)\,y(n-1) & y^2(n) & y(n)\,y(n-1) \\ y(n)\,y(n-2) & y(n)\,y(n-1) & y^2(n) \end{bmatrix}$$

This is a symmetric *Toeplitz* matrix. Thus:

$$\mathbf{\Phi}_{yy} = \begin{bmatrix} 1.35 & 0.5 & 0 \\ 0.5 & 1.35 & 0.5 \\ 0 & 0.5 & 1.35 \end{bmatrix}$$

Since the lag d is 1, the z-transform of the cross-correlation sequence is:

$$S_{yx'}(z) = z^{-1}\,(z + 0.5)$$
$$= 1 + 0.5z^{-1}$$

Hence the cross-correlation terms are: $\phi_{yx'}(0) = 1$, $\phi_{yx'}(1) = 0.5$ and $\phi_{yx'}(2) = 0$. The cross-correlation vector is:

$$\mathbf{\Phi}_{yx'} = \begin{bmatrix} 1.0 \\ 0.5 \\ 0 \end{bmatrix}$$

Solving the three simultaneous equations yields the impulse response vector of the equaliser:

$$\mathbf{h} = \begin{bmatrix} 0.69 \\ 0.13 \\ -0.05 \end{bmatrix}$$

The minimum MSE can then be obtained using equation (8.9):

$$E[e^2] = \sigma_x^2 - \mathbf{h}^T\,\mathbf{\Phi}_{yx'}$$
$$= 0.24$$

It is worth noting that the minimum MSE is larger than the noise variance of 0.1. In attempting to reverse the effect of the channel filter, the equaliser has increased the level of the noise.

Self assessment question 8.2: Repeat example 8.1 for a lag of 2 and show that the MMSE obtained is 0.14.

8.3 Algorithms for adaptive filtering

Conventional signal processing systems, for the extraction of information from a received or observed signal, operate in an open loop manner, i.e. the same processing operation is carried out in the present time interval regardless of whether that operation provided the correct result in the previous time interval. In other words, they make the assumption that the signal degradation is a known and time-invariant quantity (Figure 8.1(a)). The design of optimal linear filters or estimators, such as the Wiener FIR filter of section 8.2, requires more explicit knowledge of the signal environment in the form of correlation functions, state space models or possibly even the probability density functions of Chapter 7. In many situations, such functions are unknown and/or time-varying.

Adaptive processors are designed to approximate these estimators. They operate with a closed loop (feedback) arrangement where the processor frequency response is controlled by the feedback algorithm. This permits them to compensate for time-varying distortions and still achieve performance, close to the optimal estimator function. Adaptive filters use a programmable filter whose frequency response or transfer function is altered, or adapted, to pass without degradation the desired components of a signal and to attenuate the undesired or interfering signals, i.e. to minimise the distortions present on the input signal.

The strategy by which the impulse response of the adaptive filter is altered is the adaptive filter algorithm. An adaptive filter is thus a time-varying filter whose impulse response at a particular time is dependent on the input sequence, the training sequence and the adaptive filter algorithm. The time-varying nature of an adaptive filter gives rise to the concept of convergence. In a stationary environment, the convergence performance is a measure of how many data samples are required for the impulse response of the adaptive filter to approach that of the optimal filter. In a non-stationary environment, the convergence performance is also a measure of how closely the impulse response of the adaptive filter follows the now time-varying impulse response of the optimal filter. Adaptive filters may be both infinite impulse response (IIR) or finite impulse response (FIR) in nature. In this chapter, only adaptive FIR filters will be considered as FIR filters are inherently stable. Thus, provided the feedback loop is properly designed, then stability of the overall adaptive filter will be ensured.

The block diagram of the general adaptive FIR filter is shown in Figure 8.7. The input data sequence, $\{y(n)\}$, is convolved with the FIR sequence, $\{h_i(n)\}$, and the resulting filter output, $\{\hat{x}(n)\}$, equation (6.1), is:

$$\hat{x}(n) = \sum_{i=0}^{N-1} h_i(n-1)\, y(n-i)$$

This can also be written as a vector inner product:

$$\hat{x}(n) = \mathbf{h}^T(n-1)\,\mathbf{y}(n)$$

The difference here is that the weights of the FIR filter are adapted in a time-varying manner: i.e. $\mathbf{h}(n-1)$ rather than simply $\mathbf{h}$. The current output, $\hat{x}(n)$, is calculated using the *previous* set of weight values, $\mathbf{h}(n-1)$. The training sequence, $x(n)$, is subtracted

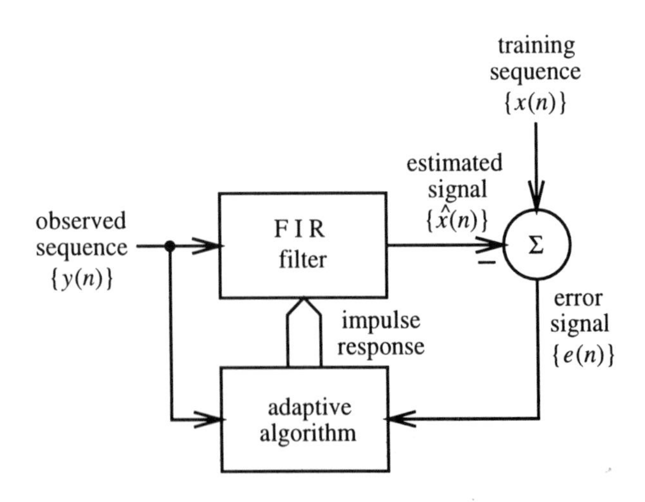

Figure 8.7 *General adaptive filter configuration.*

from the output signal to yield a scalar error signal, $e(n)$, as in equation (8.1):

$$e(n) = x(n) - \hat{x}(n)$$

The error is now used in conjunction with the input signal vector, $\mathbf{y}(n)$, to determine the next *set* of filter weight values, $\mathbf{h}(n)$. The notation $\mathbf{h}(n)$ is used because data up to and including $y(n)$ and $x(n)$ have been used to calculate $\mathbf{h}(n)$. The eventual objective of the adaptive algorithm is to estimate the optimum impulse response. Usually, 'optimum' is defined in an MSE sense, in which case the optimum filter in a stationary environment is the Wiener FIR filter of equation (8.8). The two major classes of adaptive filter control algorithm are now described. These are: recursive least squares (RLS), and the stochastic gradient or least mean squares (LMS) algorithms. Their relative convergence performance and computational complexity will also be covered later, following a discussion of adaptive filter applications.

8.3.1 Recursive least squares

To implement the Wiener FIR filter, as performed previously in section 8.2, both the autocorrelation matrix, $\mathbf{\Phi}_{yy}$, and the cross-correlation vector, $\mathbf{\Phi}_{yx}$, must be obtained from the signal samples. (Strictly speaking, the vector should be typed in lower case bold typefont, to distinguish vectors from matrices, but this is not readily available so the above simpler notation has been adopted here.) The Wiener filter is obtained by minimising the MSE cost function, ξ, which is defined again using the expectation operator, $E[.]$ of equation (8.4). Ensemble averages such as $\mathbf{\Phi}_{yy}$ and $\mathbf{\Phi}_{yx}$ are however rarely accessible in practice. Rather, what is available are portions of the data sequences themselves. One method of defining the impulse response of an adaptive filter is to choose the coefficients to minimise a sum of squared errors cost function:

$$\xi = \sum_{k=0}^{n} (x(k) - \hat{x}(k))^2 \tag{8.12}$$

When this cost function is minimised with respect to the impulse response vector, $\mathbf{h}(n)$, associated with the estimate, $\hat{x}(n)$, we obtain a least squares (LS) estimate. The impulse response vector, $\mathbf{h}(n)$, which minimises the sum of squared errors cost function, is now a function of the available data samples rather than ensemble averages. While an expression for $\mathbf{h}(n)$ could be derived by a similar method to that employed to derive the Wiener filter, a more elegant approach is to note that the two optimisation problems are similar and obtain the solution by analogy. Since the sum of squared errors cost function, equation (8.12), was obtained from the MSE cost function by replacing expectation with summation, the impulse response vector $\mathbf{h}(n)$ can be obtained from equation (8.8) by replacing expectation with summation. Thus:

$$\mathbf{R}_{yy}(n)\,\mathbf{h}(n) = \mathbf{r}_{yx}(n) \tag{8.13}$$

where:

$$\mathbf{R}_{yy}(n) = \sum_{k=0}^{n} \mathbf{y}(k)\,\mathbf{y}^T(k) \tag{8.14}$$

and:

$$\mathbf{r}_{yx}(n) = \sum_{k=0}^{n} \mathbf{y}(k)\, x(k) \tag{8.15}$$

In many applications of adaptive filters it is necessary that the impulse response be modified or updated as new data samples appear. The most direct method of meeting this requirement is to increment the upper limits of the summations of equations (8.12), (8.14) and (8.15), and completely resolve equation (8.13). However, a computationally more efficient method is to derive a time recursion for $\mathbf{h}(n)$ in terms of the previous least squares solution $\mathbf{h}(n-1)$ and the new data, $\mathbf{y}(n)$ and $x(n)$. The first step is to obtain recursions for $\mathbf{R}_{yy}(n)$ and $\mathbf{r}_{yx}(n)$. This is straightforward as they are both defined as summations in equations (8.14) and (8.15):

$$\mathbf{R}_{yy}(n) = \mathbf{R}_{yy}(n-1) + \mathbf{y}(n)\, \mathbf{y}^T(n) \tag{8.16}$$

$$\mathbf{r}_{yx}(n) = \mathbf{r}_{yx}(n-1) + \mathbf{y}(n)\, x(n) \tag{8.17}$$

Substitute for $\mathbf{r}_{yx}$ in equation (8.17) using equation (8.13):

$$\mathbf{R}_{yy}(n)\, \mathbf{h}(n) = \mathbf{R}_{yy}(n-1)\, \mathbf{h}(n-1) + \mathbf{y}(n)\, x(n)$$

Then use equation (8.16) to replace $\mathbf{R}_{yy}(n-1)$:

$$\mathbf{R}_{yy}(n)\, \mathbf{h}(n) = \left(\mathbf{R}_{yy}(n) - \mathbf{y}(n)\, \mathbf{y}^T(n) \right) \mathbf{h}(n-1) + \mathbf{y}(n)\, x(n)$$

After premultiplying by $\mathbf{R}_{yy}^{-1}(n)$, this yields:

$$\boxed{\mathbf{h}(n) = \mathbf{h}(n-1) + \mathbf{R}_{yy}^{-1}(n)\, \mathbf{y}(n)\, e(n)} \tag{8.18}$$

When the substitution:

$$e(n) = x(n) - \hat{x}(n)$$

is also included:

$$\boxed{e(n) = x(n) - \mathbf{h}^T(n-1)\, \mathbf{y}(n)} \tag{8.19}$$

A recursion for $\mathbf{R}_{yy}^{-1}(n)$ may be obtained by application of the Sherman–Morrison identity to equation (8.16):

$$\boxed{\mathbf{R}_{yy}^{-1}(n) = \mathbf{R}_{yy}^{-1}(n-1) - \frac{\mathbf{R}_{yy}^{-1}(n-1)\, \mathbf{y}(n)\, \mathbf{y}^T(n)\, \mathbf{R}_{yy}^{-1}(n-1)}{\left(1 + \mathbf{y}^T(n)\, \mathbf{R}_{yy}^{-1}(n-1)\, \mathbf{y}(n) \right)}} \tag{8.20}$$

Together, equations (8.18), (8.19), and (8.20) are known as the recursive least squares (RLS) algorithm.

The least squares approach, as described above, requires access to all the available data to form the estimate $\hat{x}(n)$. Further, it implicitly assumes that the data are valid over all time, i.e. before time 0, implying that it uses prewindowing, see Chapter 9. In many applications where the underlying processes are non-stationary, it is often more appropriate to minimise the effect of old data by progressively reducing the contribution to the squared error cost function. This is akin to assuming that the processes are stationary over short data records. A convenient technique for providing a forgetting mechanism is to replace the cost function of equation (8.12) by an exponentially weighted cost function:

$$\xi = \sum_{n=0}^{k} (x(n) - \hat{x}(n))^2 \, \alpha^{k-n} \qquad (8.21)$$

α is a positive constant parameter which is less than or equal to unity. It is often referred to as the forgetting factor, as it controls the effective length of the memory of the algorithm. As α is reduced in size, the memory of the algorithm reduces. One of the major attractions of using the exponentially weighted cost function of equation (8.21) is that the resultant RLS algorithm is similar in form and hence complexity to the standard RLS form of equations (8.18), (8.19) and (8.20). The only difference is that equation (8.20) is replaced with:

$$\mathbf{R}_{yy}^{-1}(n) = \frac{1}{\alpha}\left(\mathbf{R}_{yy}^{-1}(n-1) - \frac{\mathbf{R}_{yy}^{-1}(n-1)\,\mathbf{y}(n)\,\mathbf{y}^T(n)\,\mathbf{R}_{yy}^{-1}(n-1)}{\left(\alpha + \mathbf{y}^T(n)\,\mathbf{R}_{yy}^{-1}(n-1)\,\mathbf{y}(n)\right)}\right) \qquad (8.22)$$

Table 8.1 summarises the operations involved in the RLS algorithm.

The fact that the initial weight vector is zero is updated into the first set of weight values during the first iteration.

Self assessment question 8.3: Verify equation (8.16).

Self assessment question 8.4: Verify equation (8.17).

8.3.2 Stochastic gradient methods

The minimum mean-square error (MMSE) definitions of optimality are attractive in that, for a FIR filter, there is only one unique solution and that solution is provided by the Wiener equation (8.8). Many other optimisation problems are less attractive in that direct solutions do not necessarily exist and iterative search techniques must be applied. One of the most important iterative techniques is the method of steepest descent. The importance of this method is that it provides the basis for the stochastic gradient (SG) algorithms which are widely applied in adaptive filtering. An understanding of the steepest descent method can provide a clearer insight into the operation of the stochastic gradient algorithms.

Table 8.1 *RLS algorithm summary.*

Initialise by setting:

$$\mathbf{R}_{yy}(0) = \frac{1}{\delta} I_N, \quad \delta \text{ is a small positive number}$$

$$\mathbf{h}(0) = \mathbf{0}$$

For each time sample $n = 1, 2, \cdots$ do the following:

$$\hat{x}(n) = \mathbf{h}^T(n-1)\,\mathbf{y}(n)$$

$$e(n) = x(n) - \hat{x}(n)$$

$$\mathbf{R}_{yy}^{-1}(n) = \frac{1}{\alpha}\left(\mathbf{R}_{yy}^{-1}(n-1) - \frac{\mathbf{R}_{yy}^{-1}(n-1)\,\mathbf{y}(n)\,\mathbf{y}^T(n)\,\mathbf{R}_{yy}^{-1}(n-1)}{\left(\alpha + \mathbf{y}^T(n)\,\mathbf{R}_{yy}^{-1}(n-1)\,\mathbf{y}(n)\right)}\right)$$

$$\mathbf{h}(n) = \mathbf{h}(n-1) + \mathbf{R}_{yy}^{-1}(n)\,\mathbf{y}(n)\,e(n)$$

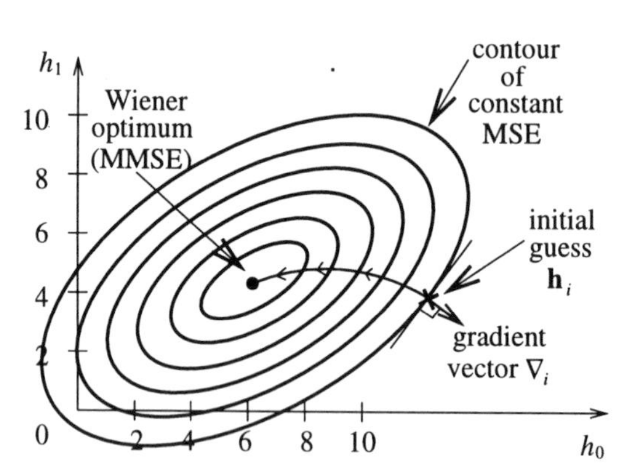

Figure 8.8 *Method of steepest descent.*

The method of steepest descent is illustrated in Figure 8.8, by examining the contours of constant MSE for a two-coefficient, h_0, h_1, FIR filtering problem. Note in particular the elliptical nature of the contours. In Figure 8.8, the optimum Wiener solution is clearly identified in the centre of the concentric contours which, on a 3-D plot, corresponds to the bottom of a bowl-shaped surface □. In the method of steepest descent, an initial guess at the weight vector, $\mathbf{h}_i$, is made. The gradient vector, ∇_i, of the MSE surface for this $\mathbf{h}_i$ value is then calculated as in equation (8.7):

$$\nabla_i = \left[\frac{\partial \xi}{\partial h_0} \; \frac{\partial \xi}{\partial h_1} \cdots \frac{\partial \xi}{\partial h_{N-1}} \right]^T_{\mathbf{h} = \mathbf{h}_i}$$

$$= 2\, \mathbf{\Phi}_{yy}\, \mathbf{h}_i - 2\, \mathbf{\Phi}_{yx} \tag{8.23}$$

The gradient vector points in the direction of the greatest rate of *increase* of the MSE cost function. The initial guess may thus be improved by moving in the opposite direction, i.e. following the line of steepest descent. Such a move is accomplished by subtracting a scaled version of the gradient, from the initial weight vector, $\mathbf{h}_i$, to give a new vector update, $\mathbf{h}_{i+1}$:

$$\mathbf{h}_{i+1} = \mathbf{h}_i - \mu\, \nabla_i \tag{8.24}$$

The small positive scaling constant μ is usually known as the step size. The optimum solution may be obtained by repeated application of equations (8.23) and (8.24) if the step size lies within a range specified by the largest eigenvalue, λ_{max} (Appendix A), associated with the autocorrelation matrix, $\mathbf{\Phi}_{yy}$:

$$0 < \mu < \frac{1}{\lambda_{max}} \tag{8.25}$$

The selection of such a magnitude for μ ensures that the weight trajectory always moves towards the MMSE optimum solution (Figure 8.8) and this process is known as convergence of the adaptive filter.

Computation of the gradient of equation (8.23) still requires access to the autocorrelation matrix and cross-correlation vector. Thus the method of steepest descent is not strictly an adaptive filter algorithm but is an indirect method of solving the linear equations summarised by equation (8.8).

To obtain the LMS stochastic gradient algorithm, the gradient of equation (8.23) is replaced by a noisy estimate of the gradient, $\hat{\nabla}$, based on the data, and the iterative step of equation (8.24) is replaced by a time recursion to give:

$$\mathbf{h}(n) = \mathbf{h}(n-1) - \mu\, \hat{\nabla}(n-1) \tag{8.26}$$

The vector $\mathbf{h}(n)$ provides an estimate of the Wiener impulse response at time sample n and the vector $\hat{\nabla}(n)$ is an associated estimate of the gradient. The exact expression for the gradient vector, $\nabla(n)$, in terms of the impulse response vector, $\mathbf{h}(n)$, can be obtained from equation (8.7):

$$\nabla(n) = -\,2\, E[\mathbf{y}(k)\,(\,x(k) - \mathbf{h}^T(n)\,\mathbf{y}(k))]$$

$$= -\,2\, E[\mathbf{y}(k)\, e(k)] \tag{8.27}$$

where $e(k)$ represents the error at data sample k of a filter with impulse response $\mathbf{h}(n)$:

$$e(k) = x(k) - \mathbf{h}^T(n)\,\mathbf{y}(k)$$

An estimate, $\hat{\nabla}(n)$, of the gradient, $\nabla(n)$, could be formed by replacing the ensemble average of equation (8.27) by a time-average operation. This would involve holding the filter impulse response constant at $\mathbf{h}(n)$ while data were processed within the filter. However, if the impulse response of the data changes at every sample point as suggested by equation (8.26), the time average reduces to a single multiplication at

$k = n$. This leads to simple expressions for the estimated gradient and the associated error:

$$\hat{\nabla}(n-1) = -2\, \mathbf{y}(n)\, e(n) \tag{8.28}$$

$$e(n) = x(n) - \mathbf{h}^T(n-1)\, \mathbf{y}(n) \tag{8.29}$$

Substitution of equation (8.28) into (8.26) yields the LMS algorithm:

$$\mathbf{h}(n) = \mathbf{h}(n-1) + 2\, \mu\, \mathbf{y}(n)\, e(n) \tag{8.30}$$

The block least mean-squares (BLMS) algorithm, where the impulse response of the adaptive filter is held constant for several data samples, represents another possible stochastic gradient technique. It provides a possibly improved estimate of the gradient. This approach is attractive for adaptive filters, particularly when the convolution operation in Figure 8.7 is realised by frequency-domain multiplication, see later (Figure 10.10), as this is also a block processing operation. The BLMS algorithm exploits the further advantage that both the filtering operation and the calculation of the gradient estimate may be implemented very efficiently using the fast Fourier transform (FFT) (see later – Figure 10.10).

An adaptive FIR filter with LMS algorithm is illustrated in Figure 8.9 where the $2\mu e(n)$ operations are clearly seen in the diagram as the updates to the individual $h(n)$ weight values stored in the recursive integrator within the structure.

Table 8.2 summarises the operations involved in the LMS algorithm.

Table 8.2 *LMS algorithm.*

Initialise by setting:
$\mathbf{h}(0) = \mathbf{0}$
For each time sample $n = 1, 2, \cdots$ do the following:
$\hat{x}(n) = \mathbf{h}^T(n-1)\, \mathbf{y}(n)$
$e(n) = x(n) - \hat{x}(n)$
$\mathbf{h}(n) = \mathbf{h}(n-1) + 2\, \mu\, \mathbf{y}(n)\, e(n)$

The simplest method of performing a theoretical analysis of the convergence of adaptive filter algorithms considers the evolution of the mean of the impulse response vector of the adaptive filter and attempts to show that, as the time index n increases, the mean will tend to the impulse response of the Wiener filter. This technique, first suggested by Widrow for the LMS algorithm, while not being strictly rigorous, does clearly highlight the major factors which control the convergence rate of the SG algorithms. Simplifications such as this are essential as convergence analysis is a very

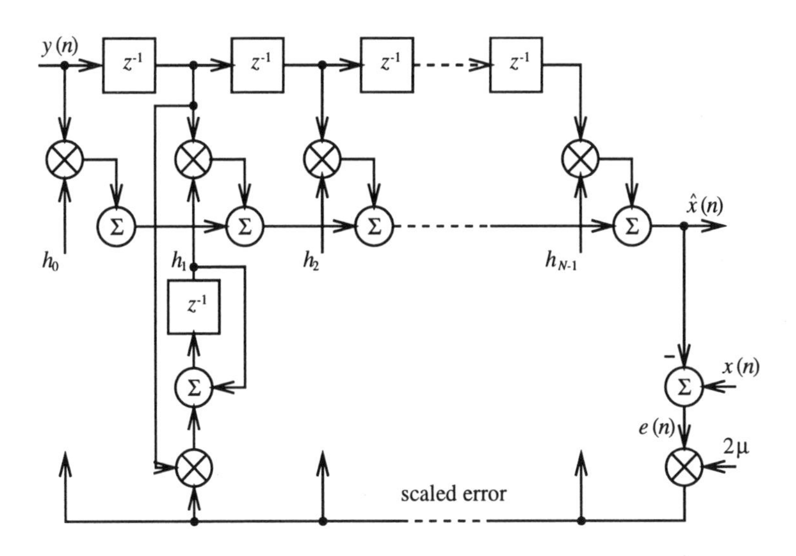

Figure 8.9 *The least mean-square (LMS) adaptive filter design.*

complicated task requiring considerable mathematical sophistication.

It can be shown that the convergence performance is limited by the modes of the filter which are in turn related to the eigenvalues, λ_i where $i = 1, \cdots, N-1$, of the input autocorrelation matrix, Φ_{yy}. A time constant, τ_j, can be defined for mode j and is given approximately as:

$$\tau_j \approx \frac{1}{2\mu\lambda_j}$$

where λ_j is the eigenvalue associated with that mode. The largest time constant, $\tau_{\max}$, is due to the smallest eigenvalue, $\lambda_{\min}$, and is thus given by:

$$\tau_{\max} \approx \frac{1}{2\mu\lambda_{\min}} \tag{8.31}$$

Combining equation (8.25) with the limiting case in equation (8.31) gives:

$$\tau_{\max} > \frac{\lambda_{\max}}{2\lambda_{\min}} \tag{8.32}$$

which suggests that the larger the eigenvalue ratio (EVR) or condition number, $\lambda_{\max}/\lambda_{\min}$, of Φ_{yy}, the longer the LMS algorithm will take to converge. The eigenvalues of the input autocorrelation matrix can be related to the power spectral components in the signal. Thus strong signal spectral components result in high loop gain in the adaptive filter and have an associated small τ value. The weak signal spectral components give rise to the $\tau_{\max}$ value.

Insight into the relationship between EVR and the convergence performance of the LMS algorithm can be gained through a consideration of the steepest descent algorithm of equation (8.24). Figure 8.10(a) illustrates how the eigenvalues ($\lambda_{\max}$, $\lambda_{\min}$) and their associated eigenvectors ($\mathbf{v}_{max}$, $\mathbf{v}_{min}$) of the autocorrelation matrix Φ_{yy} define the contours of constant MSE for a two-tap adaptive filtering problem. In Figures 8.10(b) and 8.10(c), the operation of the steepest descent algorithm is illustrated for EVRs of 2 and 4 respectively. In the latter two figures, an asterisk indicates the optimum solution. The eigenvectors of the (2×2) autocorrelation matrix Φ_{yy} are in the direction of the principal axes of the ellipses of constant MSE. The EVR is simply the ratio of the minor axis to the major axis. Since the eigenvectors are identical in Figures 8.10(b) and 8.10(c), the orientation of the ellipses in the two figures is identical. In both cases, 10 steps of the steepest descent algorithm are taken from an initial starting point of $\mathbf{h} = [4\ 5]^T$. For an EVR of 2, the algorithm is closer to the optimum solution after 10 steps than for an EVR of 4. For the higher EVR case, the contours of constant MSE are closely approximated by parallel lines in the region of the starting point. The algorithm moves initially on a trajectory parallel to $\mathbf{v}_{max}$ and only much later turns towards the optimum solution compared with the case where the EVR is 2.

In the practical situation it is difficult or impossible to determine the eigenvalue ratio and hence find an optimum value for the stepsize μ. A more practical limit is given by:

$$0 < \mu < \frac{1}{3N\sigma_y^2} \tag{8.33}$$

where N is the number of taps and $\sigma_y^2 = E[y^2(k)]$ represents the variance of the input signal. It is often possible to produce a crude estimate of σ_y^2 from knowledge of the application. Often a reasonable choice for μ is to set it to the mid-point of the region, i.e.:

$$\mu = \frac{1}{6N\sigma_y^2} \tag{8.34}$$

Self assessment question 8.5: An adaptive filter has a white noise input of variance 2. The cross-correlation vector Φ_{yx} has a value $[3\ 1]^T$. Calculate the gradient when the impulse response vector $\mathbf{h}$ of the filter is $[2.9\ \ 4.0]$.

EXAMPLE 8.2

The table below shows the input and training signal to an adaptive filter at sample numbers 4 and 5:

k	$y(k)$	$\mathbf{h}(k)$	$x(k)$
4	0.25	$[1.2\ 3.7]^T$	1.03
5	0.5		-0.27

Using the LMS algorithm, evaluate $\mathbf{h}(5)$. The step size is fixed at 0.01.

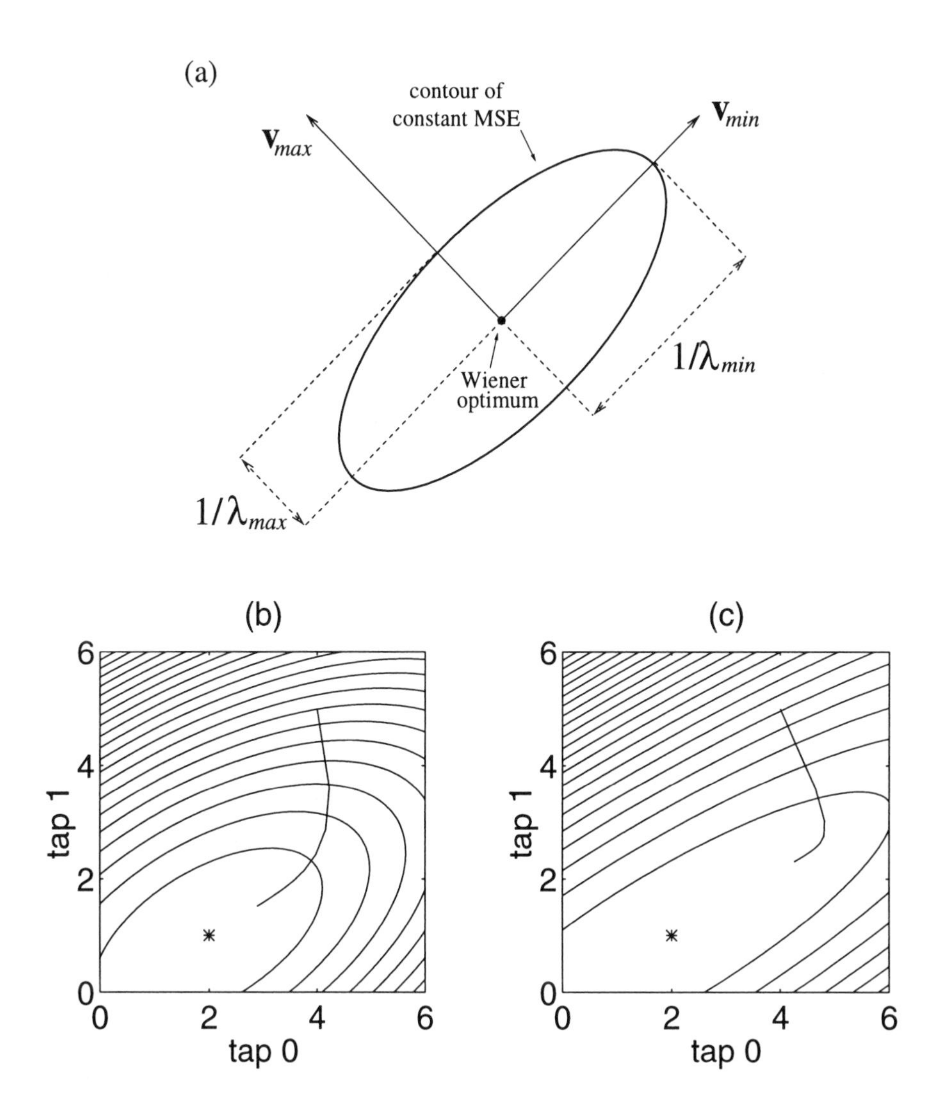

Figure 8.10 *Eigenvectors, eigenvalues and convergence: (a) the relationship between eigenvectors, eigenvalues and constant MSE contours; (b) steepest descent for EVR of 2; (c) EVR of 4.* □

Solution

Using equation (8.29):

$$e(5) = x(5) - \mathbf{h}^T(4)\,\mathbf{y}(5)$$

$$= -0.27 - [1.2\ 3.7]\begin{bmatrix} 0.5 \\ 0.25 \end{bmatrix} = -1.795$$

The impulse response is updated using (8.30):

$$\mathbf{h}(5) = \mathbf{h}(4) + 2\mu\, \mathbf{y}(5)\, e(5)$$

$$= \begin{bmatrix} 1.2 \\ 3.7 \end{bmatrix} + 2(0.01) \begin{bmatrix} 0.5 \\ 0.25 \end{bmatrix} (-1.795)$$

$$= \begin{bmatrix} 1.1821 \\ 3.6910 \end{bmatrix}$$

EXAMPLE 8.3

Using the table of example 8.2, evaluate $\mathbf{R}_{yy}^{-1}(5)$ given that:

$$\mathbf{R}_{yy}^{-1}(4) = \begin{bmatrix} 1.26 & -0.13 \\ -0.13 & 1.26 \end{bmatrix}$$

Solution

Rewriting equation (8.20) to emphasise the order in which the calculation is performed:

$$\mathbf{R}_{yy}^{-1}(5) = \mathbf{R}_{yy}^{-1}(4) - \frac{\left(\mathbf{R}_{yy}^{-1}(4)\, \mathbf{y}(5)\right)\left(\mathbf{R}_{yy}^{-1}(4)\, \mathbf{y}(5)\right)^T}{1 + \mathbf{y}^T(5)\left(\mathbf{R}_{yy}^{-1}(4)\, \mathbf{y}(5)\right)}$$

Note that $\mathbf{R}_{yy}$ is symmetric and hence $\mathbf{R}_{yy}^{-1}$ is symmetric. Therefore $(\mathbf{R}_{yy}^{-1})^T = \mathbf{R}_{yy}^{-1}$. Thus in using (8.20) it is more economical to calculate an intermediate variable vector $\mathbf{u}(5)$ where:

$$\mathbf{u}(5) = \mathbf{R}_{yy}^{-1}(4)\, \mathbf{y}(5)$$

$$= \begin{bmatrix} 1.26 & -0.13 \\ -0.13 & 1.26 \end{bmatrix} \begin{bmatrix} 0.5 \\ 0.25 \end{bmatrix} = \begin{bmatrix} 0.5975 \\ 0.25 \end{bmatrix}$$

Thus:

$$\mathbf{R}_{yy}^{-1}(5) = \begin{bmatrix} 1.26 & -0.13 \\ -0.13 & 1.26 \end{bmatrix} - \frac{\begin{bmatrix} 0.5975 \\ 0.25 \end{bmatrix} [0.5975\ 0.25]}{1 + [0.5\ 0.25] \begin{bmatrix} 0.5975 \\ 0.25 \end{bmatrix}}$$

$$= \begin{bmatrix} 1.26 & -0.13 \\ -0.13 & 1.26 \end{bmatrix} - \begin{bmatrix} 0.3570 & 0.1494 \\ 0.1494 & 0.0625 \end{bmatrix} \frac{1}{1.3613}$$

$$= \begin{bmatrix} 0.9978 & -0.2397 \\ -0.2397 & 1.2141 \end{bmatrix}$$

8.3.3 A comparison of algorithms

Simulation results are presented here to illustrate the differences in performance between the two major algorithm types that have been discussed in the previous sections. The structure used to highlight these differences is illustrated in Figure 8.11. This is an example of a system modelling mode of application where the objective is to use the adaptive filter to identify the impulse response $\mathbf{h}_{opt}$ of a unknown system from input and output data. The noise shaping filter is a useful mechanism for adjusting the characteristic of (conditioning) the input signal to the unknown system and the adaptive filter, $y(n)$, and hence to control the eigenvalue ratio. The error vector norm or deviation from the optimum weight vector, $\mathbf{h}_{opt}$, is defined as:

$$\rho(n) = E[(\mathbf{h}(n) - \mathbf{h}_{opt})^T \, (\mathbf{h}(n) - \mathbf{h}_{opt})] \tag{8.35}$$

This was chosen as the preferred measure of performance over the output MSE. The error vector norm permits the convergence plots to be plotted to demonstrate clearly the operation of the algorithm below the output noise floor, which is determined by the noise added to the training signal.

Three sets of results for a N = 16-coefficient adaptive FIR filter are illustrated in Figure 8.12. For all simulations, the additive noise was set at –60 dB. Firstly we consider the initial performance of the various algorithms under good SNR conditions. When the input signal is white (Figure 8.12(a)), the condition number or eigenvalue ratio of the input autocorrelation matrix is unity. The RLS algorithm then converges to the noise floor in approximately 30 or $2N$ data samples or iterations. The LMS algorithm reaches the noise floor within approximately 400 iterations.

If the eigenvalue ratio of the input signal is increased to 11, by altering the frequency response of the noise shaping filter, the performance of the RLS algorithm, depicted in Figure 8.12(b), is hardly altered. On the other hand, the performance of the LMS algorithm is significantly degraded, now taking greater than 1000 iterations to reach the noise floor. Finally, in Figure 8.12(c) the eigenvalue ratio is 68, which would

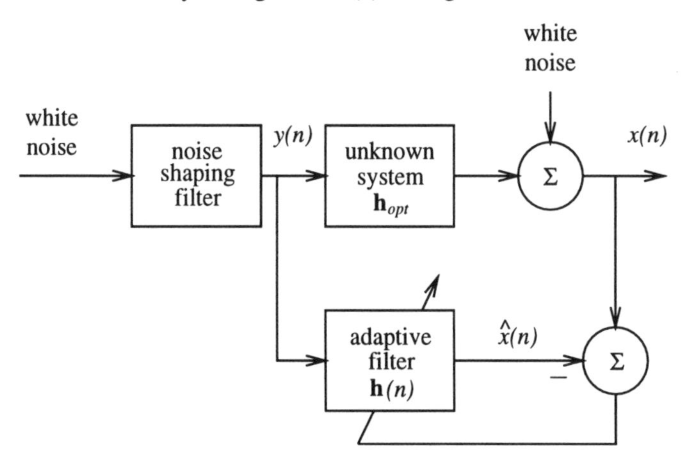

Figure 8.11 *System identification simulation.*

be considered to be extremely severe conditions. The RLS reaches the noise floor in 70 iterations while the LMS now takes approximately 2000 iterations.

When the performance goal is to achieve the noise floor level, the results can be summarised as follows. The RLS provides consistent fast convergence independent of the eigenvalue ratio. The LMS, on the other hand, provides the poorest performance, being highly dependent on the eigenvalue ratio. The second point of interest is the performance of the algorithms below the noise floor. For instance, in all three cases the convergence rate of the RLS algorithm decreases dramatically after it reaches the noise floor. However, the LMS algorithm converges to a constant level on or just below the noise floor. This is a consequence of the choice of step size, μ, that was employed. The LMS step size for all these simulations was set at $1/(6N)$ in order to ensure convergence over the wide range of input conditions. The final value of the norm achieved by the LMS algorithm could have been improved if a smaller step size had been used, but this would have resulted in a further slowing down of the overall rate of convergence.

Table 8.3 *Complexity comparison of N-point FIR filter algorithms.*

Algorithm class	*Implementation*	*Computational load*		
		multiplications	*adds/subtractions*	*divisions*
Recursive least squares	fast Kalman	$10N+1$	$9N+1$	2
Stochastic gradient	LMS	$2N$	$2N$	–
	BLMS (via FFT)	$10\log(N)+8$	$15\log(N)+30$	–

The choice of an appropriate adaptive algorithm for a particular application involves an assessment of the computational complexity as well as the convergence performance. Considerable research effort has been applied to achieve computationally efficient implementations of the various adaptive filter algorithms. Some of these are summarised in Table 8.3, where their complexity is compared in terms of computational multiplication, division and add/subtraction operations for a filter of order N. Here the RLS algorithm is based on a Kalman filter [Bozic]. Note that this also needs a division operation which presents severe practical problems for high speed implementations. Table 8.3 also includes the block LMS approach where the filter operation is performed by frequency-domain processing, see later (Figure 10.10), rather than the FIR time-domain convolution operation of Figure 6.1. The BLMS is particularly significant as the computations scale by $\log N$ rather than N and, in high complexity systems, with long FIR filters this approach will be the preferred solution.

Self assessment question 8.6: Calculate the number of multiplications associated with a 4-tap and 128-tap adaptive filter trained with the LMS algorithm; repeat the calculation for an FFT-based BLMS algorithm. What conclusions do you draw?

Self assessment question 8.7: What are the relative advantages and disadvantages of the RLS and LMS algorithms?

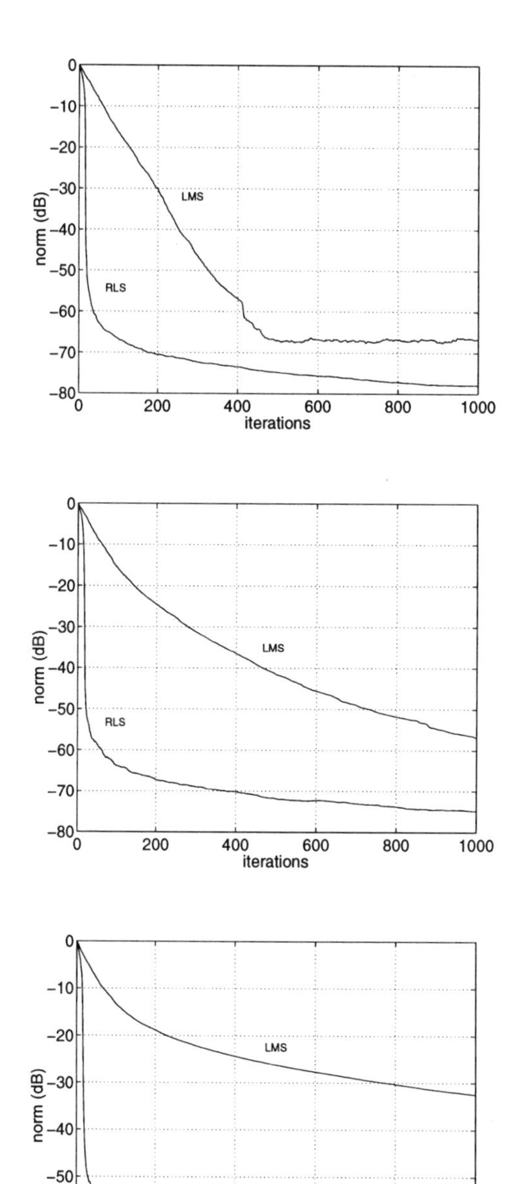

(a)

(b)

(c)

Figure 8.12 *Convergence plots for a 16-tap adaptive filter performing system identification: (a) shows results with an eigenvalue ratio (EVR) of 1, i.e. a white noise signal; (b) shows the EVR increased to 11; and (c) 68 by progressively increasing the spectral coloration of the input signal.* ☐

8.4 Applications

There are basically three modes of operation of an adaptive filter and these are illustrated in Figure 8.13:

(i) Direct system modelling (Figure 8.13(a)) which is typified by the application to echo cancellation where the echoes in the unknown system are duplicated in the adaptive filter and then cancelled in the summer.

(ii) Inverse system modelling (Figure 8.13(b)) which is what is normally implied in the communications channel equalisation application (section 8.4.5), to overcome signal distortion and bandlimiting in the transmission channel.

(iii) Prediction (Figure 8.13(c)), for instance as used in autoregressive spectral analysis and in the linear predictive coding (LPC) of speech (Figure 9.21).

Here we examine how these modes are reflected in the practical application of adaptive filters. The initial question that most people ask is, 'where does the training signal come from?' Hopefully this is answered for each of the following applications.

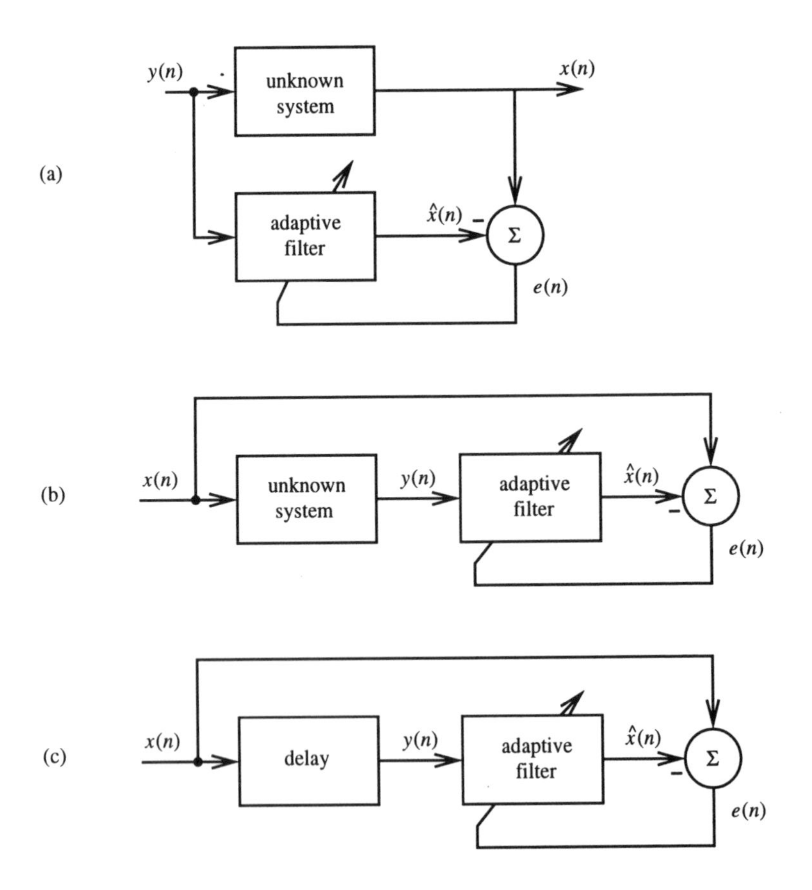

Figure 8.13 *Modes of operation of the adaptive filter: (a) direct system modelling; (b) inverse system modelling; (c) linear prediction.*

The applications cover: adaptive line enhancement; adaptive tone suppression; noise whitening; echo cancellation; and channel equalisation.

8.4.1 Adaptive line enhancement

Consider the situation illustrated in Figure 8.14(a). We have a narrow-band signal of interest centred at a frequency f_0. This signal might be a single sine wave of frequency f_0 or an amplitude modulated carrier. The signal is contaminated by adding broad-band random noise to it – the noise might be thermal in nature and generated by the amplifiers in a communications receiver. If we knew beforehand (*a priori*) both the bandwidth (B) of the signal and the centre frequency, we could design an analogue or digital bandpass filter with the same centre frequency and bandwidth. This filter would attenuate the noise components at frequencies outside its bandwidth and hence the signal-to-noise ratio at the output of the filter would be better than the signal-to-noise ratio at the input – we would have cleaned up the signal.

If however we only knew that the signal was narrow-band and did not know either its centre frequency or its bandwidth, we could not design such a filter. In cases such as this, an adaptive filter solution may be appropriate. A narrow-band signal such as a sine

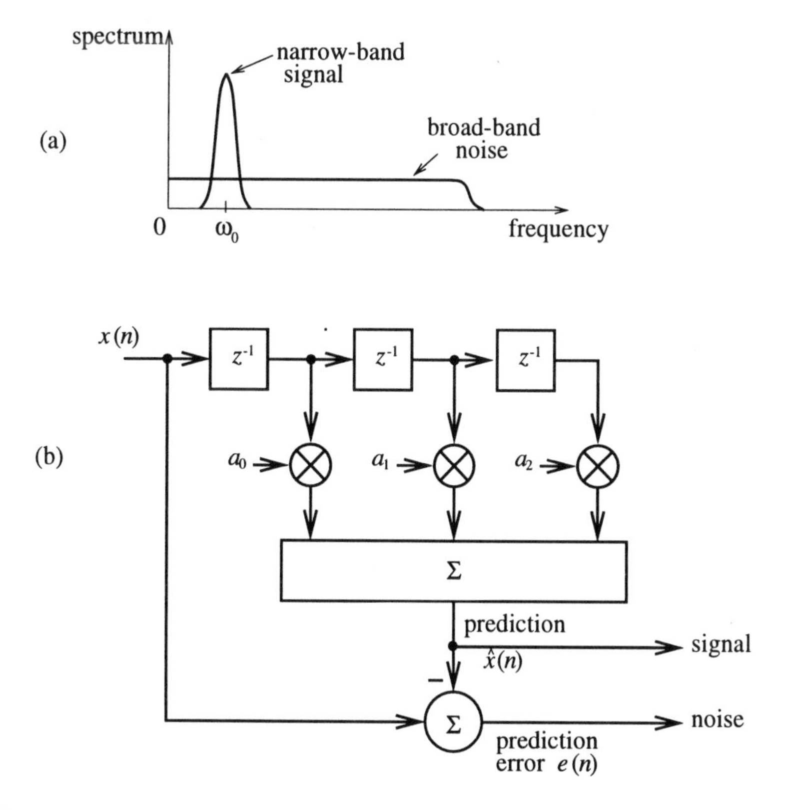

Figure 8.14 *Adaptive line enhancer: (a) signal spectrum; (b) enhancer configuration.*

wave is by its very nature predictable. If we were given 10 consecutive samples of a sine wave we could make a good guess at what the next sample is likely to be. A sine wave is a highly correlated signal. Broad-band noise, on the other hand, is highly uncorrelated from sample to sample. If we were given 10 samples of broad-band noise, we would have very little idea what the next sample would be. It is this difference in predicability between the narrow-band signal and the broad-band noise that we can exploit to separate the signal from the noise.

Figure 8.14(b) illustrates a 3-tap predictor to which we apply $x(n)$ which contains both the signal and the noise. $x(n)$ is delayed by one sample and we use older samples such as $x(n-1)$, $x(n-2)$, $x(n-3)$ to estimate the current input $x(n)$. The estimate $\hat{x}(n)$ is thus a prediction of $x(n)$ based on the samples $x(n-1)$, $x(n-2)$, $x(n-3)$. Because the narrow-band signal is more 'predictable' than the broad-band noise, the prediction $\hat{x}(n)$ will contain less noise than $x(n)$.

To train the predictor we want to make $\hat{x}(n)$ as close an approximation to $x(n)$ as possible. To do this we could use either of the adaptive filtering algorithms presented earlier in this chapter. The training signal is $x(n)$, while the input to the adaptive filter is $x(n-1)$ which is equivalent to $y(n)$ in the general adaptive filter structure of Figure 8.7. When the adaptive filter has been trained, $\hat{x}(n)$ will contain mostly narrow-band signal and the error $e(n)$ by implication will contain mostly the broad-band noise. In

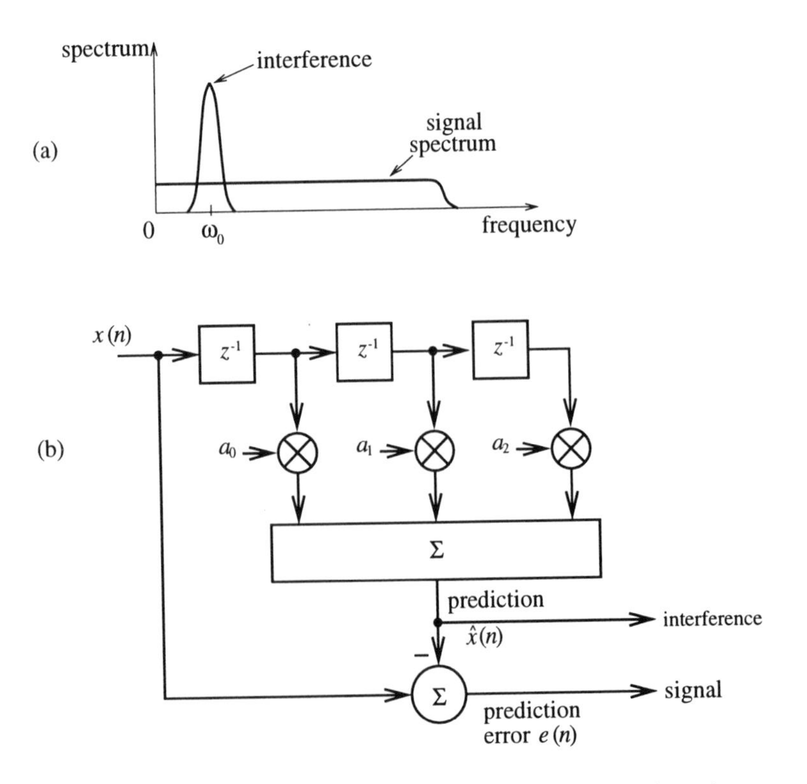

Figure 8.15 *Prediction error filter: (a) signal spectrum; (b) predictor configuration.*

this signal enhancement application, the output of interest is $\hat{x}(n)$ as shown in Figure 8.14(b). In summary, only one input signal is required. It can be used to provide both the input and the training signal to the adaptive filter!

The filter in Figure 8.14(b) with input $x(n-1)$ and output $\hat{x}(n)$ is the prediction filter and it has transfer function: $a_0 + a_1 z^{-1} + a_2 z^{-2}$. In the above example we would expect the frequency response of this filter to be bandpass with a centre frequency ω_0 – after training of course. Thus $\hat{x}(n)$ is the required output for the earlier adaptive line enhancement application.

The filter with input $x(n)$ and error output $e(n)$ is the prediction error filter and has transfer function: $1 - a_0 z^{-1} - a_1 z^{-2} - a_2 z^{-3}$. In the above example we would expect the frequency response of this filter to be bandstop with the notch centred at angular frequency ω_0 – after training.

8.4.2 Adaptive tone suppression

Another very similar application of adaptive filtering in a prediction mode is illustrated in Figure 8.15(a). In this example the signal of interest is broad band in nature – it might be the signal transmitted by a wide-band communications system such as spread spectrum. This time the interference is narrow band in nature – it might be coming from a large rotating machine or fan in the same room as the receiver. The adaptive filter solution is almost identical to the previous example except that we are interested in 'cleaning up' the unpredictable part – the broad-band signal. Thus the estimate $\hat{x}(n)$ contains mostly narrow-band interference and the error $e(n)$ from the prediction error filter contains the recovered cleaned-up signal providing the overall output of the system.

8.4.3 Noise whitening

Figure 8.16(a) illustrates the power spectral density of a random signal. The spectrum is not flat and hence the random signal is coloured or correlated. In applications such as linear predictive coding (LPC) of speech or autoregressive (AR) spectral analysis (Chapter 9), we need a filter which will whiten such a signal. If we apply the correlated random signal to the whitening filter, the output will have a white spectrum. The design of the whitening filter is obviously dependent on the signal so an adaptive solution can again be used.

Figure 8.16(b) illustrates the prediction structure. When the adaptive filter has been trained, $\hat{x}(n)$ is the predictable part of $x(n)$. Thus $e(n)$ is the unpredictable part and must have an approximately white spectrum. The prediction error filter is thus the required whitening filter.

Self assessment question 8.8: White noise is applied to a bandpass filter. The bandwidth of the filter is decreased at the beginning of each day. Is the signal at the output of the filter more or less predictable at the end of the week?

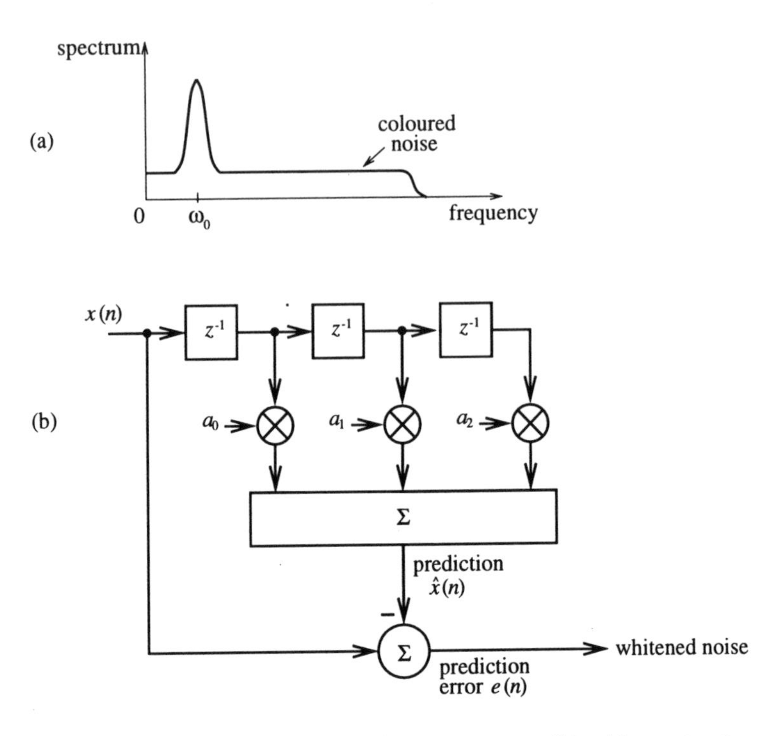

Figure 8.16 *Noise whitening filter: (a) input spectrum; (b) whitener structure.*

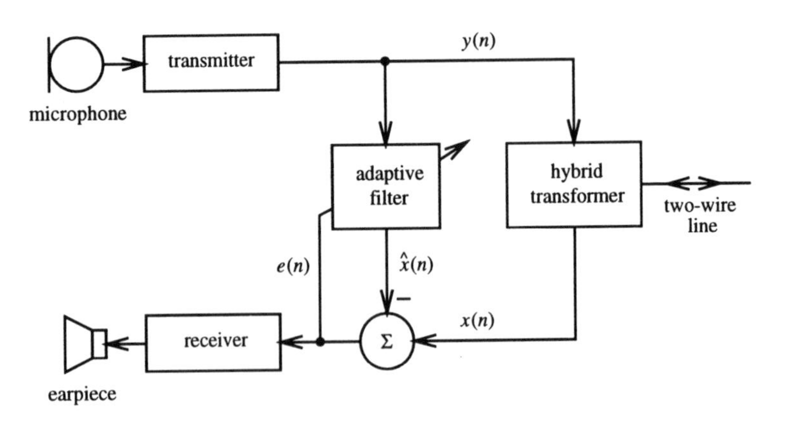

Figure 8.17 *Application of adaptive echo cancellation in a telephone handset.*

8.4.4 Echo cancellation

The essential elements of a telephone handset are illustrated in Figure 8.17. We speak into the microphone which, in combination with the transmitter, converts the sound energy to electrical energy to give the signal $y(n)$. This signal is applied to a hybrid transformer which allows most of the energy to pass up the two-wire line to the local exchange with the subsequent connection to the person at the other end. The hybrid transformer is not perfect and some of the transmitted energy leaks across it into the receiver (ignore the adaptive filter for the moment). We can view the transformer as a filter with a frequency response or transfer function relating its input $y(n)$ to its output $x(n)$.

While we are speaking into the handset we can hear a slight 'echo' of our own voice in the earpiece. For speech this is a good thing since with this 'feedback' we can adjust the level of our voice appropriately. When someone is speaking to us from the far end, the signal travels up the two-wire line from the exchange and is routed by the hybrid transformer into the receiver and our earpiece. The receiver amplifier is quite sensitive because it has to amplify the signal after the signal has passed all the way up the line from the exchange.

Unfortunately, when we wish to use the handset as a modem for transmitting binary data, the 'echo' across the hybrid becomes a major problem. The binary data we transmit through the mouth piece transmitter leak across the hybrid and corrupt the received data coming from the exchange.

If we knew the impulse response or transfer function of the hybrid we could build a filter with similar characteristics and apply the signal $y(n)$ to its input. The output of this filter $\hat{x}(n)$ would be a good estimate of the 'echo' component of the signal coming from the hybrid. If we then subtract $\hat{x}(n)$ from $x(n)$, the resultant signal $e(n)$ would have no component of the 'echo' in it. Hybrid transformers are manufactured on a production line. Their characteristics vary widely from one to the other because the manufacturer wants to keep the price of a handset as low as possible. When we wish to use a particular handset with our modem, we have no idea what the impulse response of the hybrid transformer will be – thus we need an adaptive solution.

Figure 8.17 illustrates an adaptive echo canceller for use with a modem. It is an adaptive filter in a system identification mode of operation. When a modem is switched on, it goes through a long initialisation procedure with the exchange. During part of the initialisation, the exchange does not transmit to the handset and the handset transmits a random binary sequence. This random binary sequence forms the input to both the adaptive filter and the hybrid. The output from the hybrid is the training sequence for the adaptive algorithm. As more random data are passed through the hybrid, the adaptive filter learns its impulse response. After the adaptive filter has reduced the MSE to a suitable level, the random binary sequence is stopped, the adaptive algorithm stops updating the weights and the modem is ready to send and receive data.

8.4.5 Channel equalisation

If we transmit a binary data sequence over a radio communications channel or a telephone channel, the signal can be very severely distorted. Looking at the signal at the output of the A/D convertor in the receiver we would have no idea what digital data had been transmitted.

A communications channel including: transmitter filter; multipath effects; and receive and filter – can be accurately modelled as a finite impulse response filter (Figure

6.1). Usually the transmitted data look like a random binary sequence where $x(n)$ is either +1 or −1 depending on whether we wish to transmit 'logic one' or 'logic zero'. Gaussian white noise (receiver amplifier noise) is added to the output of the FIR filter to produce the received samples. In its simplest case all the coefficients of the FIR filter would be zero except h_0. The received signal samples would then be:

$$y(n) = \pm h_0 + \text{noise}$$

and we could tell what data were being transmitted by simply testing whether $y(n)$ was greater than or less than zero, i.e.:

$$\text{if } y(n) \geq 0 \qquad \text{then } x(n) = +1 \quad \text{else } x(n) = -1$$

If however the other coefficients were not zero we could not use such a simple test, e.g.:

$$y(n) = \sum_{i=0}^{2} h_i \; x(n-i) + \text{noise} \tag{8.36}$$

Here the received sample $y(n)$ is a function of 3 transmitter bits (or symbols). This is called intersymbol interference (ISI).

Each channel has a particular frequency response. If we knew this frequency response we could build a filter in the receiver with the 'opposite' or inverse frequency response, as in Figure 7.13. Everywhere the channel had a peak in its frequency response, the inverse filter would have a trough and vice versa. The frequency response of the channel in cascade with the inverse filter would have a flat amplitude response and a linear phase response, i.e. as far as the data were concerned, the cascade of the channel and the inverse filter together would look like a simple delay. Effectively we would have 'equalised' the frequency response characteristics of the channel.

In practice, when we switch on our digital mobile radio or use a modem to transmit data down a telephone line we have no idea what the frequency response of the channel between the transmitter and the receiver will be. What is worse is that every phone call will produce a new channel. We again need an adaptive solution.

Figure 8.18 illustrates an adaptive equalisation scheme. There is a transversal filter or FIR filter in the receiver which we wish to train adaptively to form the inverse to the

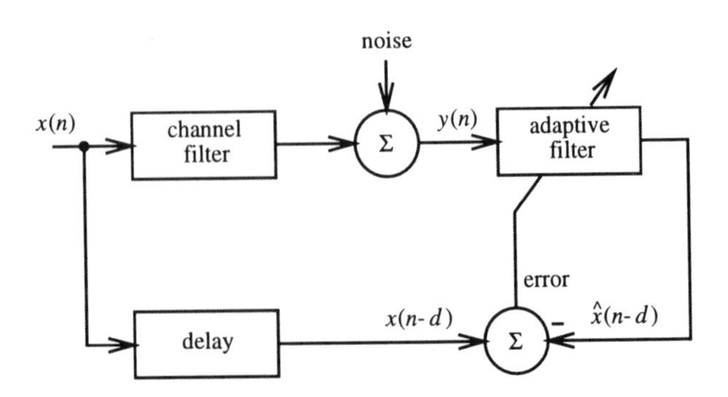

Figure 8.18 *The adaptive equaliser.*

channel filter. The problem is finding a training signal. Fortunately, we design the transmitter as well as the receiver. When the transmitter is switched on, it sends a random binary sequence or training sequence before the actual data are sent. This binary sequence can be stored in the receiver in memory and used as a training signal for the adaptive filter. The objective is to make the output $\hat{x}(n)$ as close to the transmitted data as possible, in which case the adaptive algorithm would have 'learned' the impulse response of the inverse filter. We can then switch off the adaptive algorithm and start transmitting actual data traffic.

On telephone lines the channel filter does not change with time. So once we have trained the equaliser, we can switch off the adaptive algorithm until we dial up a new line. In digital mobile radio this is not the case. The transmitter may be in a car and, as the car moves, the channel between the transmitter and receiver will change with time. Thus the optimum inverse filter needs to change with time. The receiver needs to track changes in the channel. If the channel is changing while actual data are being transmitted, where do we get a training signal from?

One way of doing this is to use what is called 'decision directed' operation (Figure 8.19). If the equaliser is working well then $\hat{x}(n)$ will just be the binary data plus noise. We can remove the noise in a decision circuit which simply tests whether $\hat{x}(n)$ is greater or less than zero. So if the equaliser is working well, the output of the decision circuit is the transmitted data. We could use such data to continue training the adaptive filter by putting the switch in the upper position (Figure 8.19). Here the error is the difference between the binary data and the filter output. This 'decision directed' system will work well, provided the equaliser continues to make correct decisions on the received data samples. If it makes an incorrect decision the training signal will be incorrect, and the adaptive filter will learn the wrong inverse filter, which will lead to more incorrect decisions, i.e. error propagation will result. The decision directed equaliser thus works well in slowly changing channels where the adaptive filter feedback loop can track these changes. If the filter cannot track the signal, then the switch in Figure 8.19 must be moved to the lower position and the actual transmitted data used again to train the filter, as previously shown in Figure 8.18.

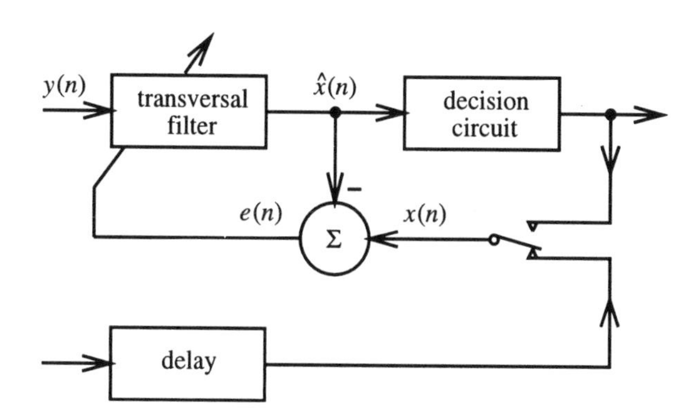

Figure 8.19 *Decision directed equaliser design.*

8.5 Summary

The Wiener filter is optimum in a mean-square error (MSE) sense, i.e. it is the filter, of all possible linear filters, which provides the smallest mean-square error. If the filter is restricted to an N-tap FIR filter, the optimum choice is found by setting the gradient of the MSE (with respect to the filter impulse response vector) to zero. This leads to the Wiener equation which identifies the optimum filter as the solution to a set of N linear simultaneous equations. These equations are defined in terms of the autocorrelation matrix $\mathbf{\Phi}_{yy}$ associated with the vector $\mathbf{y}(n)$ of signal samples in the filter, and the cross-correlation vector $\mathbf{\Phi}_{yx}$ associated with the vector $\mathbf{y}(n)$ and the desired scalar output of the filter $x(n)$.

The design of the Wiener filter requires knowledge of the associated autocorrelation and cross-correlation functions, and is rarely used in practical application. However the importance of the Wiener FIR filter lies in its role in the design of adaptive FIR filters where it can be used as a benchmark against which performance can be measured.

Adaptive filters provide the method for implementing optimal filters when there is insufficient *a priori* information to permit the design of an optimal filter. The required information is obtained from the actual data by the adaptive algorithm through the use of an additional input known as the training signal. A major advantage of adaptive filters lies in their ability to operate in hostile non-stationary or time-varying environments.

This chapter has described the operation of the two major control algorithms, the LMS and RLS, which are used to realise practical adaptive filters. Simulations have been presented to show the superior performance of the RLS algorithm, but its increased computational demand and requirement for more accurate arithmetic are major drawbacks over the simpler LMS algorithm.

At first glance, the requirement for a training signal may appear to limit the application of these adaptive filters. However, further discussion of the many practical applications for these filters in spectral line enhancement, signal prediction, noise whitening, echo cancellation and equalisation shows that the requirement for a separate training signal is not a severe one.

8.6 Problems

8.1 A communications channel has transfer function: $1.0 + 0.5z^{-1}$. The signal $x(n)$ is zero-mean and white with variance 2.0. The additive noise is zero-mean and white with variance 0.2 and is uncorrelated with the signal. Design a 3-tap Wiener equaliser with a lag of 0 and calculate the minimum MSE that it achieves. [$\mathbf{h}_{opt} = [0.88 \; -0.38 \; 0.14]^T$; MMSE $= 0.24$]

8.2 For the two-filter system of problem 7.2, design a 3-tap Wiener filter to estimate the output of filter 2 from the output of filter 1 with an estimation lag of 1.
[$\mathbf{h}_{opt} = [-0.21 \;\; 0.36 \;\; -0.31\,]^T$; MMSE $= 0.21$]

8.3 The following input sequence, $\{y(n)\}$, is applied to an unknown 2-coefficient FIR filter:

$$y(1) = 1.0,\; y(2) = 1.0,\; y(3) = -1.0,\; y(4) = -1.0,\; y(5) = 1.0$$

The associated output sequence is $\{x(n)\}$ where:

$$x(1) = 0.898,\ x(2) = 1.346,\ x(3) = -0.415,\ x(4) = -1.236,\ x(5) = 0.467$$

Use the LMS algorithm to estimate the impulse response of the FIR filter. For the purposes of this calculation, use a step size of 0.25 and assume $y(0) = 0.0$. [$0.852 + 0.385z^{-1}$]

8.4 An electronic diagnosis system is proposed to monitor patient health in an intensive care unit of a busy hospital. Three symptoms are recorded every hour. At the nth hour these are: the pulse rate $y_0(n)$; the temperature $y_1(n)$; the blood pressure $y_2(n)$. A linear estimate, $\hat{x}(n)$, of the patient health, $x(n)$, is formed using a set of coefficients h_0, h_1 and h_2:

$$\hat{x}(n) = h_0(n)\ y_0(n) + h_1(n)\ y_1(n) + h_2(n)\ y_2(n)$$

Show from first principles that the estimate will be optimal in an MSE sense if the following conditions apply:

$$E[y_0\ e(n)] = 0$$

$$E[y_1\ e(n)] = 0$$

$$E[y_2\ e(n)] = 0$$

The error, $e(n)$, is the difference between the actual patient health and the estimate:

$$e(n) = x(n) - \hat{x}(n)$$

In order to implement the system, a senior medical doctor makes himself available as the expert and provides the training signal for the first couple of days. The coefficients are calculated using the LMS algorithm during the training period. Draw a diagram showing how the individual symptoms are combined to form the estimate and how the error signal is generated. Write down equations to show explicitly how the coefficient, h_0, at the nth hour is calculated from the coefficients at the previous hour.

8.5. It is necessary to measure accurately the frequency response of a 20 dB audio amplifier which has a specified 3 dB bandwidth of 100 Hz to 5 kHz and operates with input signal amplitudes of ± 100 mV.

(a) Discuss how to measure the frequency response using only a sine wave oscillator and an oscilloscope.

(b) Explain how to use adaptive signal processing techniques to obtain the amplifier frequency response. Sketch the block diagrams of the test equipment used, indicating the required sampling rates, pulse widths and filter lengths.

(c) Finally, discuss the relative advantages and disadvantages of methods (a) and (b) and select the more appropriate solution for a production line environment.

CHAPTER 9

The Fourier transform and spectral analysis

9.1 Development of the discrete Fourier transform

This chapter develops the mathematical relationship behind the discrete Fourier transform (DFT) operation and highlights the limitations of the finite data record on the DFT performance analysis. This is then developed into the classical and modern techniques which are widely used for estimation of the power spectrum from signal sample data records. The power spectral density of a signal was defined previously in section 7.3 via the autocorrelation sequence. Here we extend this to the case of a discretely sampled sequence, $x(n)$, which, in most practical situations, is restricted to a finite and limited number of signal samples.

9.1.1 The continuous Fourier transform

Consider first an analogue deterministic signal, $x(t)$, with Fourier transform $X(\omega)$, equation (1.16), i.e.:

$$X(\omega) = \int_{-\infty}^{\infty} x(t) \exp(-j\omega t)\, dt \tag{9.1}$$

In the practical situation we observe $x(t)$ and wish to calculate and estimate its Fourier transform. Usually we are supplied with a *finite* number of samples of the signal. The sampling of the signal may not cause loss of information provided we sample fast enough (see the sampling theorem, Chapter 4). However the finite nature of the data record is such that we cannot perform the integration in equation (9.1) from the beginning of time (i.e. $t = -\infty$) to the end of time (i.e. $t = \infty$). Thus we cannot expect to calculate the exact Fourier transform in a practical situation. There will be some loss or error associated with truncating the integration. While we might accept such an error, it is important to understand its nature so that we do not draw false conclusions after performing this estimate of the Fourier transform.

To put this idea on a more formal basis, let us assume that we have sampled the analogue signal, $x(t)$, every Δt s with the impulse train $\delta_T(t)$ of section 4.2 to yield $x_c(t)$ as in Figure 4.4:

$$x_c(t) = \sum_{n=-\infty}^{\infty} x(n\Delta t)\,\delta(t - n\Delta t) \tag{9.2}$$

The Fourier transform of the sampled signal is, by definition:

$$X_c(\omega) = \int_{-\infty}^{\infty} x_c(t) \exp(-j\omega t)\,dt \tag{9.3}$$

If we substitute equation (9.2) into (9.3) we get:

$$X_c(\omega) = \int_{-\infty}^{\infty} \left[\sum_{n=-\infty}^{\infty} x(n\Delta t)\,\delta(t - n\Delta t) \right] \exp(-j\omega t)\,dt \tag{9.4}$$

Changing the order of the integration and summation:

$$X_c(\omega) = \sum_{n=-\infty}^{\infty} x(n\Delta t) \left[\int_{-\infty}^{\infty} \delta(t - n\Delta t) \exp(-j\omega t)\,dt \right] \tag{9.5}$$

The integration then simplifies, because of the properties of the impulse, into:

$$X_c(\omega) = \sum_{n=-\infty}^{\infty} x(n\Delta t)\; \exp(-j\omega n\Delta t) \tag{9.6(a)}$$

Equation (9.6(b)) is an expression for the Fourier transform of a sampled signal. It is known as the *discrete-time Fourier transform* (DTFT) to distinguish it from the Fourier transform of equation (9.1). To calculate the DTFT we still need an infinite number of time samples from the signal.

9.1.2 Fourier transform of a finite length data record

If we only have a data record comprising the *finite* number of data samples from $n = 0$ to $n = N - 1$ then we might approximate the infinite summation of equation (9.6(a)) with a finite summation to produce an estimate $\hat{X}_c(\omega)$ of the actual Fourier transform $X_c(\omega)$:

$$\hat{X}_c(\omega) = \sum_{n=0}^{N-1} x(n\Delta t)\; \exp(-j\omega n\Delta t) \tag{9.6(b)}$$

It is fairly natural to ask: how good an estimate is $\hat{X}_c(\omega)$? One way of answering this question is to introduce the concept of a time-domain window as used previously in Chapter 6 for FIR filters. Truncating the sampled signal into the N-sample record is equivalent to multiplying it by a rectangular window $w_T(t)$ of width $T = N\Delta t$ s:

$$\begin{aligned} \hat{x}_c(t) &= \sum_{n=0}^{N-1} x(n\Delta t)\,\delta(t - n\Delta t) \\ &= x_c(t)\, w_T(t) \end{aligned} \tag{9.7}$$

Multiplication in the time domain is equivalent to convolution in the frequency domain, hence:

$$\hat{X}_c(\omega) = \frac{1}{2\pi} X_c(\omega) * W_T(\omega) \tag{9.8}$$

and the Fourier transform of the rectangular window, $W_T(\omega)$, now distorts the actual Fourier transform to produce the estimated Fourier transform, as shown in Figure 6.6.

EXAMPLE 9.1

An example of these processes is illustrated in Figure 9.1 for the case where $x(t)$ is a sine wave or cosine wave of frequency ω_0 rad/s. Figure 9.1(a) shows the analogue signal and its Fourier transform. As indicated in Chapter 1, its Fourier transform consists of impulses at ω_0 and $-\omega_0$. Sampling this waveform is akin to multiplying it by an infinite train of impulses at the sample rate ω_s and, since we are multiplying in the time domain, we can obtain the Fourier transform of the sampled signal by convolving the transform of the analogue signal with the transform of the impulse train (as described in Chapter 4). This produced replicas in the frequency-domain transform of the analogue signal at multiples of the sampling frequency ($\omega_s = 2\pi/\Delta t$ rad/s), as shown in Figure 4.8. If we are merely presented with a finite block of N data samples such as $\hat{x}_c(t)$, we can evaluate its Fourier transform using the time-domain window concept developed in Chapter 6. To get $\hat{x}_c(t)$ from the infinite data sequence $x_c(t)$, we multiply the latter by a rectangular window $w_T(t)$ of width just less than T s (Figures 9.1(b) and (d)).

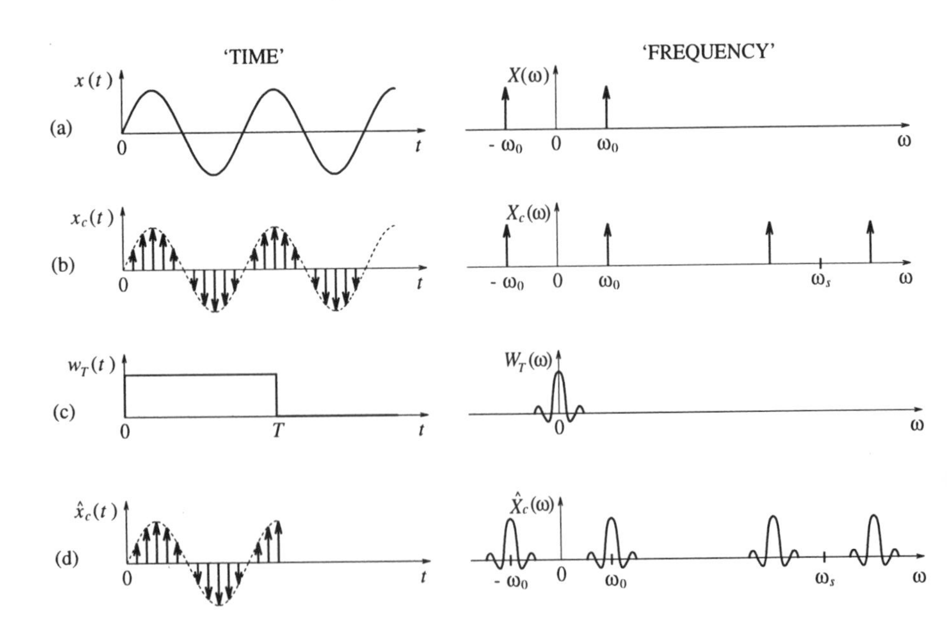

Figure 9.1 *Estimating the Fourier transform of a finite set of data samples: (a) analogue signal; (b) sampled signal; (c) rectangular window; (d) finite data record.*

The transform of the finite data record is then the convolution of the transform of the sampled analogue signal with the transform of the rectangular window. It is obvious from Figure 9.1 that while $\hat{X}_c(\omega)$ is related to $X(\omega)$, they are not the same. The Fourier transform of the finite data record is a distorted version of the desired Fourier transform. The major contributor to this distortion is the Fourier transform of the rectangular window. We cannot avoid using a window of some sort, since we can never have access to an infinite data record. We can, however, in common with the design of FIR filters, choose an appropriately tapered window other than a rectangular one, see equations (6.10) to (6.13) – this will be discussed further in section 9.3.

The Fourier transform of any sampled data signal is a continuous function of frequency, i.e. $X_c(\omega)$ or $\hat{X}_c(\omega)$, which is defined at every possible value of ω from 0 to the sampling frequency. If however we are working with actual data then we can only calculate it numerically at specific values of ω. It is usual to calculate it at N equally spaced frequencies between 0 and the sampling frequency ω_s, i.e. at frequencies spaced by:

$$\Delta\omega = \frac{2\pi}{(\Delta t N)} \tag{9.9}$$

Thus for a given sample rate, $\Delta\omega$ can be arbitrarily decreased by increasing the transform length N, as $T = N\Delta t$ in Figure 9.1. We can now modify equation (9.6(b)) to highlight this 'sampling' in the frequency domain by replacing ω by $k\Delta\omega$:

$$\hat{X}_c(k\Delta\omega) = \sum_{n=0}^{N-1} x(n\Delta t)\ \exp(-jk\Delta\omega n\Delta t) \tag{9.10}$$

and we can remove explicit reference to Δt and $\Delta\omega$ by substituting equation (9.9):

$$\hat{X}_c(k) = \sum_{n=0}^{N-1} x(n)\ \exp\left(\frac{-jnk2\pi}{N}\right) \tag{9.11}$$

This is the usual expression for the discrete Fourier transform (DFT). It is more usual to simply write $\hat{X}_c(k)$ as $X(k)$.

9.1.3 Definition of the DFT

The discrete Fourier transform can thus now be defined as:

$$X(k) = \sum_{n=0}^{N-1} x(n)\ \exp\left(\frac{-jnk2\pi}{N}\right) \tag{9.12}$$

where $0 \leq n < N-1$ and $0 \leq k < N-1$.

Thus in the DFT we take N consecutive samples of the analogue signal and calculate N equally spaced values of $X_c(\omega)$ from 0 to ω_s. The integer k, often referred to as the 'bin' number, gives the DFT value $X(k)$ corresponding to a particular output frequency component ω_k.

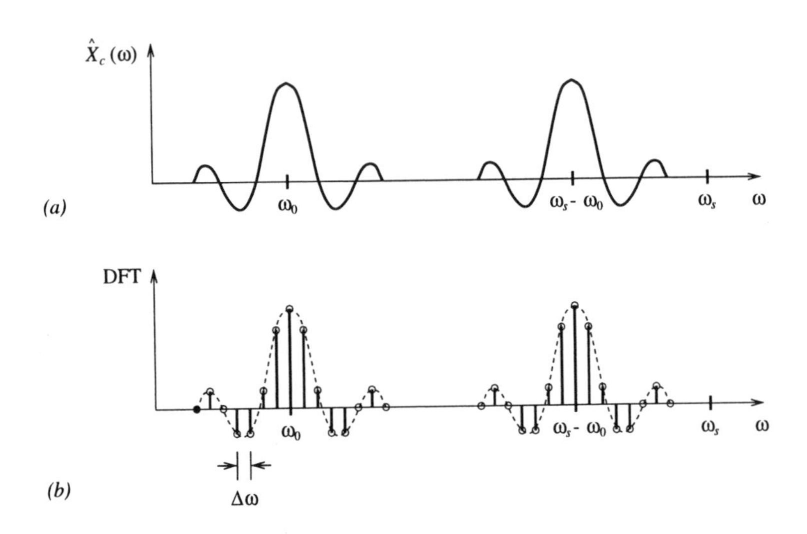

Figure 9.2 *Discrete Fourier transform: (a) expanded view of Figure 9.1(d); (b) discrete frequency samples.*

An expanded view of Figure 9.1(d) and the corresponding DFT, $X(k)$, are shown in Figure 9.2. Note in particular that while there is only one frequency present in the signal and one impulse at ω_0 in $X_c(\omega)$, the energy has smeared over all the bins of the DFT. This smearing is due to the rectangular window, which selects the N sample record from the infinite time series. If we were only presented with Figure 9.2(b) and told nothing about the signal, it might be difficult to come to definite conclusions about the frequency content of the signal other than 'there appears to be a lot of energy in the vicinity of bin number $k = \omega_0$'. These issues will be considered further in section 9.3.

As with the conventional Fourier transform it is possible to define an inverse DFT (IDFT) which relates samples in the frequency domain to time-domain samples:

$$x(n) = \frac{1}{N} \sum_{k=0}^{N-1} X(k) \exp\left(\frac{jnk2\pi}{N} \right) \tag{9.13}$$

where again $0 \leq n < N - 1$ and $0 \leq k < N - 1$. Note here that the processing operation is similar to that of the DFT except that the phase of the exponential is reversed.

The IDFT is of particular importance in deriving numerically efficient (fast) convolution algorithms. These exploit the property that convolution in the time domain is equivalent to multiplication in the frequency domain. Hence the convolution operation of equation (6.1) can be implemented by DFT processing of the input time-domain signal followed by multiplication of the transformed signal samples and then an inverse DFT operation to return to the time domain, see later (Figure 11.9), to obtain the conventional time-domain FIR filter output response.

The four classes of Fourier transform, i.e. the Fourier series, the Fourier transform, the discrete-time Fourier transform and the discrete Fourier transform, are summarised in Appendix B.

9.1.4 Properties of the DFT

There are certain shift and symmetry properties that apply to the DFT of a sampled data sequence $x(n)$. The symmetry property implies that the magnitude DFT of a real (non-complex) data sequence is symmetrical about the DC or 0 Hz component. For complex input data then, the complex conjugate symmetry property applies and the phase response is skew symmetric about 0 Hz. Hence the phase at $+\omega$ is equal to the negative of the phase at $-\omega$. This is shown below by substituting $-k$ into equation (9.12):

$$\begin{aligned} X(-k) &= \sum_{n=0}^{N-1} x(n)\ \exp\left(\frac{-jn-k2\pi}{N}\right) \\ &= \sum_{n=0}^{N-1} x(n)\ \exp\left(\frac{jnk2\pi}{N}\right) \\ &= X^*(k) \end{aligned} \tag{9.14}$$

This symmetry for real signals gives rise to the aliased repeat at ω_s. The modulus of the DFT components from zero up to $\omega_s/2$ thus repeat from $\omega_s/2$ up to ω_s in a mirror image fashion (see Figures 9.2 and 10.8). This can be analysed for complex data by substituting $k = N - r$, where $r < N/2$, into the DFT equation (9.12):

$$\begin{aligned} X(N-r) &= \sum_{n=0}^{N-1} x(n)\ \exp\left(\frac{-jn(N-r)2\pi}{N}\right) \\ &= \sum_{n=0}^{N-1} x(n)\ \exp\left(\frac{jnr2\pi}{N}\right)\exp(-jn2\pi) \\ &= \sum_{n=0}^{N-1} x(n)\ \exp\left(\frac{jnr2\pi}{N}\right) \\ X(N-r) &= X^*(r) \end{aligned} \tag{9.15}$$

The DFT also repeats after N samples and thus a similar analysis to the above equations (9.15) shows that $X(N+k) = X(k)$.

The shift property implies that the magnitude DFT of a sampled signal record is fixed or constant, irrespective of the precise timing of the signal record within the DFT data block. The timing information is contained in the DFT output phase components.

9.2 Computation of the discrete Fourier transform

9.2.1 DFT matrix coefficient values

It is conventional and convenient to adopt the short-hand notation $W_N^k = \exp(-j2\pi k/N)$ for the sampled complex phasor in equation (9.12). The term W_N^1 is the fundamental Nth root of unity (representing the first of the N, Nth roots). These roots are complex numbers which, when raised to the power N, all equal 1.

EXAMPLE 9.2

The eight, eighth roots of unity, W_8^0 to W_8^7, are illustrated in Figure 9.3 in the complex plane. Note that, as the exponential is negative, the rotation is clockwise. These are separated by 360°/N or 360°/8, i.e. 45°. W_8^4 is thus a 180° rotated version of W_8^0 so that $W_8^4 = -W_8^0$. Similarly $W_8^6 = -W_8^2$. Note that $W_8^0 = 1$, $W_8^4 = -1$, $W_8^6 = j$, $W_8^2 = -j$, etc.

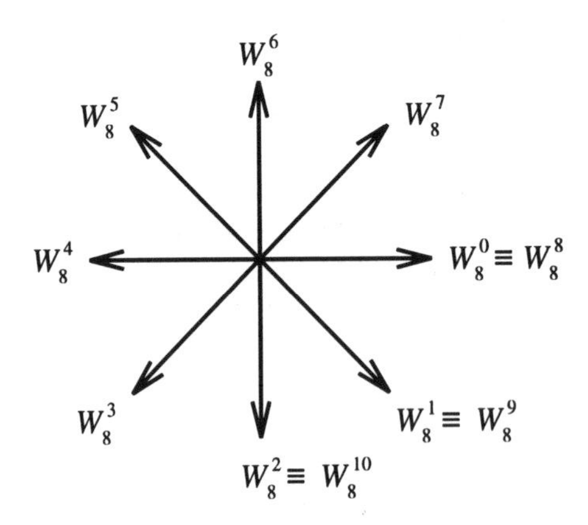

Figure 9.3 *Complex phasor representation of the eight, eighth roots of unity.*

Anticlockwise rotation by 45°, W_8^{-1}, is the same as clockwise rotation through the larger 315° sector, W_8^7. Also if we change the subscript from W_8 to W_4 then we reduce the figure to the four 90° separated roots only, but these are the same as before, so $W_8^2 = W_4^1$.

Self assessment question 9.1: Write out the eight roots of unity in complex $(a + jb)$ form.

9.2.2 Matrix formulation of the DFT

$X(k)$ in equation (9.12) represents the kth component of the frequency-domain representation of $x(n)$. This equation may be expanded horizontally as a function of sample number n from $n = 0$ to $N - 1$, and vertically as a function of output frequency k from $k = 0$ to $N - 1$, giving the set of equations below, with W_N^0 replaced by unity:

$$\begin{array}{lllllll}
X(0) & = & x(0) & + & x(1) & + \;.\;.\;+ & x(N-1) \\
X(1) & = & x(0) & + & x(1)\,W_N^1 & + \;.\;.\;+ & x(N-1)\,W_N^{N-1} \\
X(2) & = & x(0) & + & x(1)\,W_N^2 & + \;.\;.\;+ & x(N-1)\,W_N^{N-2} \\
\cdot & & \cdot & & \cdot & & \cdot \\
\cdot & & \cdot & & \cdot & & \cdot \\
X(N-1) & = & x(0) & + & x(1)\,W_N^{N-1} & + \;.\;.\;+ & x(N-1)\,W_N^1
\end{array} \tag{9.16}$$

Equation (9.16) can also be expressed in vector/matrix form as illustrated below:

$$\begin{bmatrix} X(0) \\ X(1) \\ X(2) \\ \cdot \\ \cdot \\ X(N-1) \end{bmatrix} = \begin{bmatrix} 1 & 1 & 1 & . \; . & 1 \\ 1 & W_N^1 & W_N^2 & . \; . & W_N^{N-1} \\ 1 & W_N^2 & W_N^4 & . \; . & W_N^{N-2} \\ \cdot & \cdot & \cdot & & \cdot \\ \cdot & \cdot & \cdot & & \cdot \\ 1 & W_N^{N-1} & W_N^{N-2} & . \; . & W_N^1 \end{bmatrix} \begin{bmatrix} x(0) \\ x(1) \\ x(2) \\ \cdot \\ \cdot \\ x(N-1) \end{bmatrix} \tag{9.17}$$

In the above DFT matrix, the individual W_N^{nk} terms from equation (9.13) represent the set of N, Nth roots of unity. In the second row of the matrix, $k = 1$, and the sampled complex phasor values are used in the order W_8^1, W_8^2, W_8^3, etc. In the third row, $k = 2$ and thus the sampled phasors are used twice as fast in the order W_8^2, W_8^4, W_8^6, etc.

Each row of this matrix represents a sampled complex phasor. The top row is at zero frequency so it is the fixed unity phasor. The second row consists of samples from a phasor with one cycle over the complete N samples, as it uses all N roots in Figure 9.3 in sequence with one rotation round the circle. The third row, in comparison, represents a phasor at two cycles per N samples (or block length), etc.

A pictorial representation of the contents of the DFT matrix may be obtained by replacing the elements of equation (9.17) with representations as phasors, as shown in Figure 9.4, for $N = 8$. This clearly shows the speed of rotation increasing as we move down, from row to row, and access the sampled complex phasors in order in the DFT matrix.

The DFT fundamentally performs complex processing operations on a complex input signal which is often the case when the signal is obtained from a radar or communications receiver with 'I' and 'Q' (or 'real' and 'imaginary' channels). In the DFT operation the vector of time-domain samples, $[x(0)\ x(1) \cdots x(N-1)]^T$, is also complex. Alternatively, if the signal is coming from a baseband source (e.g. from a microphone), then it is 'real' and its 'imaginary' component is zero. In this case the input array holding the imaginary input components must be set to zero if a valid transform is to be obtained (Figure 10.8). In fact, for real inputs we can perform two separate N-point real transforms within an N-point complex DFT, as explained later in section 10.5.3.

EXAMPLE 9.3

Explain the operation of the 8-point DFT by a complex input phasor with an integer numbers of cycles within the 8-sample input data sequence.

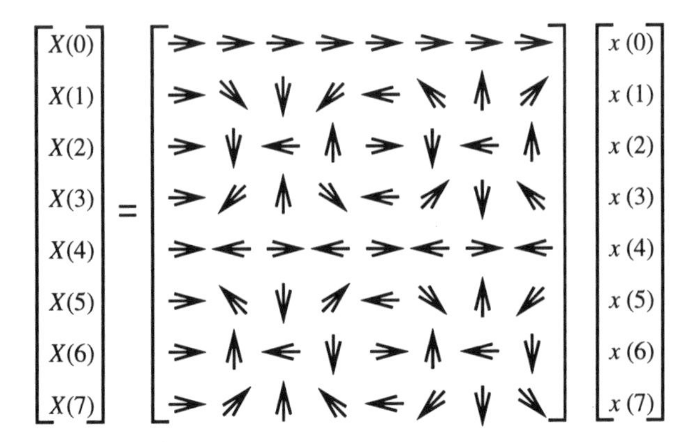

Figure 9.4 *Phasor representation of the 8-point DFT matrix coefficients.*

Solution

If $N = 2$ or 4 then the DFT can be performed without recourse to any complex multiply operations, as shown later in section 10.5.2. Hence the term 'degenerate' can be applied to these two sizes. The 8-point DFT is the smallest non-degenerate transform which is a power of two.

The 8-point DFT matrix contains the eight, eighth roots of unity as shown in Figure 9.3. An 8th root of unity is a complex number which when raised to the power 8 equals unity. Progressing round the roots in increasing order implies clockwise rotation round Figure 9.3. The input signal phasor, on the other hand, rotates in an anticlockwise direction in the complex plane (cosine leads sine by $\pi/2$ radians). Each of the phasors in the DFT matrix seeks to derotate the component in the input signal at the frequency corresponding to that specific matrix row, as shown in Figure 9.4. The resulting stationary component will build up through the transform and appear in the corresponding frequency output. This will be further explained later in this example.

The integer powers 0 to $N - 1$ produce distinct roots in the primary region (0 to 2π). Integer powers of N and above do not produce more roots, but repeat one of the existing roots by an 'aliasing' like effect. Note that the final row in Figure 9.4, which rotates $N - 1$ cycles in the N-point samples, appears as a single-cycle rotation in the opposite direction after calculation, modulo 8. This arises directly from the aliasing of the $(N - 1)$-cycle rotation with the N-cycle sampling frequency and is further pictorial evidence of equation (9.15).

When the N-element input time-domain vector or data sequence $\mathbf{x}(n)$ is multiplied by the DFT matrix it produces a corresponding N element frequency-domain output vector. The first row of the DFT matrix (row zero) finds the DC component in the signal or its average value. The second row finds the component at one cycle per block length, etc. Each successive frequency-domain output is referred to as residing in a frequency bin. For an N-point transform there are N equally spaced bins from DC up to close to the sample rate. The sample rate itself is not a valid DFT output as it aliases to DC and is represented by the $X(0)$ term.

This may all be illustrated by applying a simple input signal, a sampled complex phasor with a period of 8 samples:

$$x(n) = \exp(j2\pi n/8) = \cos(2\pi n/8) + j\sin(2\pi n/8) \tag{9.18}$$

to the 8-point DFT.

We take 8 equally spaced (complex) samples, $x(0)$ to $x(7)$, from this signal. These samples are shown in phasor form in Figure 9.5 and they rotate in an anticlockwise direction for the positive exponential. The starting phase has been chosen to be zero and the input frequency has been chosen to have only one cycle within the complete block length. These assumptions are very restrictive and the problem is hence a very artificial one, but they do allow the mechanism of the transformation to be seen more clearly. We will remove both restrictions later in turn, explaining the consequences at each step.

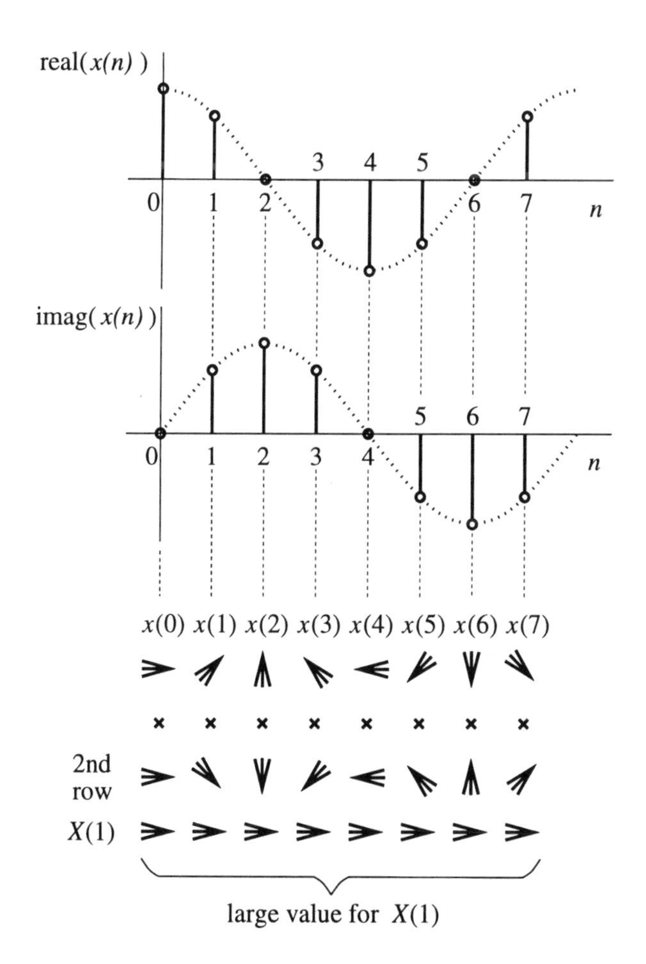

Figure 9.5 *Calculation of the second DFT component for an input signal with one cycle in the block length.*

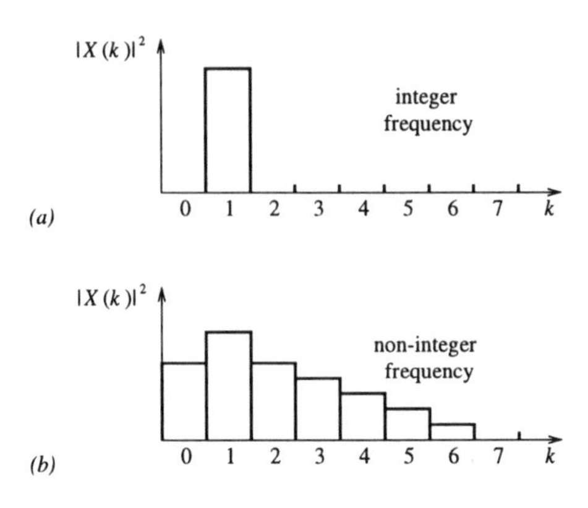

Figure 9.6 *Illustration of leakage at the DFT outputs when analysing a sinusoid with (a) integer and (b) non-integer number of cycles in the block length.*

The term-by-term multiplication of the complex input vector with the second row of the DFT matrix is also illustrated in Figure 9.5. Recall that this row represents a filter matched to a centre frequency corresponding to a component at one rotational cycle of the phasor per block length. (This is also the only component in this particular input signal.) Since, in this case, each element of this row of the DFT matrix is being multiplied by its complex conjugate value, then the result of each multiplication is real and all eight vectors add co-linearly to give a large output in the second frequency bin, indicating a large component at one cycle per block length, i.e. $X(1)$.

If this signal is multiplied by the third row of the DFT matrix, the input signal is still rotating once anticlockwise per transform block while the phasor in this third row of the DFT matrix is rotating twice in a clockwise direction. Considering the term-by-term multiplication as a mixing operation will reveal that the resultant will rotate once in a clockwise direction, giving a net zero result. Repetition of the above for each of the other rows of the DFT matrix reveals zero output from each of them. The resultant output vector is thus $[0\ 1\ 0\ 0\ 0\ 0\ 0\ 0]^T$, as the output appears at bin $k = 1$ only (Figure 9.6(a)), for the chosen complex input signal.

The case of an arbitrary starting phase θ is now considered. The terms of $X(1)$ build up linearly as before, indicating a component at one cycle per block length. The magnitude of the resultant is the same as previously but the phase component now has a value θ, corresponding to the input signal phase in equation (9.18). Thus phase information is not lost in performing the complex DFT operation, and upon performing the inverse DFT (equation (9.13)), the signal will be recovered with the correct starting phase.

Self assessment question 9.2: Repeat the calculation of the input vector in Figure 9.5 with the third row of the DFT matrix and verify that the sum is still zero.

Since many people are only interested in signal power as a function of frequency, they take the squared magnitude of each complex frequency component ($|X(\omega)|^2$) (or, more correctly, for the sampled DFT output ($|X(k)|^2$)) and, in the process, they discard the phase information. It is interesting to note that if the inverse DFT is performed on the resulting power spectrum, an estimate of the autocorrelation function of the original signal is obtained, see later (section 9.5). The autocorrelation function of a signal is also independent of the starting phase of the signal.

Note in Figure 9.5 that if the input was only the real signal $\cos(\theta)$, with the same rotational frequency as before, then this would give the same magnitude of summed output when evaluating the vector products for the second and eighth rows in the DFT matrix of Figure 9.4. These two rows actually give complex conjugate solutions! When squaring and summing the output to obtain the final magnitude (power spectral) values, both $X(1)$ and $X(7)$ have identical and significant magnitude values. We would expect these results from the aliased components of the real input signal at ω_s/N being reflected about ω_s, as discussed in equations (9.15).

9.2.3 Analogies for the DFT

The most revealing analogy for a DFT is the filterbank, as illustrated in Figures 9.7 and 9.8. Each row of the DFT matrix in equation (9.17) is equivalent to performing a convolution or FIR filter operation on the input signal samples with a set of sampled complex phasors representing one row in equation (9.17). These filters are set on progressively increasing centre frequencies, ω_c (Figure 9.7), corresponding to each of the DFT outputs $X(k)$. This can be seen by examining again the rows in Figure 9.4. However in the DFT only one sample is output for each k value over each block of input data! Thus the outputs are downsampled (see later in Chapter 11) by N, the transform block size which means that all the $N \times N$ FIR multiplies do not actually need to be performed at the input sample rate.

This output sample rate reduction is possible because each DFT channel realises a narrow-band filter, and the reduced bandwidth output signal can be easily accommodated within a lower output sample rate. The 8-point DFT can thus be thought of as being equivalent to a contiguous bank of 8 bandpass filters, spaced apart by $\Delta\omega = \omega_s/N = \omega_s/8$, with each output from the DFT being considered equivalent to the output of the corresponding filter. This is shown for $N = 8$ in Figure 9.8. The parallel channel filterbank concept is dealt with further in Figure 11.10, and its application to signal coding is explored briefly in Chapter 11.

Self assessment question 9.3: Use the Nyquist criterion of Figure 4.10(c) to confirm in Figure 9.7 that the downsampling rate is indeed N for each channel in this filterbank design.

Thus the DFT of equation (9.12) performs the filtering operation:

$$X(k) = \sum_{n=0}^{N-1} x(n)\left[\cos\left(\frac{2\pi nk}{N}\right) + j\sin\left(\frac{2\pi nk}{N}\right)\right] \tag{9.19}$$

by summing over N samples, which is inherent in the vector/matrix multiplication in equation (9.17). Although we think of the DFT outputs as comprising frequency-domain samples they are, in fact, time-domain signals representing the amplitude, in the reduced output bandwidth, of each separate DFT channel in Figure 9.7.

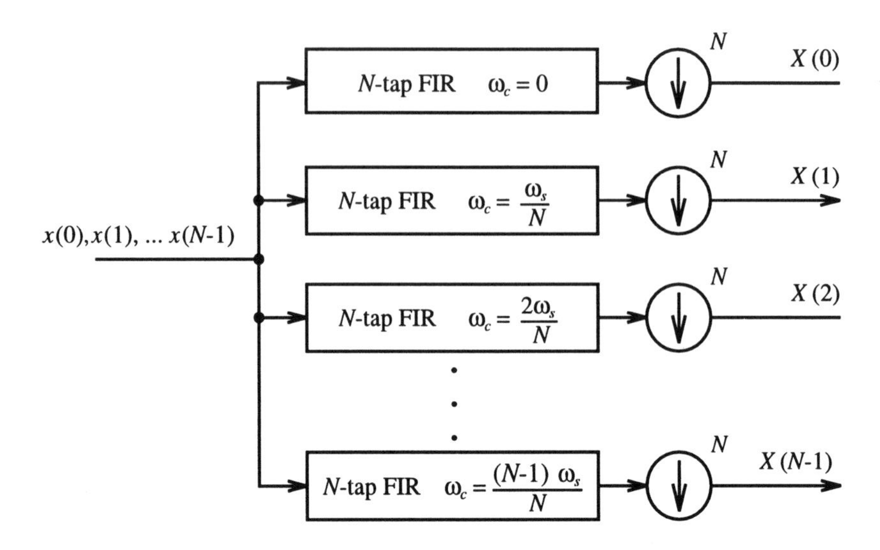

Figure 9.7 *Alternative schematic for N-point DFT.*

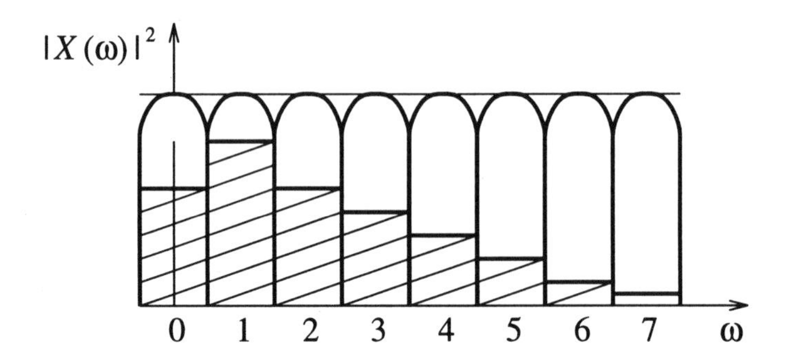

Figure 9.8 *Filterbank analogy for an 16-point DFT.*

The problem with direct implementations, such as filters or the DFT, is that they effectively perform a set of parallel convolution operations involving N^2 complex multiplications for an N-point transform. This is computationally intensive and was the main restriction or inhibitor to the widespread adoption of computer-based frequency-domain signal analysis techniques. This situation changed dramatically with the discovery of the FFT algorithm for rapid computation of the DFT (Chapter 10).

9.3 Resolution and window responses

9.3.1 Resolution

The filterbank analogy in Figure 9.8 also indicates the resolution of the transform processor. When processing N signal samples, the transformed output has a sinc x shape (Figure 9.2), which is controlled by the N-sample data record. If the input signal comprises a sinusoid with n complete cycles within the N samples, then the output occurs on bin $k = n$ only. In fact, this output has a sinc x shape which peaks on bin n with the zeros lying on all the other output bins, as shown in Figure 9.6(a). Resolution is the ability to distinguish between two or more closely spaced sinusoidal inputs without error. One can think of this as one signal falling on the centre of one of the filters in the bank, and the other sinusoid(s) falling entirely within the bandwidth of the other filters. Clearly, minimum resolution occurs when one signal falls at the centre of one of the bins and the other signal falls on the adjacent bin. The smallest resolution, $\Delta\omega$, is thus equal to the filter spacing, ω_s/N (equation (9.9)), which is related to the sample rate and transform size. Note also that the resolution is also constrained to be the inverse of the duration of the transformed data record, $\Delta\omega = 2\pi/T = 2\pi/(N\Delta t)$ in Figure 9.1(c).

Self assessment question 9.4: Verify, for an 8-sample signal vector, sampled at 1 kHz rate, that the 8-sample basis vectors do indeed have zero values on the DFT bin frequencies.

Self assessment question 9.5: What is the resolution of a 1024-point transform with a 512 kHz input sample rate?

9.3.2 Leakage effects

The assumption that the signal comprises an integer number of cycles per block length will now be removed. Recall that the continuous Fourier transform was valid if the integration was performed over the interval $-\infty$ to $+\infty$ or over an integer number of cycles of the waveform. If we now attempt to perform a DFT over a non-integer number of cycles of the input signal we expect in some way to invalidate the transform. This indeed happens and the problems manifest themselves in the form of 'leakage' of energy from any given bin, particularly into the surrounding bins (Figure 9.6).

The leakage problem arises because it is only possible to work with a time-series of *finite* length. A series with N samples corresponds to an infinite series multiplied by a rectangular window of width $T = N\Delta t$ s. This is equivalent to convolving the FT of the continuous signal with the function $T\text{sa}(\omega T/2)$. The convolution smears, or smooths, $X(\omega)$ on a frequency scale of approximately $2/T$ Hz, as shown in Figure 6.6. This leakage and its control are of major importance in practical DFT processor designs. Tapered windows are applied to weight the input data samples to control leakage and to minimise its effect on reducing the dynamic range capability of the transform output.

The worst case leakage occurs when there are $n + ½$ cycles in the input data block length. The resultant output $X(n)$ is now no longer stationary but is rotating at ½ a cycle per block length. As a result, $X(n)$ is diminished in amplitude. The missing energy is spread into the other bins as there is a level of 'match' between the non-integer input signal component and the basis vectors corresponding to the other rows in the DFT matrix. In fact there is equal magnitude for bins $X(n)$ and $X(n+1)$ and this is reduced, from the earlier integer sinusoid case, by the scalloping loss. Figure 9.6(b) illustrates this 'leakage' of energy for such a non-integer input frequency. This will now be shown for the DFT of real data signals. The leakage values can be calculated by examining the sinc x spectrum of the input data record and then superimposing on to this the presence of the actual DFT frequency bin values $0, 1, 2, \cdots, k, \cdots, N-1$.

9.3.3 The rectangular window

Every DFT process is windowed, since a finite time-series is obtained by effectively multiplying an infinite duration time-series by a rectangular (sampling) window equal to the data block length, $N\Delta t$. Recall that multiplication in the time domain is equivalent to convolution in the frequency domain. For the case of the rectangular window the DFT obtained is the convolution of the DFT of the signal itself, with the DFT of a rectangle. The result is a set of sinc ($(\sin x)/x$) functions, each raised on the corresponding unwindowed frequency-domain signal component (Figure 9.1). The problem of leakage and its control is now illustrated with a carefully chosen test signal comprising two tones, as shown in Figure 9.9. Here, a large sine wave of amplitude 1 is shown with a normalised frequency of 10 cycles per block length, i.e. $10\omega_s/N$. Signal 2 has 16 cycles per block length and a relative amplitude of 0.01. Thus both of these signals have an integer number of cycles over a block length and will exhibit no leakage, as the DFT samples lie at the zeros on the sidelobes of the sinc x responses for both the input signals. The DFT output in Figure 9.9 plots $|X(k)|$ on a logarithmic vertical scale in dB.

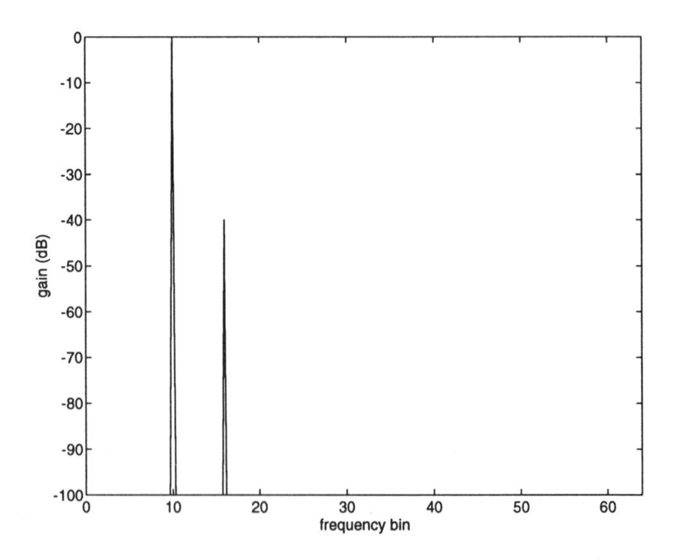

Figure 9.9 *128-point DFT of sinusoids, with integer number of cycles in the block length. The displayed outputs $k = 0$ to 63 correspond to frequencies from 0 to $\omega_s/2$.* □

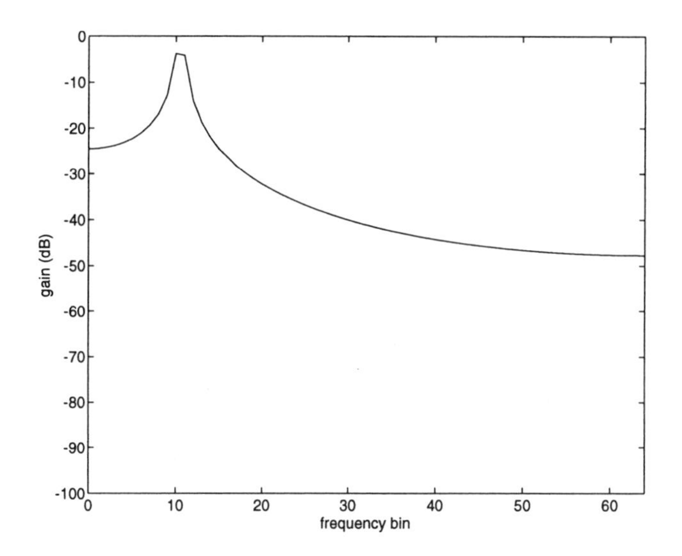

Figure 9.10 *As Figure 9.9 except that the large amplitude sinusoid now falls exactly in between the DFT output bins.* □

In contrast, Figure 9.10 shows what happens if the frequency of the large signal is increased by half a resolution cell in the DFT to $\omega = 10.5\omega_s/N$, so that it falls in between bins 10 and 11. Thus there are 10.5 cycles of this signal within the 128 samples which are being processed to obtain this DFT output. A worse case leakage problem now results. First, as the signal is represented by separate outputs in the two bins 10, i.e. $X(10)$, and 11, i.e. $X(11)$, the magnitudes in these bins are less than that of the single bin 10 case in Figure 9.9. Leakage from the non-integer signal also spreads through the complete transform, and masks the small signal, which is now well below leakage response from the large signal. Thus, the dynamic range of the processor or analyser is severely reduced, particularly where close to large signals. This is unacceptable since we have a much larger dynamic range capability in the DFT processor.

Self assessment question 9.6: What is the dynamic range for a DFT processor using 16-bit fixed-point arithmetic?

The problem is that the leakage here follows the envelope of the sinc function shown in Figure 9.11 and, in the previous smaller sample set in Figure 9.10, the output samples in the display fell exactly on the peaks of these sinc x lobes. Figure 9.11 is developed from Figure 9.10, extending the data record from 128 to 1024 samples, by adding 896 zero elements and by plotting the squared magnitude of this highly zero-padded 1024-point DFT. The zero padding, see later (section 10.6), creates an interpolation effect which allows plotting of the fine detailed structure of the sidelobes in these sinc x responses. The leakage has arisen because the output sampling grid in the 128-point transform is no longer sampling at the zeros of the sinc x function, as occurred previously in Figure 9.9.

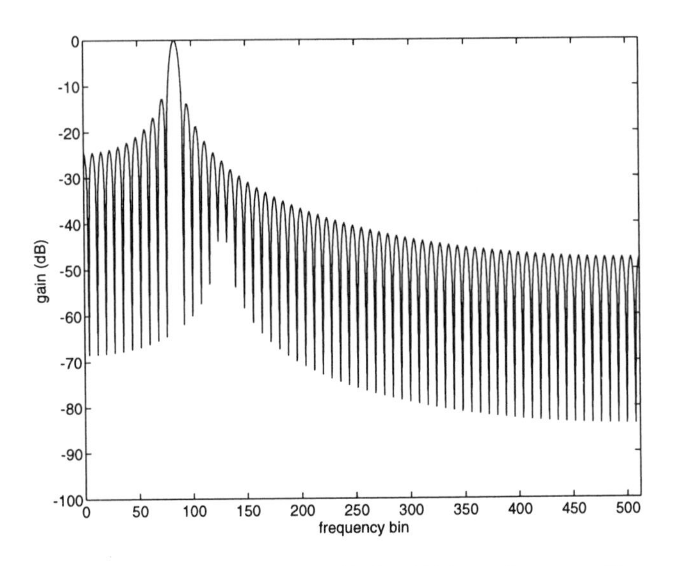

Figure 9.11(a) *As Figure 9.10 but zero-padded to 1024 samples to provide more detail.* □

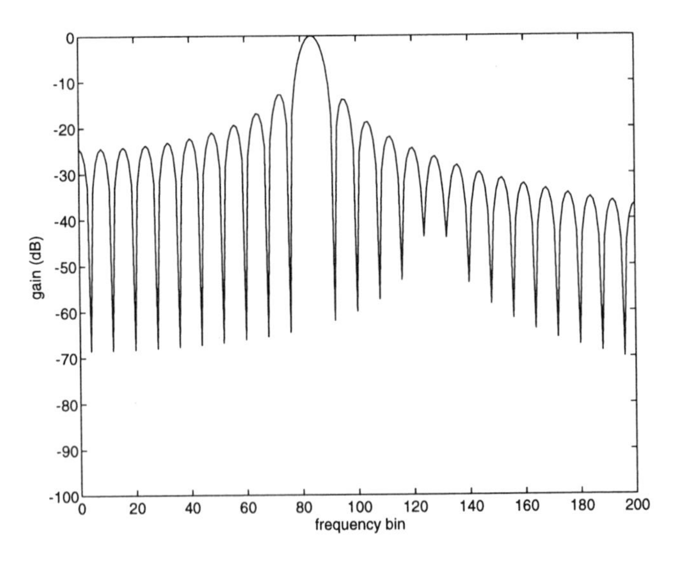

Figure 9.11(b) *As Figure 9.11(a) but only displaying output bins 0 to 200.* □

Self assessment question 9.7: Calculate, for this case, the precise level of leakage in bin 16 for the sinusoidal signal centred exactly in between bins 10 and 11?

Self assessment question 9.8: If you know in advance that, for the 128-point DFT sampled at 12.8 kHz, the input signal does lie exactly at bin 10.5, how can this be made to fall precisely on one of the DFT output bins?

Self assessment question 9.9: For a 256-point DFT sampled at 51.2 kHz, with a sinusoidal input signal of 63 Hz at +10 dBm power level, what is the absolute leakage level in dBm at bin $k = 7$?

Figure 9.10 is obtained by plotting only the reduced set of 64 output $|X(k)|$ samples from the signal in Figure 9.11. Note that the resolution or ability to distinguish between two sinusoidal signals is unaltered by the zero padding operation, but the spectrum is simply interpolated with more output samples.

This leakage may now be reduced by use of a 'tapered' window, as introduced in Chapter 6. Almost any tapered window will be better than the rectangular window. The main families of these tapered windows will now be considered.

9.3.4 Hanning window

The Hanning window is a raised cosine window, as defined previously in Chapter 6, comprising a cosine function on a pedestal of height 0.5, as shown in Figure 9.12. Equation (6.10) is restated here for the N-point DFT with the input sample index $n = 0, \cdots, N-1$ as:

$$w_n = 0.5 - 0.5\cos\left(\frac{2\pi n}{N}\right) \tag{9.20}$$

This window function is used to weight the input sample sequence $\{x(n)\}$ prior to processing in the DFT:

$$x'(n) = x(n) \times w_n$$

The sequence $\{x'(n)\}$ is then input to the DFT processor.

Self assessment question 9.10: For the Hanning window with $N = 64$, verify that $w_0 = 0$, $w_{32} = 1$, $w_{64} = 0$. Also calculate the precise values for w_{16} and w_{48}.

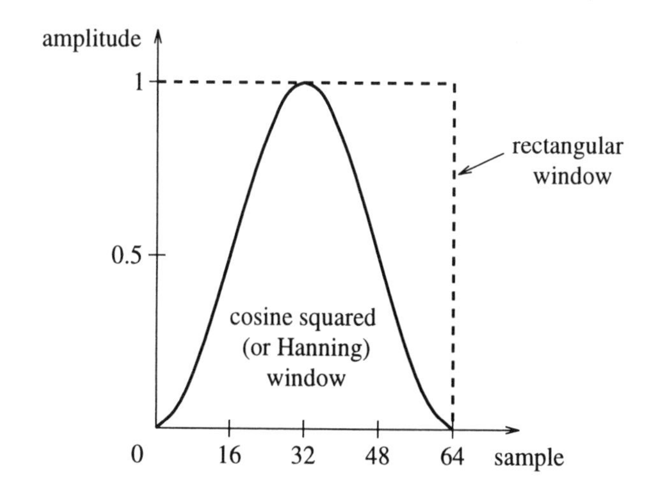

Figure 9.12 *Example of rectangular and Hanning window functions.*

Figure 9.13(a) illustrates the effect of this window on the previous test signal for the case of the large signal halfway between bins 10 and 11, and the small signal in bin 16. The leakage is well controlled in comparison with the rectangular window, with the small signal now being just visible on the side of the reduced leakage from the large signal. The highest sidelobe level in this window is –32 dB and it rolls off rapidly to

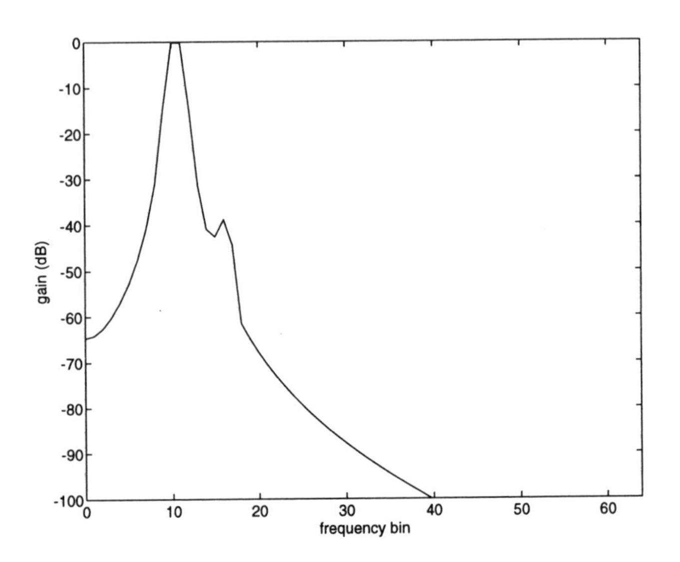

Figure 9.13(a) *As Figure 9.10 but with Hanning-windowed input data.*

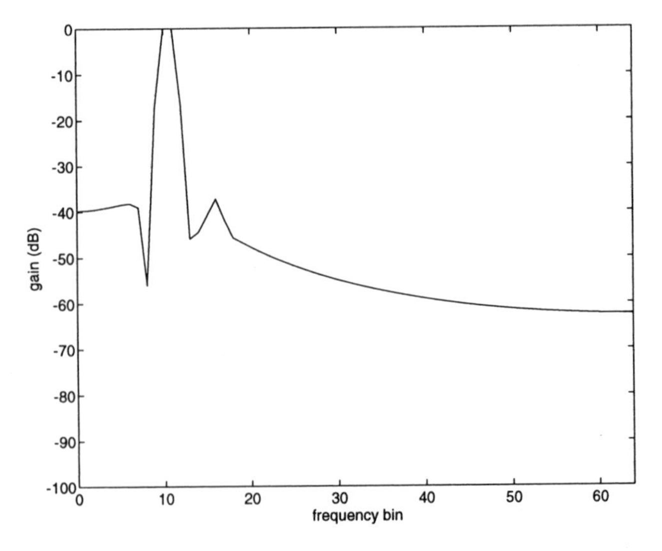

Figure 9.13(b) *As Figure 9.10 but with Hamming-windowed input data.*

give very good visibility well away from the mainlobe.

Although the window reduces the leakage, this is not obtained without a penalty. If you closely examine the width of the responses in Figure 9.13(a) and compare these with Figure 9.10, then you will see that the base width of the windowed responses is widened. This degrades the resolution of the transform, making it more difficult to distinguish between closely spaced sinusoidal inputs. This is further verification of the results quoted earlier in Table 6.1. There is also a loss in signal energy following the window operation, compared with Figure 9.10, but this has been ignored and the display, in Figure 9.13, has been rescaled to 0 dB for the peak output values.

9.3.5 Hamming window

The Hamming window is a slight modification of the Hanning window and is redefined here for the N-point DFT with the index $n = 0, \cdots, N-1$ as:

$$w_n = 0.54 - 0.46 \cos\left(\frac{2\pi n}{N}\right) \tag{9.21}$$

Self assessment question 9.11: Confirm for the Hamming window with $N = 64$ the values for w_0, w_{32} and w_{64}.

The Hamming window has a maximum sidelobe level of –43 dB, which is 11 dB lower than the corresponding maximum of the Hanning window, but the trade-off here is that, far away from the main lobe, the sidelobes do not decrease as rapidly, as with the Hanning window. They 'bottom out' at about 50 dB below the main lobe. This means that, by symmetry, leakage from a signal in the position of the main lobe will come through in bins which are far away from that bin with an attenuation of 50 dB, and will mask smaller signals that may lie in those bins. Figure 9.13(b) shows the effect of this Hamming window on the test signal. The small signal is more visible with the Hamming window, which possesses a superior resolution compared with the Hanning window.

9.3.6 Dolph–Chebyshev window

Another important family of DFT windows is the Dolph–Chebyshev family given by the formula [Brigham]:

$$w_n = \frac{N-1}{N-n-1} \sum_{k=0}^{M} \binom{n-1}{k} \binom{N-n-1}{k+1} \beta^{k+1}, \quad n \neq 0 \text{ or } N-1 \tag{9.22(a)}$$

where:

$$M = \begin{cases} n-1, & n \leq N/2 \\ N-n-2, & n \geq N/2 \end{cases}$$

and:

$$w_0 = w_{N-1} = 1$$

Alternatively this window can also be defined in the frequency domain [Harris] as:

$$W(k) = \frac{(-1)^k \cos\left\{N \cos^{-1}\left[\beta \cos\left(\frac{k\pi}{N}\right)\right]\right\}}{\cosh[N \cosh^{-1}(\beta)]}, \qquad 0 \le |k| \le N-1 \qquad (9.22(b))$$

where:

$$\beta = \cosh\left[\frac{1}{N} \cosh^{-1}(10^{\alpha})\right]$$

and:

$$\cos^{-1}(X) = \begin{cases} \frac{\pi}{2} - \tan^{-1}\left(\frac{X}{\sqrt{1.0 - X^2}}\right), & |X| \le 1.0 \\ \ln(X + \sqrt{X^2 - 1.0}) & |X| \ge 1.0 \end{cases}$$

The α parameter controls the taper or severity of the window. α represents the log of the sidelobe level. Thus $\alpha = 2.0$ represents two decades down from the main lobe or –40 dB. The leakage properties of a typical member of this family, $\alpha = 3.0$, are shown in Figure 9.14. The feature of these windows is the fact that the background leakage from the main lobe is at a constant amplitude across the transformed output. The window is optimum in that, for a given level of leakage, the width of the mainlobe will be a minimum. The main attraction of this window family is for small transforms where there only are a small number of available input signal samples. This window provides very good visibility close to the mainlobe. The sidelobe levels for these windows are adjustable from –50 dB to –70 dB, with $\alpha = 3.0$ in Figure 9.14(a) giving –60 dB. Figure 9.14(b) shows the same response when zero-padded to 1024 samples.

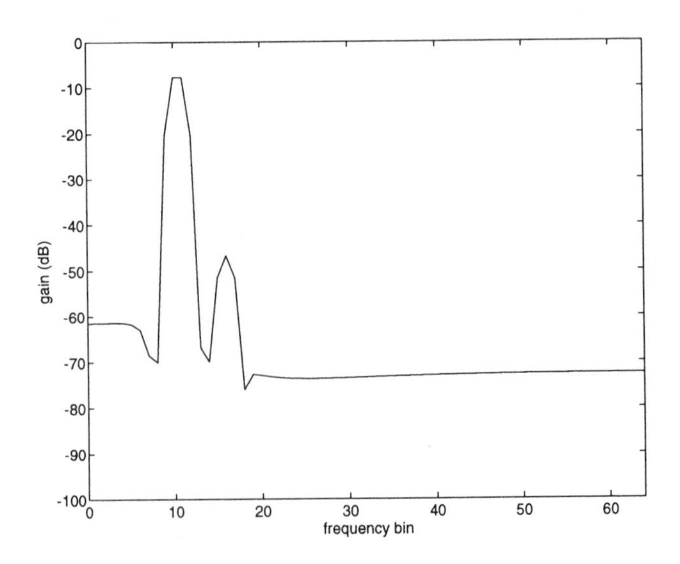

Figure 9.14(a) *As Figures 9.10 and 9.13 but with Dolph–Chebyshev-windowed data.*

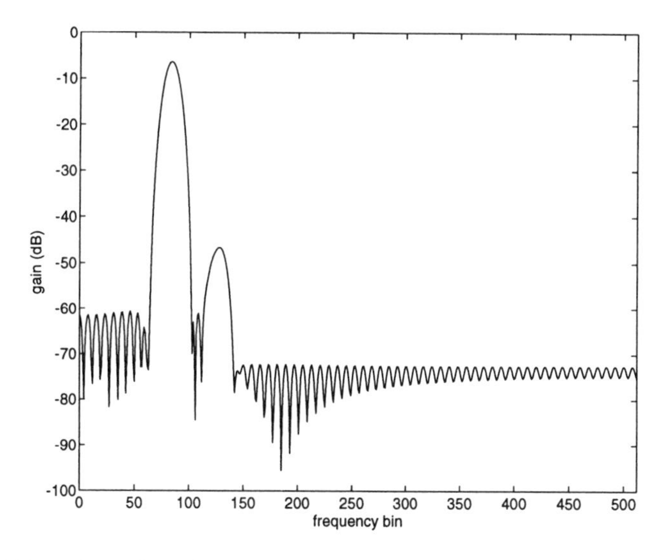

Figure 9.14(b) *Dolph–Chebyshev-windowed zero-padded 1024 point DFT, for comparison with the unwindowed responses of Figure 9.10.*

This simply adds extra points to the Figure 9.14(a) display, giving more *apparent* detail on the transformed output. Obviously the lower the sidelobe level the better, but what prevents α being made very large and the sidelobe level arbitrarily low? As the window becomes more strongly tapered, resolution is lost owing to a progressive widening of the mainlobe. Hence visibility is achieved at the expense of resolution and it depends on the particular application where the optimum trade-off lies. The Dolph–Chebyshev window does, however, provide the minimum mainlobe width for a given amount of sidelobe leakage. Note in Figure 9.14(a) that the energy or processing loss in the window operation has been included in this output display. The maximum output values are thus now somewhat smaller than those in the unwindowed case of Figure 9.10.

9.3.7 Window comparisons

Table 9.1, derived from Brigham and from Harris, compares the various window functions described here and includes other important functions. This table shows the increase in basewidth of the various windows, as a –3 dB bandwidth in Hz for transform length N, at a sample time of Δt s. It also defines the processing or SNR loss due to applying each of these windows. The signal loss was shown previously in Figure 9.14(b) (which was not autoscaled) but there is also a consequent noise reduction by applying the window, so the processing or SNR loss (Table 9.1) is consequently smaller than the amplitude loss in the plotted DFT outputs. The largest sidelobe leakage level is defined for each window, and also the roll-off rate in the sidelobes, defined for the upper four table entries, was shown previously in Figures 9.11(a), 9.13(a), 9.13(b) and 9.14.

Table 9.1 *Typical DFT window characteristics.*

Window	−3 dB *bandwidth* (Hz)	*Processing loss* (dB)	*Peak sidelobe level* (dB)	*Asymptotic roll-off* (dB/octave)
Rectangular	$\frac{0.89}{N\Delta t}$	0	−13	−6
Hanning	$\frac{1.4}{N\Delta t}$	4	−32	−18
Hamming	$\frac{1.3}{N\Delta t}$	2.7	−43	−6
Dolph–Chebyshev	$\frac{1.44}{N\Delta t}$	3.2	−60	0
Blackman	$\frac{1.52}{N\Delta t}$	3.4	−51	−6
Kaiser–Bessel	$\frac{1.57}{N\Delta t}$	3.6	−57	−6

Self assessment question 9.12: Select appropriate window functions for spectrum analysers with < -40 dB and < -55 dB peak sidelobe levels and justify your selection of window function in each case.

9.4 Fundamentals of spectral analysis

Spectral analysis is a large and much researched subject. Spectral analysis involves estimating the amplitude of the harmonics of a periodic signal from a finite set of data samples, i.e. making an estimate of the energy in each Fourier component. Since practical signals are rarely periodic, it is more useful to replace the Fourier series with the Fourier transform and resort to energy density rather than energy, to measure the frequency content of a signal. Finally, if the signal is to some degree random and thus of infinite energy, a more appropriate measure is power. Thus, in order to deal with the widest class of signal, which would include periodic, non-periodic and stochastic processes, spectral analysis is most often defined as an estimate of the signal power spectral density (PSD).

Spectral analysis can be visualised conveniently as a bank of contiguous bandpass filters (BPF), as shown in Figure 9.7. If the bandwidth of each filter is extremely narrow, then an estimate of the PSD at a particular frequency can be formed by squaring the output of a filter and averaging it over M samples (Figure 9.15), which extends the previous Figure 7.7. This visualisation can also be used to highlight the important concepts of spectral resolution and quality.

Spectral resolution is a measure of how closely spaced in frequency two sinusoids can become before they merge into one. Thus in Figure 9.15, the resolution (section

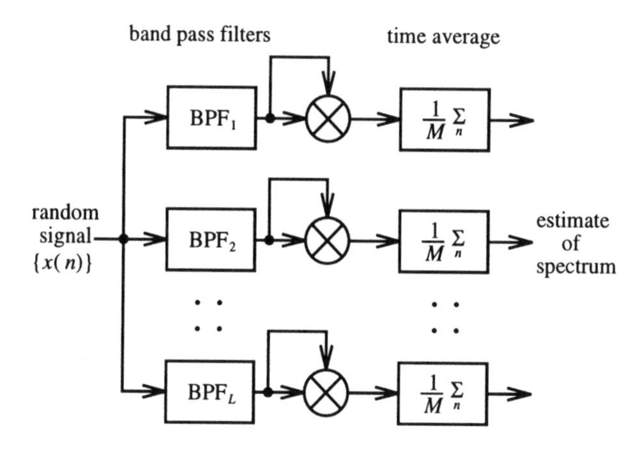

Figure 9.15 *Spectrum analysis with bandpass filterbank.*

9.3.1) was shown to be directly related to the bandwidth of the BPFs. The narrower the bandwidth of the BPFs, the better the spectral resolution will be. However, remember that as the bandwidth is narrowed then the impulse response of the filter becomes longer and hence the time duration of the signal record must be appropriately increased. This is intuitively obvious as, for a deterministic (sinusoidal) signal, the time bandwidth product is unity and hence the resolution is inversely proportional to the time duration of the data record.

Spectral quality, on the other hand, is a measure of how good the spectral estimate is in a statistical sense, e.g. in terms of its mean and variance (Chapter 7). In order to improve the quality of a spectral estimate at a particular frequency, the number of samples, M, over which the time average is performed must be increased.

However, both the quality and resolution of a spectral estimate cannot be improved simultaneously for a given limited number of data samples, M. If, in attempting to improve resolution, the bandwidth of the BPFs is reduced, then the impulse response of these filters will become longer and, consequently, it will take longer for the output of a filter to achieve a steady-state response. The time-averaging processor can only form an unbiased estimate of the spectral density if the signal at the output of a BPF is stationary, i.e. it is not time varying. Thus, in an attempt to improve resolution, two alternatives are open. Either:

(i) Average over all M time samples and accept a spectral estimate which becomes increasingly degraded or biased as the resolution improves; or
(ii) Allow enough time for the natural response of the BPF to decay to zero and average over the remaining data samples with the result that the variance of the spectral estimate and hence the quality become degraded with increasing resolution.

Before discussing spectral estimation, power spectral density must be more rigorously defined. First the autocorrelation of $x(n)$ – equation (7.7) – is redefined here:

$$\phi_{xx}(m) = E[x(n)\, x(n+m)] \tag{9.23}$$

Here $E[.]$ is the statistical expectation operator, as defined previously in section 7.3. For a wide sense stationary random process, $\{x(n)\}$, the power spectral density, $S_{xx}(\omega)$, is defined as the discrete-time Fourier transform (DTFT) of the autocorrelation sequence, $\{\phi_{xx}(m)\}$ – see equation (7.10) again:

$$S_{xx}(\omega) = \sum_{m=-\infty}^{\infty} \phi_{xx}(m) \exp(-j\omega m \Delta t) \tag{9.24}$$

The autocorrelation function can thus be obtained from the spectral density, which is a periodic function of ω, by evaluating the inverse Fourier transform of $S_{xx}(\omega)$ by using equation (1.11) as in equation (7.11):

$$\phi_{xx}(m) = \frac{\Delta t}{2\pi} \int_{0}^{2\pi/\Delta t} S_{xx}(\omega) \exp(j\omega m \Delta t)\, d\omega \tag{9.25}$$

The topic of spectral analysis is developed further in the next two sections. This is approached first from the classical or DFT based approach, and then the newer modern approach is introduced. This latter technique overcomes some of the resolution restrictions of the classical approach and permits the examination of time-varying and non-stationary signals, such as those experienced in speech signal analysis.

Self assessment question 9.13: How does the power spectral density differ from the power spectrum measurement?

9.5 Classical spectral analysis

Power spectral density computation, using equation (9.24), requires an infinite set of autocorrelation coefficients, $\{\phi_{xx}(m)\}$. All that is usually available in a practical situation is the finite set of N signal samples, $x(0), x(1), \cdots, x(n), \cdots, x(N-1)$.

Spectral analysis thus aims to estimate the power spectral density from the finite set of data samples. The oldest and most popular approach is to form an N-point DFT of the finite sequence (equation (9.12)). An estimate of the spectral density at the points $k = 0, 1, 2, \cdots, N-1$ is then formed by dividing the energy, i.e. squared amplitude, at each frequency by the number of data samples:

$$\hat{S}_{xx}(k) = \frac{\left| \sum_{n=0}^{N-1} x(n) \exp\left(\frac{-jnk2\pi}{N}\right) \right|^2}{N} \tag{9.26}$$

where $k = 0, 1, 2, \cdots, N-1$. This technique is the simplest form of what is known as a periodogram, the structure of which is illustrated in Figure 9.16, as an extension of Figure 7.7 where the DFT implements the narrow-band filterbank. This is appropriate for the analysis of random signals, where their time averages are equivalent to their statistical averages. The popularity of the approach lies in the ready availability of the FFT to perform the DFT computation.

Another approach utilises the transform relationship of equation (9.25). If the signal $x(n)$ is stationary, or more strictly ergodic (see section 7.2), then the expectation operator of equation (9.23) can be replaced with a time average. This is approximated by a finite summation over the available data to form an estimate, $\hat{\phi}_{xx}(m)$, of the autocorrelation coefficient, $\phi_{xx}(m)$, as in equation (7.5):

$$\hat{\phi}_{xx}(m) = \frac{1}{N} \sum_{n=0}^{N-1} x(n)\, x(n+m) \tag{9.27}$$

For reliable estimates of the autocorrelation sequence it is usual to restrict the range of m such that: $|m| < N/10$. Equation (9.24) can then be used to obtain an estimate of the PSD:

$$\hat{S}_{xx}(\omega) = \sum_{m=-M}^{M} \hat{\phi}_{xx}(m) \exp(-j\omega m \Delta t)$$

where $0 < M < N/10$. As in section 9.1.2, the DTFT is only evaluated at uniformly spaced discrete frequencies, using a FFT algorithm (Chapter 10). Thus:

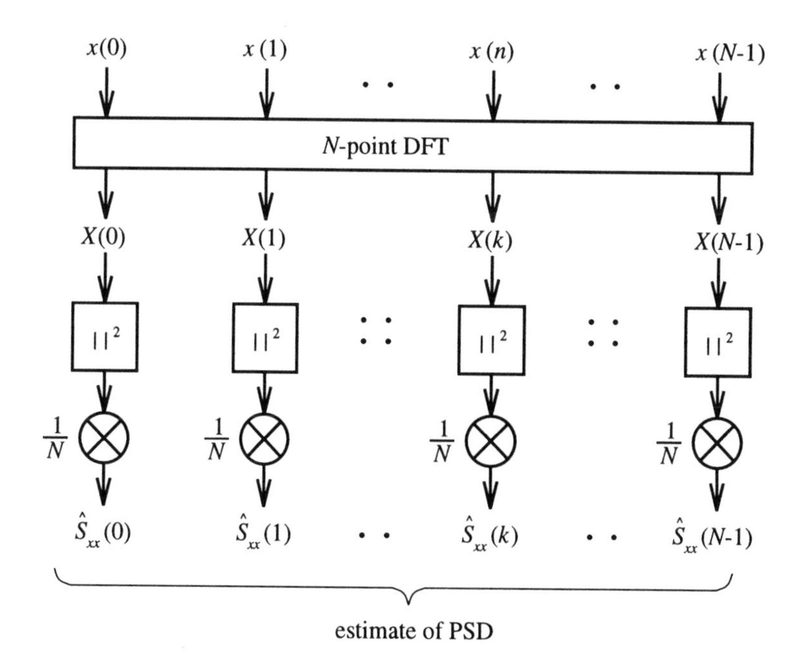

Figure 9.16 *Periodogram power spectrum estimator.*

$$\hat{S}_{xx}(k) = \sum_{m=-M}^{M} \hat{\phi}_{xx}(m) \exp\left(-\frac{j2\pi mk}{N_T}\right) \tag{9.28}$$

Care must be taken with this however as the autocorrelation sequence is symmetric with respect to m and hence the transform size N_T must be an odd number. The symmetry can be simply accommodated by shifting the sequence to the right – delay and hence phase shift do not affect the PSD. A suitable input vector to an DFT or FFT algorithm can then constructed by padding the autocorrelation sequence with zeros. Thus, for example, if $M = 2$ the input vector to a 8-point DFT would be:

$$[\hat{\phi}_{xx}(-2)\ \hat{\phi}_{xx}(-1)\ \hat{\phi}_{xx}(0)\ \hat{\phi}_{xx}(1)\ \hat{\phi}_{xx}(2)\ 0\ 0\ 0]^T$$

This form of spectral estimate is known as a correlogram (Figure 9.17), and its function is equivalent to the earlier structure of Figure 9.16.

Classical spectral analysis is based on either the correlogram or periodogram techniques or hybrids of these. They are robust in that their performance is not usually affected by the signal environment. They have the additional advantages that they are well understood because of their age and can be made computationally efficient through the use of FFT techniques (Chapter 10). On the negative side, they often require long data records with good SNR to achieve high resolution. Thus they usually provide a poor compromise between spectral resolution and quality. These drawbacks have spurred the development of the modern analysis techniques.

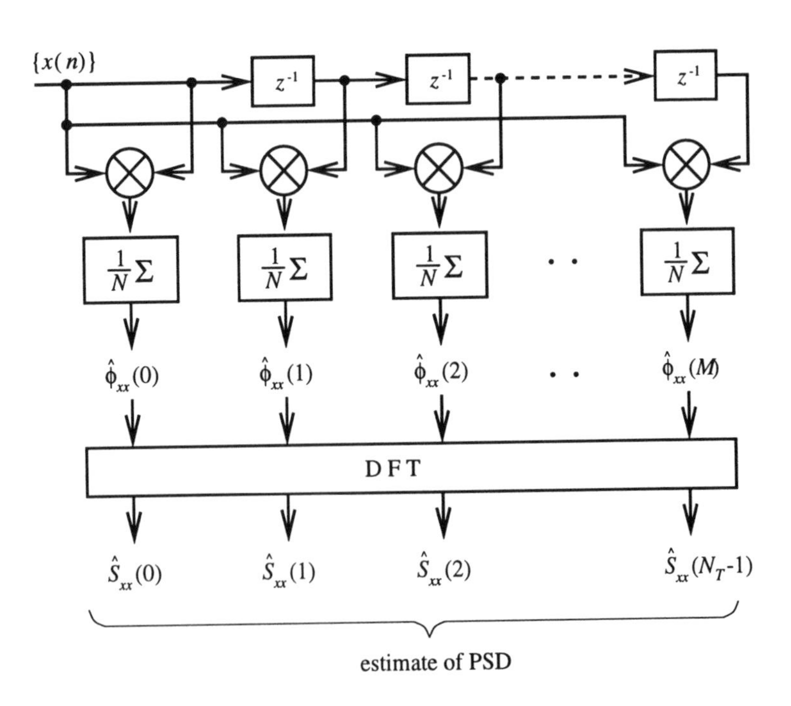

Figure 9.17 *Correlogram power spectrum estimator.*

9.6 Modern spectral analysis

9.6.1 Introduction to parametric techniques

These modern spectral analysis techniques assume that the observed signal is produced directly by a filtering operation on a white noise sequence. The flat PSD of the white noise is thus shaped by the implied filter which generates the signal that we call the observed signal representing the input signal to our spectrum analyser. The most frequently used form of parametric spectral analysis technique is the autoregressive (AR) process. Many of the available techniques permit us to estimate the parameters or AR coefficients which define the input signal PSD. The most notable of these are the least squares (LS) algorithm and the Burg algorithm, and these are closely related to the adaptive filter concepts considered in Chapter 8. The theoretical foundation to this model is provided by equation (7.12) which relates the PSD, $S_{xx}(\omega)$, at the output of a digital filter with the transfer function, $H(z)$, to the PSD at the input, $S_{uu}(\omega)$:

$$S_{xx}(\omega) = | H(\omega) |^2 S_{uu}(\omega) \tag{9.29}$$

For a spectrally white input signal, $S_{uu}(\omega)$ has a constant variance, independent of frequency, which is equal to σ_u^2. The output PSD will then simplify to:

$$S_{xx}(\omega) = | H(\omega) |^2 \sigma_u^2$$

and the PSD will be completely characterised by the amplitude response of the filter and the variance of the white noise. If the filter can be defined by a finite set of parameters then the PSD is also defined by these parameters.

Recall that, as the process with PSD $S_{uu}(\omega)$ is not accessible to us, the filter $H(z)$ cannot be readily identified. However an alternative approach is possible. If an inverse filter $H^{-1}(z)$ of Figure 7.14 can be found which whitens the process $x(n)$ then, from equation (7.13), it will also completely parameterise the PSD of the signal. If the whitened signal $\hat{S}_{uu}(\omega)$ has a variance σ_e^2 then this is given by:

$$\hat{S}_{uu}(\omega) = \sigma_e^2 = | H^{-1}(\omega) |^2 S_{xx}(\omega)$$

and by mathematical manipulation the PSD of the process or signal $x(n)$ is given by:

$$S_{xx}(\omega) = \frac{\sigma_e^2}{| H^{-1}(\omega) |^2} \tag{9.30}$$

The parametric spectral analysis calculation thus involves two stages:

(i) Estimation of the coefficients of the whitening filter, $H^{-1}(z)$, and the noise variance, σ_e^2.

(ii) Making an estimate of the PSD using equation (9.30).

The complete signal modelling and analysis structure is illustrated in Figure 9.18. The key attraction in this approach is that it delivers a high resolution estimate, without incurring the window degradations of the DFT.

9.6.2 Autoregressive spectral analysis

A simplistic approach to achieve parametric spectral estimation is to assume that the signal generating filter, $H(z)$, takes an autoregressive (AR) form, or that it is based on an IIR filter operation (Chapter 5), i.e.:

$$x(n) = u(n) + \sum_{i=1}^{P} a_i \, x(n-i)$$

$$H(z) = \frac{1}{1 - \sum_{i=1}^{P} a_i \, z^{-i}}$$

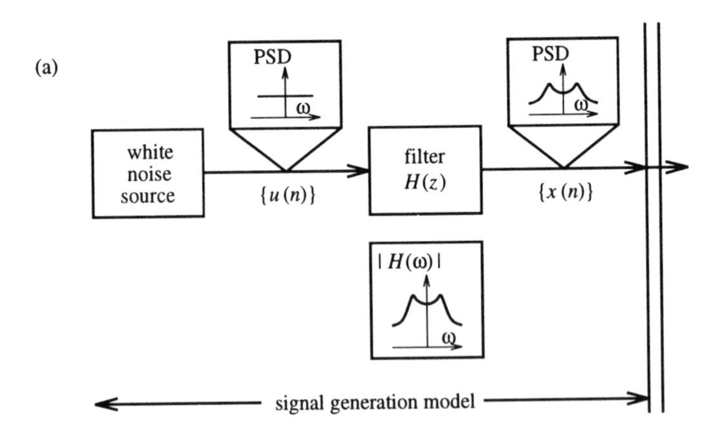

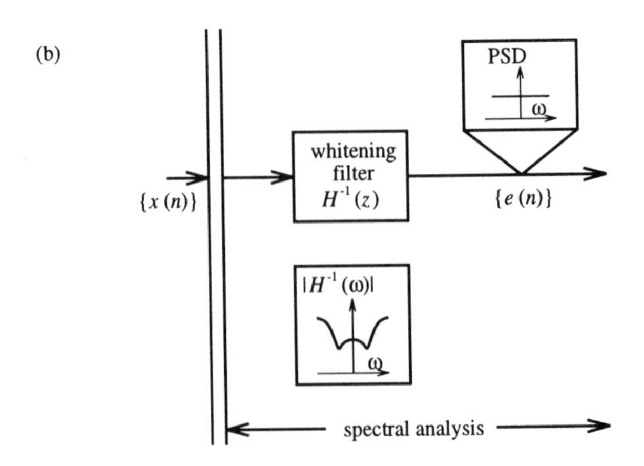

Figure 9.18 *Parametric spectral analysis: (a) generation model; (b) analysis filter.*

where $u(n)$ represents the white noise process and the summation is performed over the P previous output samples. Although other approximations are also possible, the AR model assumes that the signal comprises a set of resonances or sinusoids which is an appropriate model for most practical signals. Given an order-P generation model, the whitening filter, $H^{-1}(z)$, is then an order-P moving average (MA) or FIR filter (Chapter 6):

$$H^{-1}(z) = 1 - \sum_{i=1}^{P} a_i \, z^{-i} \tag{9.31}$$

Figure 9.19 shows the order-P AR filter and the corresponding MA whitening filter. The attractiveness of this approach lies in the fact that the whitening filter is FIR and hence its a_i coefficients may be estimated conveniently from a finite data set by the adaptive signal processing techniques of Chapter 8. Thus, although the AR filter might seem an unlikely choice of signal generation model, it leads *directly* to a relatively simple and stable spectral analysis process.

The whitening filter structure of Figure 9.19 relies on the linear predictor or forward prediction error filter structure of Figure 8.15. A prediction, $\hat{x}(n)$, of the current signal sample, $x(n)$, is formed from the vector $\mathbf{x}(n-1)$, of the P previous signal samples, $[x(n-1)\; x(n-2) \cdots x(n-P)]^T$:

$$\hat{x}(n) = \sum_{i=1}^{P} a_i \, x(n-i)$$

The whitening filter output, $e(n)$, forms the difference between the current signal sample and the prediction, $e(n) = x(n) - \hat{x}(n)$, see equation (8.1). This linear estimation problem is solved using the Wiener FIR filter where the coefficients are chosen to minimise the MSE, ξ, as defined in equation (8.2):

$$\xi = E[e^2(n)]$$

Using the results obtained in Chapter 8 for the Wiener FIR filter, the optimum tap vector $\mathbf{a}$ is again given by equation (8.8) as:

$$\boldsymbol{\Phi}_{yy} \, \mathbf{a} = \boldsymbol{\Phi}_{yx} \tag{9.32}$$

or:

$$E[\mathbf{x}(n-1)\, \mathbf{x}^T(n-1)] \times [a_1 \; a_2 \cdots a_P]^T = E[x(n)\, \mathbf{x}(n-1)]$$

Figure 9.19 shows the cascaded signal generation and whitening filter processes. Here the input to the P-tap FIR filter, formerly $y(n)$ in Chapter 8, is replaced with a delayed version, $x(n-1)$, of the signal sequence, $x(n)$. The optimum filter, $\mathbf{a}$, of equation (9.32) is obtained by simply replacing $y(n)$ with $x(n-1)$. The minimum MSE, σ_e^2, is given by:

$$\min \xi = \sigma_e^2 = E[x^2(n)] - \mathbf{a}^T \boldsymbol{\Phi}_{yx} \tag{9.33}$$

The Wiener equation (9.32) provides a suitable method of calculating the P AR coefficients from the $(P+1)$ autocorrelation coefficients, $\phi_{xx}(m)$, used to construct $\boldsymbol{\Phi}_{yy}$ and $\boldsymbol{\Phi}_{yx}$. In common with the adaptive filter techniques of Chapter 8, and the classical

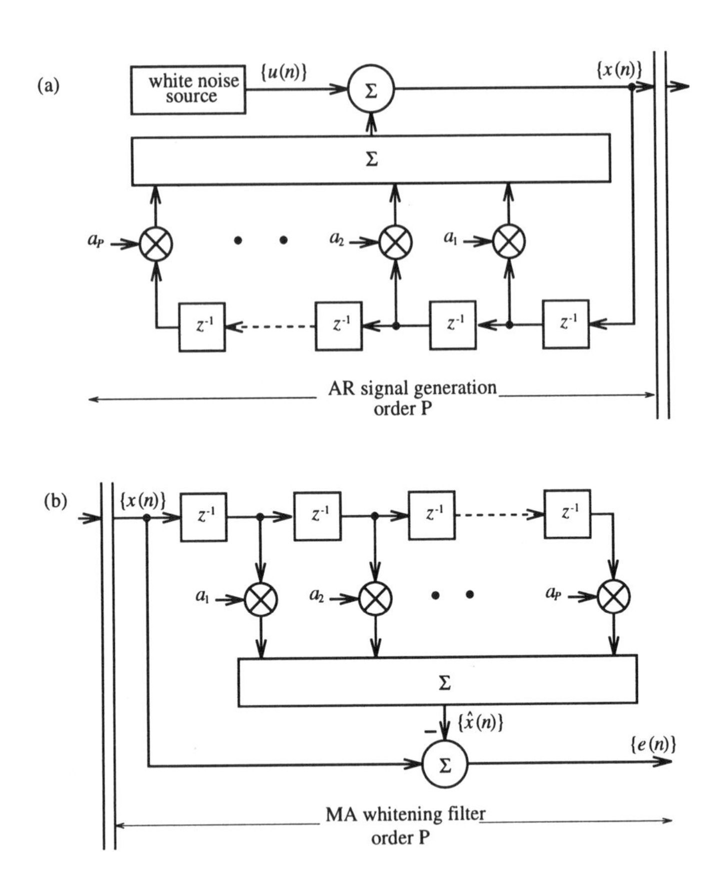

Figure 9.19 *Autoregressive spectral analysis: (a) AR order-P generation model; (b) MA whitening filter.*

methods of spectral analysis discussed in section 9.5, AR analysis is usually performed on signal data itself rather than on autocorrelation coefficients which are generally unknown. Again, in common with adaptive filtering in Chapter 8, one method to circumvent this problem is to replace the MSE cost function with a sum of squared error cost function, ρ:

$$\rho = \sum_n (x(n) - \mathbf{a}^T \mathbf{x}(n-1))^2 \tag{9.34}$$

The vector **a** is adjusted to minimise the cost function, in a MSE sense, and the least squares solution to this is provided by the deterministic form of the Wiener equation (8.8):

$$\mathbf{R}_{yy}\,\mathbf{a} = \mathbf{r}_{yx} \tag{9.35}$$

or:

$$\sum_n \mathbf{x}(n-1)\,\mathbf{x}^T(n-1) \times [\, a_1 \; a_2 \cdots a_P \,]^T = \sum_n x(n)\,\mathbf{x}(n-1)$$

Substituting equation (9.35) into equation (9.34) achieves the minimum value of the cost function:

$$\rho = \sum_{n} x^2(n) - \mathbf{r}_{yx}^T \mathbf{a} \tag{9.36}$$

The range, over which the cost function summation is evaluated, has not yet been defined. Selecting limits on this summation implies assumptions about the signal before the first signal sample $x(0)$ and after the last signal sample $x(M-1)$. These windowing assumptions control the accuracy of both the estimate of the AR coefficients provided by equation (9.35) and hence the accuracy of the spectral estimate itself. To illustrate this, assume the summation starts at $n = 0$. The error associated with this signal sample must first be computed. The summation usually starts at $n = 0$ or $n = P$, and elements which lie outside the window of available data are assumed to be zero. Similar arguments hold for the other end of the data block.

There are four possible variations of the AR coefficient process which lead to four possible LS estimates, depending on the start and finish points in the summation. These variations are summarised in Table 9.2.

Table 9.2 *Least squares windows for AR coefficient estimation.*

lower limit	*upper limit* $\sum^{M+P}$	$\sum^{M-1}$
$\sum_{n=0}$	autocorrelation	pre-windowed
$\sum_{n=P}$	post-windowed	covariance

The autocorrelation form has been widely used in signal analysis for speech coders because the least squares matrix $\mathbf{R}_{yy}$ is Toeplitz [Broyden] and the Levinson recursion can be used to solve equation (9.35) at a lower computational cost than with Gaussian elimination techniques. Further, the Toeplitz property also ensures that the AR filter has all its poles within the unit circle in the z-plane, resulting in a stable solution. This is important in the linear predictive coding (LPC) of speech where the AR filter must be constructed in order to regenerate the speech in the receiver (section 9.8). However, since the autocorrelation form embodies assumptions about the data at both ends of the data block, it provides the poorest spectral estimates.

The other forms of coefficient estimation in Table 9.2 have non-Toeplitz least squares matrix solutions or more accurately a solution comprising the product of two Toeplitz matrices. This near to Toeplitz property has facilitated the development of efficient fast algorithms for the calculation of the AR parameters. The covariance form is at the other extreme from the autocorrelation form, embodying no assumptions about the data outside those samples that are available. Thus it is to be preferred as it produces the best spectral estimates of the four variants. Finally, the minimum of the sum of squared error cost functions ρ provides a mechanism for estimating the minimum MSE, σ_e^2. Taking the covariance form as an example, an estimate, $\hat{\sigma}_e^2$, of the minimum MSE can be formed by dividing the minimum of the sum of squared errors,

ρ, by the number of errors, $M-1-P$:

$$\hat{\sigma}_e^2 = \frac{\sum_{n=P}^{M-1} x^2(n) - \mathbf{r}_{yx}^T \mathbf{a}}{(M-1-P)} \tag{9.37}$$

Finally the covariance-form order-P AR spectral estimate is given by equation (9.30) as:

$$\hat{S}_{xx}(\omega) = \frac{\hat{\sigma}_e^2}{\left| 1 - \sum_{n=1}^{P} a_n \exp(-jn\omega\Delta t) \right|^2} \tag{9.38}$$

The major difficulty in AR spectral analysis is the definition of the model order, P. Intuitively one can increase the order until the MSE drops below a specified threshold. However the MSE decreases monotonically with increasing order and a clear decision point is not always available. The choice of P is a compromise between the degree of spectral detail required, the accuracy and the computational requirement. A good estimate is to make P equal to half or one-third of the number of samples processed. Also remember that we need two poles per sinusoid to model the signal generation mechanism. If too few poles are chosen then the output spectrum becomes highly smoothed, as two adjacent sinusoids can only be modelled by a single resonance.

9.7 Comparison of spectral analysis techniques

The relative performance achieved by these spectral analysis algorithms is illustrated via a simple example in Figure 9.20.

EXAMPLE 9.4

A test signal consists of three sinusoids with relative power levels of 0 dB, –10 dB and –10 dB in additive noise. The noise is white with a PSD of –40 dB. Compare the PSD representations, obtained from 64 samples of this signal, using DFT and AR spectral estimates.

Solution

The frequency spacing of the sinusoids in the test signal was deliberately chosen so that it is difficult to resolve them with classical techniques on this short data record. The correlogram is obtained with DFT processing which has been zero-padded to 512 points to provide more spectral detail. Clearly, the three sinusoids cannot be distinguished.

Finally, the results for an order-8 AR spectral estimate are shown as a continuous line. This provides the best resolution as the three sinusoids can be clearly identified. However, there is a spurious peak due to the background noise and the height of the two peaks associated with the sinusoids at –10 dB are different. As the model order is $P=8$ then the total number of peaks displayed is $P/2$. Thus good resolution has been

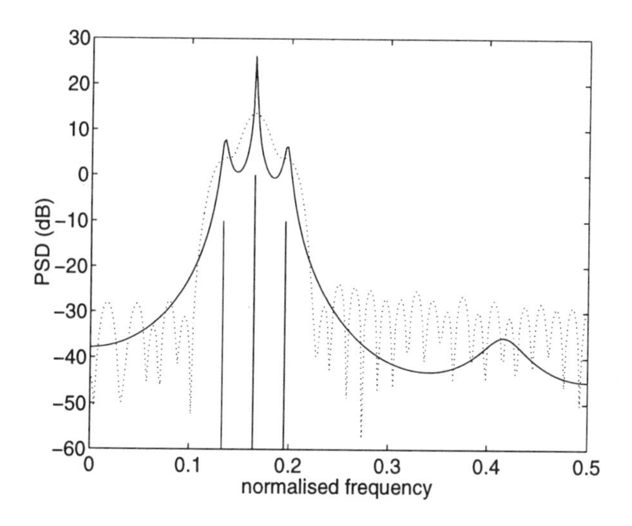

Figure 9.20 *Comparison of spectral estimates:* 64 *real data samples – three sine waves, random phases, normalised frequencies of* 0.13, 0.16 *and* 0.19 *and relative powers of* −10, 0 *and* 10 *dB respectively (vertical lines); periodogram – zero-padded* 512*-point FFT with Hamming window (dotted line); covariance form AR spectral estimate –* $P = 8$ *(solid line).* □

obtained but detailed information on the signal power levels is not so readily available from the AR spectral representation.

Self assessment question 9.14: Now restate the advantages and disadvantages of classical spectrum analysis compared with modern analysis techniques.

9.8 Application of AR techniques in speech coders

Speech basically comprises four format frequencies and the coder analyses the input information to find how the position and magnitude of these varies with time. Vocoders are simplified coders which extract, in an efficient manner, the significant spectral components in the signal waveform, to exploit the redundancy in the speech waveform and achieve very low-bit-rate (< 2.4 kbit/s) transmission. A major vocoder design is the linear predictive coder (LPC).

The LPC design is usually implemented as a cascade of linear prediction error filters (Figure 8.15). The prediction error filters model the vocal tract speech production mechanism as an all-pole filter. In the LPC vocoder (Figure 9.21), the analyser and encoder normally process the signal in 20-ms frames and subsequently transmit the coarse spectral information via the filter coefficients. LPC performs a full model of the input spectrum and transmits the model parameters rather than the residual error samples, as in DPCM speech coders [Glover and Grant]. The residual error is used instead to provide an estimate of the input power level which is sent, along with the pitch information and a binary decision as to whether the input is voiced or

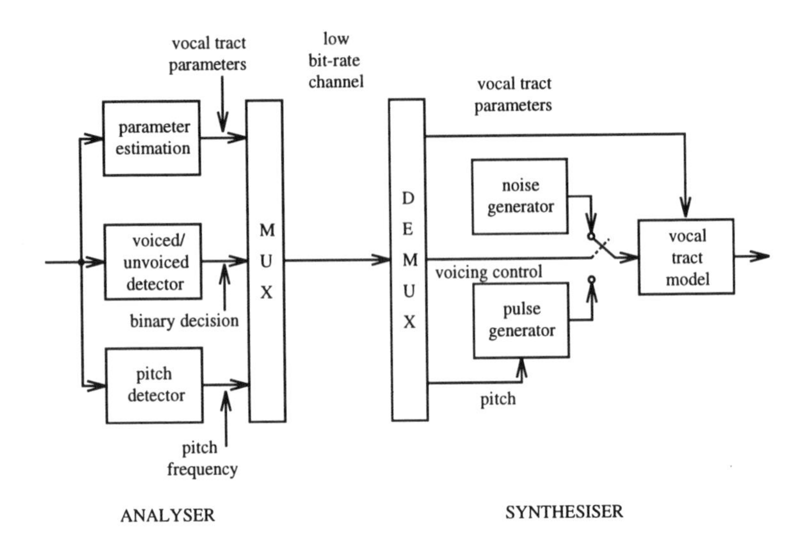

Figure 9.21 *Linear predictive vocoder simplified schematic.*

unvoiced (Figure 9.21). The decoder and synthesiser apply the received filter coefficients to a synthesising filter which is excited with impulses at the pitch frequency if voiced, or white noise if unvoiced. The excitation amplitude is controlled by the input power estimate information.

With delays in the vocal tract of about 1 ms and typical speech sample rates of 8 to 10 kHz, the number of predictor stages is normally in the range 8 to 12, with 10 being the number adopted in the integrated NATO LPC vocoder standard (LPC-10) [also US Federal Standard FS 1015], which transmits at a 2.4 kbit/s rate for a complexity which is approximately 50 times that of pulse code modulation.

Much current research is in progress to improve the quality of these vocoders while still retaining low bit rates, for use in mobile cellular communications, which have to make efficient use of the limited radio spectrum. This is providing the thrust into hybrid coder systems, such as LPC excited from a codebook of possible vectors (CELP), or achieving excitation with multiple pulses (MPE) rather than the simple pitch

Table 9.3 *Performance and rate comparison for selected speech coder designs.*

Class	*Technique*	*Bit rate kbit/s*	*Speech quality*	*Relative complexity*
Waveform coders	PCM G.711 ADPCM G.721 DM	64 32 16	4.3 4.1 3.0	1 10 0.3
Enhanced source coder	CELP/MPE FS 1016	4.8	3.2	30–100
Source coder	LPC-10 FS 1015	2.4	2-2.5	50

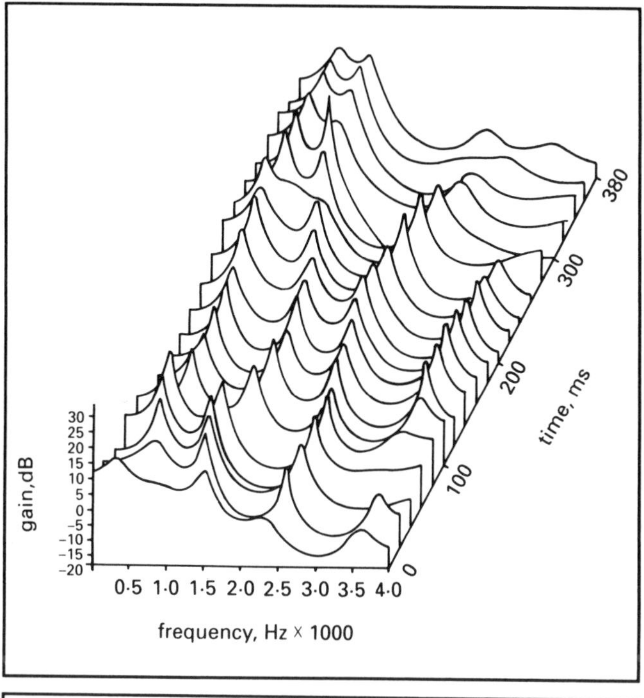

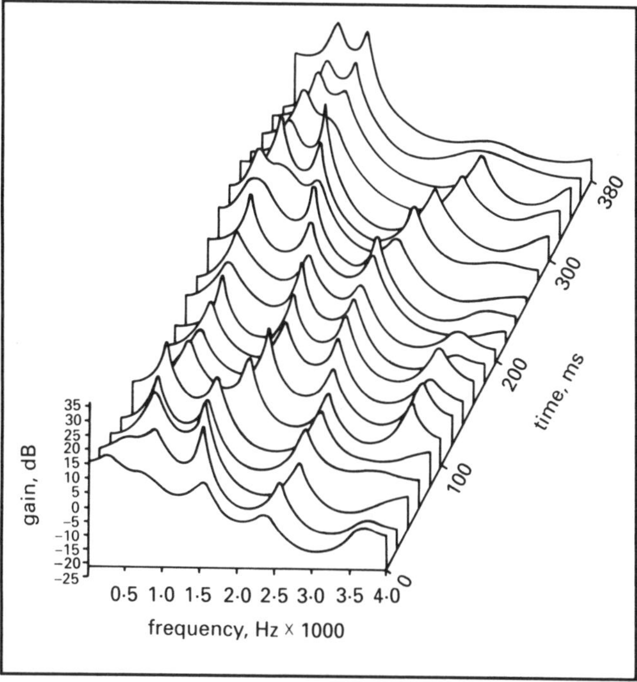

Figure 9.22 *Spectral plots of speech signal: (upper plot) original signal; (lower) decoded output from 8 kbit/s CELP coder. (Source: Boyd, 1992, see Bibliography, reproduced with the permission of the IEE).*

information of Figure 9.21. US Federal Standard FS 1016 defines a 4.8 kbit/s CELP implementation for secure systems. In developing these systems, low coder delay of 2 ms is also a requirement for mobile cellular systems where there is fast fading due to Doppler and multipath effects. Table 9.3 compares LPC and CELP with conventional waveform coders and assesses their subjective quality on the MOS scale of 0 to 5. For higher quality transmission, ITU G.728 defines a 16 kbit/s low delay CELP implementation. Figure 9.22 shows how well the spectral representation of a time-varying speech signal is reproduced within a 8 kbit/s CELP coder.

9.9 Summary

This chapter has described the development of the DFT, carefully explained its operation and shown its equivalence as a filterbank. It has covered the concepts of resolution, leakage and windowing of the data with a tapered envelope to reduce these leakage effects. Finally it has been shown how the DFT can be applied to measure the power spectrum of signals.

Further, the modern techniques for spectral analysis, recently developed to improve upon the resolution of the classical DFT based techniques, were shown to offer this performance enhancement without a significant loss in quality. The most notable of these modern techniques is AR spectral analysis, which is parametric in that the observed signal is assumed to have been generated by passing white noise through an AR filter. While AR analysis can provide excellent results when the underlying process is approximately AR (e.g. LPC speech coders), the results can be less than satisfactory when the process is not AR or the order is chosen incorrectly. Moreover, reliable amplitude information is difficult to obtain from the AR spectral analysis technique. The improved performance that can be achieved with the modern techniques over the traditional classical techniques usually involves a significant increase in computational complexity, but this is not a severe penalty with current real-time DSP products.

9.10 Problems

9.1 Calculate the DFT, $X(k)$, of the sequence $x(n) = \frac{1}{4}, \frac{1}{2}, \frac{1}{4}$.
From the DFT spectrum of $x(n)$, determine the magnitude and phase of the 3-point DFT, $X(k)$, corresponding to the original sequence, $x(n)$.

9.2 A finite-duration sequence of length L is given by:

$$x(n) = \begin{cases} 1, & 0 < n < L - 1 \\ 0, & \text{otherwise} \end{cases}$$

Determine the DFT of N samples of this sequence where $N > L$. From the result, determine an expression for an N-point DFT of this sequence.

9.3 A continuous time signal is known to contain two dominant frequencies at 1605 Hz and 1645 Hz and it is to be sampled at 10 kHz. Given 200 samples, is it possible to resolve the two frequencies by a DFT operation? If they cannot be measured, what frequency values would these two signals have to be altered to, to enable the DFT to adequately resolve the two signals? [No]

9.4 Work out, by hand, $X(0)$ and $X(1)$, the 1st two frequency bin outputs for an 8-point DFT, assuming the complex input to be a unit amplitude phasor which rotates 1 cycle during the transform length and has zero starting phase.

9.5 Given that you are required to analyse a signal comprising the following 8 sample values: $j4, -2\sqrt{2}+j2\sqrt{2}, -4, -2\sqrt{2}-j2\sqrt{2}, -j4, 2\sqrt{2}-j2\sqrt{2}, 4, 2\sqrt{2}+j2\sqrt{2}$, calculate the first four components of the Fourier transform, $X(0), \cdots, X(3)$, and then sketch the full output for the 8-point DFT analyser. [0, $32j$, 0, 0, etc.]

9.6 The first three autocorrelation coefficients of a stationary random sequence are: $\phi(0) = 1.03$; $\phi(1) = 0.31$; and $\phi(2) = -0.81$. Using a second order autoregressive estimate, calculate spectral estimates of the signal at 6 equally spaced frequencies, starting at 0 Hz and finishing at half the sampling frequency. Sketch your results. [-14.6, -11.8, 16.7, -13.4, -18.6, -20.0, all in dB]

9.7 The following set of samples is all that is available from a sequence $x(n)$:

$$3.0, 2.0, -0.5, 1.5, -1.0$$

A third order autoregressive spectral estimate is required and it is proposed that the least squares (LS) technique should be used to calculate the AR coefficients. Calculate the 4 separate forms of the LS matrix $\mathbf{R}_{xx}$.

9.8 For problem 9.7, calculate the linear prediction coefficients for the covariance case of a 2nd order autoregressive spectral estimator. [$a_0 = -0.758$, $a_1 = 0.394$]

9.9 Use the LS technique to identify the system from the data supplied in problem 8.3. [$h_0 = 0.866$, $h_1 = 0.425$]

CHAPTER 10

The fast Fourier transform

10.1 Introduction

The DFT of Chapter 9 is widely used in signal processing for spectral analysis and for filter implementations. One problem with large transform sizes, >128-point, is that the DFT processing becomes extensive and, with limited speed multiply and accumulate hardware, then the bandwidth of the DFT processor is restricted. For a real multiply and accumulate (MAC) rate of 50 MHz and a corresponding cycle time of 20 ns, the sample rate of a 1024-point DFT processor is 12.5 kHz as it requires 1024^2 complex operations. This chapter overcomes some of this restriction by developing a simplification in the DFT processing which recognises that many of the operations are being needlessly repeated when calculating all the $k = N$ output transformed values. The chapter thus develops the simplified decimation-in-time fast Fourier transform (FFT) algorithm which makes for much easier computation of the transformed values. The FFT is widely applied in the practical implementation of large-N DFT processors.

10.2 Partitioning of the DFT into two half-size matrices

We start by repeating equation (9.12) for the kth frequency-domain output from the DFT:

$$X(k) = \sum_{n=0}^{N-1} x(n)\, W_N^{nk} \tag{10.1}$$

Evaluation of $X(k)$ requires the vector multiplication (dot or inner product) of the kth row of the N-point DFT matrix with the input vector, as shown in equations (9.16) and (9.17). A total of N^2 complex multiply operations are involved when performing the calculations for the complete N-point transform operation to evaluate all of the output sample values $X(0), X(1), \cdots, X(N-1)$.

The FFT is based on the observation that there are many symmetries in the DFT matrix and that many multiplications were being needlessly repeated. This can clearly be seen by looking carefully at equation (9.17) and the phasor representation in Figure 9.4. Equation (9.17) is repeated here for convenience:

$$\begin{bmatrix} X(0) \\ X(1) \\ X(2) \\ \cdot \\ \cdot \\ X(N-1) \end{bmatrix} = \begin{bmatrix} 1 & 1 & 1 & \cdots & 1 \\ 1 & W_N^1 & W_N^2 & \cdots & W_N^{N-1} \\ 1 & W_N^2 & W_N^4 & \cdots & W_N^{N-2} \\ \cdot & \cdot & \cdot & & \cdot \\ \cdot & \cdot & \cdot & & \cdot \\ 1 & W_N^{N-1} & W_N^{N-2} & \cdots & W_N^1 \end{bmatrix} \begin{bmatrix} x(0) \\ x(1) \\ x(2) \\ \cdot \\ \cdot \\ x(N-1) \end{bmatrix} \tag{10.2}$$

It is clear from the even columns (0, 2, 4, etc.) that many of the $x(n_{even})W_N^{nk}$ complex operations are being repeated when calculating all the $X(k)$ outputs.

Formal derivation of the FFT begins by recognising these symmetries and splitting the single summation over N samples in equation (10.1) into two summations, each over $N/2$ samples, as shown in equation (10.3):

$$X(k) = \sum_{\substack{n=0 \\ n \text{ even}}}^{N-1} x(n)\, W_N^{nk} + \sum_{\substack{n=0 \\ n \text{ odd}}}^{N-1} x(n)\, W_N^{nk} \tag{10.3}$$

The partition here is into two separate summations – one over the even samples and another over the odd samples. Now make the substitution of variable m in place of n:

$$m = \begin{cases} \dfrac{n}{2}, & n \text{ even} \\ \dfrac{n-1}{2}, & n \text{ odd} \end{cases} \tag{10.4}$$

to clarify the two summations, each over the reduced range $N/2$:

$$X(k) = \sum_{m=0}^{\frac{N}{2}-1} x(2m)\, W_N^{2mk} + \sum_{m=0}^{\frac{N}{2}-1} x(2m+1)\, W_N^{(2m+1)k} \tag{10.5}$$

This step expresses the single N-point DFT as two half-size $N/2$ point DFTs. This is shown in Figure 10.1. When halving the size of the transform we also need to halve the subscript in the sampled complex phasors, W_N^k. From Figure 9.3 and section 9.2.1 we can see that $W_N^2 = W_{N/2}^1$, etc. Also the even data, $x(2m)$, can now be denoted by the sub-sequence $x_1(m)$ and the odd data, $x(2m+1)$, by the sub-sequence $x_2(m)$ (i.e. $x_1(m)$ is the mth sample from the even sequence). These substitutions are made into equation (10.5) to obtain:

$$X(k) = \sum_{m=0}^{\frac{N}{2}-1} x_1(m)\, W_{N/2}^{mk} + W_N^k \sum_{m=0}^{\frac{N}{2}-1} x_2(m)\, W_{N/2}^{mk} \tag{10.6(a)}$$

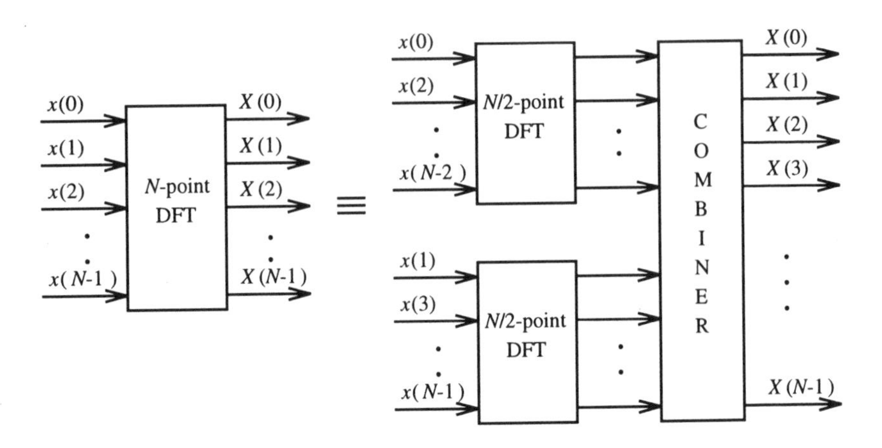

Figure 10.1 *N-point DFT realised with two $N/2$-point DFTs and combining.*

In the case of the second transform, the component W_N^k has been taken outside the summation as it is a constant with respect to the variable m. This component is referred to in the literature as a 'twiddle factor' and is a rotational operator, in that it is complex, has unit magnitude and thus rotates the vector representation of any complex number which it multiplies. In equation (10.6(a)) we can identify the two separate transforms, $X_1(k)$ and $X_2(k)$, each of half-size, i.e. $N/2$. Thus the outputs of the full N-point transform ($X(k)$, $k = 0$ to $N - 1$) may be obtained by suitable combination (Figure 10.1) of the outputs of two $N/2$ point transforms, one processing the odd input data samples and the other the even input data samples:

$$X(k) = X_1(k) + W_N^k\ X_2(k) \tag{10.6(b)}$$

It is also important to note that, from each $N/2$-point transform, the first $N/2$ final output sample values, $X(0)$ to $X(N/2 - 1)$, are obtained by summing the weighted outputs of $X_1(k)$ and $X_2(k)$. Owing to the rotation of the phasors in Figure 9.3, $W_N^{k+N/2} = -\,W_N^k$ and thus the second $N/2$ output sample values, $X(N/2)$ to $X(N - 1)$, are obtained by differencing the weighted outputs. This can be represented as follows:

$$X(k + N/2) = X_1(k) - W_N^k\ X_2(k) \tag{10.6(c)}$$

We now consider this in more detail for the specific case of the 8-point example.

EXAMPLE 10.1

Simplify the DFT matrix by using the FFT processing substitutions to highlight the redundancy.

Solution

For the 8-point DFT we can simplify equation (10.5) to:

$$X(k) = \sum_{m=0}^{3} x(2m)W_8^{2mk} + \sum_{m=0}^{3} x(2m+1)W_8^{(2m+1)k} \tag{10.7}$$

This is shown in Figure 10.2 for the 8-point example. Now each of these 4-point transforms can be split again into 2-point transforms. If we retain the same m index values this gives:

$$X(k) = \sum_{m=0}^{1} x(4m)W_8^{4mk} + \sum_{m=0}^{1} x(4m+2)W_8^{(4m+2)k} + \sum_{m=0}^{1} x(4m+1)W_8^{(4m+1)k}$$
$$+ \sum_{m=0}^{1} x(4m+3)W_8^{(4m+3)k} \tag{10.8}$$

Note that these summations are now occurring over the original input sample pairs 0 & 4, 2 & 6, 1 & 5 and 3 & 7. This is clearly shown in Figure 10.3 but the combining operations are still not fully defined at this stage. If we now drop the 8-point subscripts and expand equation (10.8) for increasing values of k starting at $k = 0$, we obtain the individual transformed output samples as a re-ordering of previous equation (9.16):

$$X(0) = x(0)\,W^0 + x(4)\,W^0 + x(2)\,W^0 + x(6)\,W^0 + x(1)\,W^0 + \text{etc.}$$
$$X(1) = x(0)\,W^0 + x(4)\,W^4 + x(2)\,W^2 + x(6)\,W^6 + x(1)\,W^1 + \text{etc.}$$
$$X(2) = x(0)\,W^0 + x(4)\,W^0 + x(2)\,W^4 + x(6)\,W^4 + x(1)\,W^2 + \text{etc.}$$
$$\text{etc.} \tag{10.9}$$

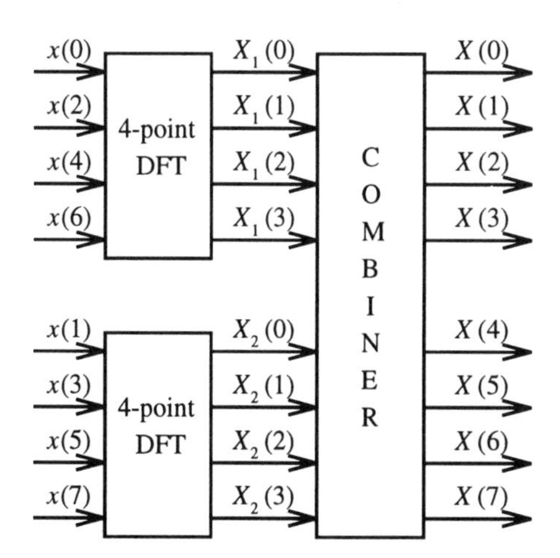

Figure 10.2 *Eight-point DFT realised with two 4-point DFTs.*

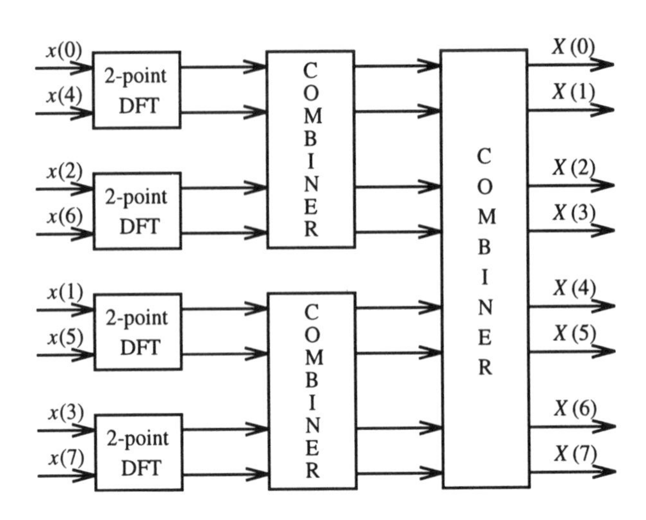

Figure 10.3 *Eight-point DFT realised by 2-point DFTs with combining operations.*

This mathematical explanation indicates how we obtain the output samples but it does not clearly show the power of the FFT. An alternative representation is obtained by returning to the earlier equation (10.2) and representing this in matrix form for the 8-point example, with the *input* data samples now re-ordered, as in equations (10.8) and (10.9), corresponding to Figure 10.3. This adjusts the right-hand vector in equation (10.10) as follows:

$$\begin{bmatrix} X(0) \\ X(1) \\ X(2) \\ X(3) \\ X(4) \\ X(5) \\ X(6) \\ X(7) \end{bmatrix} = \begin{bmatrix} W^0 & W^0 & W^0 & W^0 & W^0 & W^0 & W^0 & W^0 \\ W^0 & W^4 & W^2 & W^6 & W^1 & W^5 & W^3 & W^7 \\ W^0 & W^0 & W^4 & W^4 & W^2 & W^2 & W^6 & W^6 \\ W^0 & W^4 & W^6 & W^2 & W^3 & W^7 & W^1 & W^5 \\ W^0 & W^0 & W^0 & W^0 & W^4 & W^4 & W^4 & W^4 \\ W^0 & W^4 & W^2 & W^6 & W^5 & W^1 & W^7 & W^3 \\ W^0 & W^0 & W^4 & W^4 & W^6 & W^6 & W^2 & W^2 \\ W^0 & W^4 & W^6 & W^2 & W^7 & W^3 & W^5 & W^1 \end{bmatrix} \begin{bmatrix} x(0) \\ x(4) \\ x(2) \\ x(6) \\ x(1) \\ x(5) \\ x(3) \\ x(7) \end{bmatrix} \qquad (10.10)$$

The above re-ordering of the input data vector results in the DFT matrix of equation (10.10) having the corresponding columns re-ordered compared with equation (9.17). The next step is to simplify the W notation, taking account of the 180° rotational property to express W^7 as $-W^3$, W^6 as $-W^2$, W^5 as $-W^1$, and W^4 as $-W^0$, as in Figure 9.3. This simplifies the matrix in equation (10.10) to include only the sampled complex phasors within the range $W^0, \cdots, W^3$:

$$\begin{bmatrix} X(0) \\ X(1) \\ X(2) \\ X(3) \\ X(4) \\ X(5) \\ X(6) \\ X(7) \end{bmatrix} = \begin{bmatrix} W^0 & W^0 & W^0 & W^0 & W^0 & W^0 & W^0 & W^0 \\ W^0 & -W^0 & W^2 & -W^2 & W^1 & -W^1 & W^3 & -W^3 \\ W^0 & W^0 & -W^0 & -W^0 & W^2 & W^2 & -W^2 & -W^2 \\ W^0 & -W^0 & -W^2 & W^2 & W^3 & -W^3 & W^1 & -W^1 \\ W^0 & W^0 & W^0 & W^0 & -W^0 & -W^0 & -W^0 & -W^0 \\ W^0 & -W^0 & W^2 & -W^2 & -W^1 & W^1 & -W^3 & W^3 \\ W^0 & W^0 & -W^0 & -W^0 & -W^2 & -W^2 & W^2 & W^2 \\ W^0 & -W^0 & -W^2 & W^2 & -W^3 & W^3 & -W^1 & W^1 \end{bmatrix} \begin{bmatrix} x(0) \\ x(4) \\ x(2) \\ x(6) \\ x(1) \\ x(5) \\ x(3) \\ x(7) \end{bmatrix} \tag{10.11}$$

Now it is seen that the DFT matrix can be partitioned into quarters where the upper and lower left-hand quarters are in fact *identical.* These similarities are also present in the right-hand quarters and they can be exposed by expanding each of the right-hand sampled complex phasors into a *product* of two terms and identifying the sample dependent *post-multipliers* $W^0, \cdots, W^3$ as the initial term in the products:

$$\begin{bmatrix} X(0) \\ X(1) \\ X(2) \\ X(3) \\ X(4) \\ X(5) \\ X(6) \\ X(7) \end{bmatrix} = \begin{bmatrix} W^0 & W^0 & W^0 & W^0 & W^0 \times W^0 & W^0 \times W^0 & W^0 \times W^0 & W^0 \times W^0 \\ W^0 & -W^0 & W^2 & -W^2 & W^1 \times W^0 & W^1 \times -W^0 & W^1 \times W^2 & W^1 \times -W^2 \\ W^0 & W^0 & -W^0 & -W^0 & W^2 \times W^0 & W^2 \times W^0 & W^2 \times -W^0 & W^2 \times -W^0 \\ W^0 & -W^0 & -W^2 & W^2 & W^3 \times W^0 & W^3 \times -W^0 & W^3 \times -W^2 & W^3 \times W^2 \\ W^0 & W^0 & W^0 & W^0 & -W^0 \times W^0 & -W^0 \times W^0 & -W^0 \times W^0 & -W^0 \times W^0 \\ W^0 & -W^0 & W^2 & -W^2 & -W^1 \times W^0 & -W^1 \times -W^0 & -W^1 \times W^2 & -W^1 \times -W^2 \\ W^0 & W^0 & -W^0 & -W^0 & -W^2 \times W^0 & -W^2 \times W^0 & -W^2 \times -W^0 & -W^2 \times -W^0 \\ W^0 & -W^0 & -W^2 & W^2 & -W^3 \times W^0 & -W^3 \times -W^0 & -W^3 \times -W^2 & -W^3 \times W^2 \end{bmatrix} \begin{bmatrix} x(0) \\ x(4) \\ x(2) \\ x(6) \\ x(1) \\ x(5) \\ x(3) \\ x(7) \end{bmatrix} \tag{10.12}$$

Each of the segmented quarters of this DFT matrix represents the 4-point DFTs in Figure 10.2. Here the final four W sampled complex phasor values in each row are given by the *product* of the phasor pairs within the matrix. The top left-hand quarter of the matrix operates on the even input sample set $x(0), \cdots, x(6)$. Note that the lower left-hand quarter of the matrix is *identical* and, as it also operates on the even input sample set, then this duplicate calculation can be avoided.

The top right-hand quarter of the matrix, which operates on the odd sample set, $x(1), \cdots, x(7)$, is also identical to the left-hand quarter except that it includes row multipliers $W^0, \cdots, W^3$ which are shown in front of the phasor products. The lower right-hand quarter of the matrix is also identical to the right-hand upper quarter except that it includes a negative sign in front of the multipliers $W^0, \cdots, W^3$. This negative sign and the multipliers $W^0, \cdots, W^3$ are in fact a full representation, in matrix form, of equations (10.6(b)) and (10.6(c)) for the 8-point transform operation.

This DFT processing thus requires the data to be partitioned into odd and even samples, these are processed in separate 4-point DFTs and the results are combined with the final combiner weighting as shown in Figure 10.2.

The final combining weights, incorporated in Figure 10.2, will be shown later to be the sample dependent post-multipliers $W^0, \cdots, W^3$ which are shown in front of the right-hand sampled phasors in equation (10.12).

10.3 Radix-2 FFT

10.3.1 Decimation-in-time algorithm

Each 4×4 4-point DFT matrix in equation (10.12) and Figure 10.2 can be further simplified, using equation (10.8), into two 2-point DFTs with another product operation (Figure 10.3), this time with W_4^0 and W_4^2:

$$\begin{bmatrix} W^0 & W^0 & W^0 & W^0 \\ W^0 & -W^0 & W^2 & -W^2 \\ W^0 & W^0 & -W^0 & -W^0 \\ W^0 & -W^0 & -W^2 & W^2 \end{bmatrix} = \begin{bmatrix} W^0 & W^0 & W^0 \times W^0 & W^0 \times W^0 \\ W^0 & -W^0 & W^2 \times W^0 & W^2 \times -W^0 \\ W^0 & W^0 & -W^0 \times W^0 & -W^0 \times W^0 \\ W^0 & -W^0 & W^2 \times -W^0 & W^2 \times W^0 \end{bmatrix} \quad (10.13)$$

Now there are two stages of combining. One is associated with equation (10.13) and the other with equation (10.12). Figure 10.4 exposes the simplest element, the 2-point transform, which is the basic building block for the radix-2 FFT. This has two inputs, $x(0)$ and $x(1)$, which are respectively added and subtracted to provide the two outputs, $X(0)$ and $X(1)$. This is a direct implementation, in simplified form, of equations (10.6(b)) and (10.6(c)). This flowgraph can also be expressed in matrix form as below, when the simplification $W_4^2 = -W_4^0$ is incorperated:

$$\begin{bmatrix} X(0) \\ X(1) \end{bmatrix} = \begin{bmatrix} W^0 & W^0 \\ W^0 & -W^0 \end{bmatrix} \begin{bmatrix} x(0) \\ x(1) \end{bmatrix} \quad (10.14(a))$$

or as $W^0 = 1$, this can be further simplified to:

$$\begin{bmatrix} X(0) \\ X(1) \end{bmatrix} = \begin{bmatrix} 1 & 1 \\ 1 & -1 \end{bmatrix} \begin{bmatrix} x(0) \\ x(1) \end{bmatrix} \quad (10.14(b))$$

This basic add/subtract operation is referred to as a 'butterfly' operation. As this is a two-point example, the sampled complex phasors are only 1 and -1. The 4-point transform in equation (10.13) also involved + and $-$ W_8^2 or W_4^1, the fourth root of unity, i.e. $+j$ and $-j$, as in Figure 9.3 – see later (section 10.5).

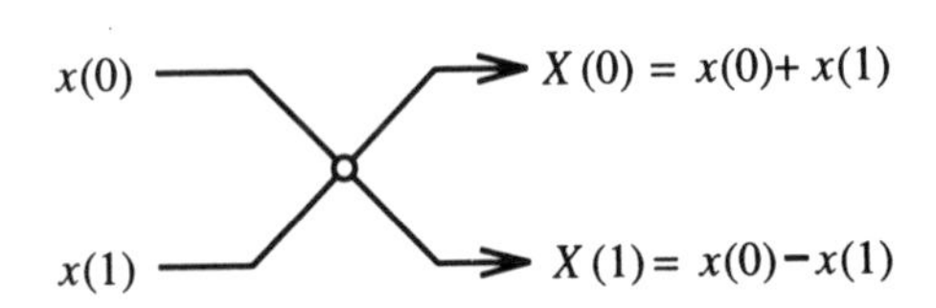

Figure 10.4 *Two-point transform or butterfly operation.*

EXAMPLE 10.2
Derive the full flowchart for the 8-point FFT.

Solution
When Figures 10.3 and 10.4 are combined, then the full 8-point FFT of Figure 10.5 results. Note here that the 4×4 matrices of equation (10.13), which are the symmetrical quadrants within the matrix of equation (10.12), are represented by the two 4-point transforms which are performed on the even input sample set, $x(0), \cdots, x(6)$, and the odd sample set, $x(1), \cdots, x(7)$. Figure 10.5 also partitions the flowgraph so that it exposes the simplest element, the 2-point transform of Figure 10.4, which is the basic building block for the radix-2 FFT.

Figure 10.5 thus comprises an interconnected set of 2-point transforms. The multipliers, which were not detailed in the combiners in Figures 10.1 and 10.3, are now

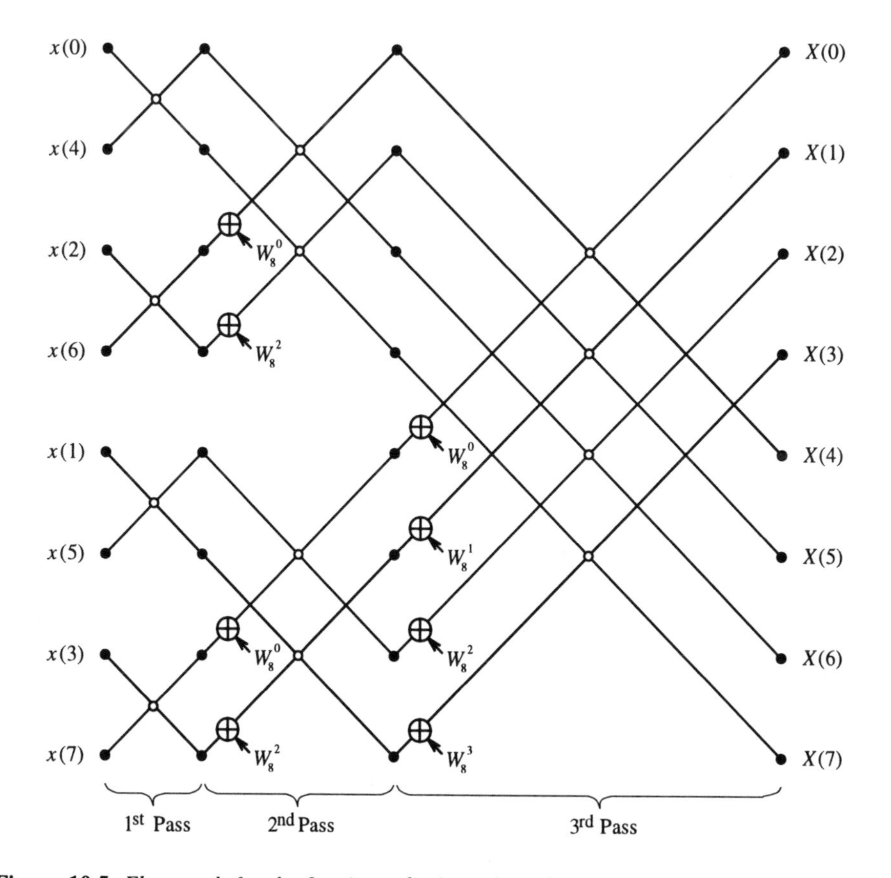

Figure 10.5 *Flowgraph for the 8-point radix-2 in-place decimation-in-time (DIT) FFT operations.*

included in Figure 10.5 on the lower *input* sample to each butterfly operation. The final set of combiners have the highest resolution, using the W^0, W^1, W^2, W^3 product multipliers within the matrix in equation (10.12). These are the same as the previous W_N^k factors in equation (10.6). Note in Figure 10.5 that, in the final butterflies of Figure 10.4, the first 4 outputs, $X(0), \cdots, X(3)$, result from the sum operation in the upper output branch while the latter 4 outputs, $X(4), \cdots, X(7)$, come from the difference operation or lower output branch in the butterflies. Thus the negative signs on the sample dependent multipliers, $W^0, \cdots, W^3$ in equation (10.12), come directly from the butterfly operation of Figure 10.4.

The prior smaller combiners in Figure 10.3 use the lower resolution W^0 and W^2 combiner post-multipliers in equation (10.13), while the 2-point butterflies at the input uses only the sum and difference operations of equation (10.14). There is in fact a weighting by W^0 on each input but, as this equals unity, it has been omitted from Figures 10.4 and 10.5. Figure 10.5 also repeats the normal convention for multipliers, as used for FIR and IIR filters (Chapters 5 and 6), but usually these are simply represented by arrowheads in most FFT textbooks.

The above operations are commonly referred to as the Cooley–Tukey, radix-2, in-place, decimation-in-time (DIT) FFT algorithm. As each of these operations requires us to take two data numbers from *a pair of* memory locations, modify them by multiplying the lower sample by a sampled complex phasor W_N^k, then perform addition and subtraction operations as in equation (10.14(b)), and return the two resultant answers to the *same two* memory locations, this is commonly referred to as the *in-place* algorithm. This minimises the memory requirement which was an important feature of the early development of this FFT algorithm. Now memory is cheap and dense and, for some applications, the in-place algorithm has been superseded by other variants. The term decimation-in-time comes from the successive splitting (decimation) of the time-domain (input) data into even and odd sub-sequences until we have $N/2$ pairs of numbers on which to perform 2-point transforms. This matrix analysis of the DFT matrix to develop the FFT is further reported in [Brigham].

The term 'pass' is used to refer to the complete cycle whereby all the data flow once through the arithmetic unit. In the case of the 8-point DIT transform there are 3 passes in the radix-2 implementation of Figures 10.3 and 10.5. In general there are $\log_2(N)$ passes for an N-point transform. Figure 10.5 is the normal full flowgraph representation of the 8-point transform. Note in the second and third passes that we successively invoke the 2-point butterfly operations of Figure 10.4 on sample pairs to generate the partially computed data results. Thus the N-point transform can be expressed as repeated applications of 2-point transforms, the outputs of which are combined with rotational multiplier operators to produce the full N-point result. On each pass the FFT controller must select the appropriate pairs of data samples, as shown in Figure 10.5, and this requires sophisticated memory addressing.

Self assessment question 10.1: For an $N = 256$ point radix-2 FFT, how many passes are there in the processor?

The particular 'shuffle' or re-ordering of the input data in equations (10.10) to (10.12) leads to the time-domain (input) data being accessed in 'bit reversed' order on the flowgraph in Figure 10.5. Thus data re-ordering for a DIT FFT processor is easily achieved by simply adjusting the address decoder.

Self assessment question 10.2: For the eight input samples $x(0), \cdots, x(7)$, show that bit reversing the 3-bit binary address does achieve the required FFT data re-ordering.

Figure 10.5 is not the only radix-2 DIT implementation. One can manipulate the schematic to obtain naturally ordered inputs and bit reversed outputs while retaining an in-place design. This requires the big crossovers of the third pass in Figure 10.5 to be used on the first pass to access the necessary sample pairs [Lynn and Fuerst]. Alternatively, both the inputs and outputs can occur in natural order but, with this arrangement, the inplace property is lost [Lynn and Fuerst]. For DIT all these different forms commonly have the 'twiddle factor' weight on one of the butterfly *inputs* and hence the products appear on both outputs.

The FFT reduces considerably the number of DFT complex multiplications required for DFT computation. Further examination of Figure 10.5 shows that there are three passes. In the general N-point FFT, where N is a power of 2, then the required number of passes is $\log_2 N$. In Figure 10.5 each pass generally involves four butterfly computations, and in the N-point FFT the number of butterfly computations per pass is $N/2$. Now each butterfly involves one complex multiply, on the lower of the two inputs (ignoring the first pass which in the DIT FFT is free of multiply operations), and so the total number of complex multiplies for an N-point FFT is given by:

$$(N/2)\log_2 N$$

Further, comparing the DFT and FFT processors up to $N = 1024$ gives the result in Figure 10.6.

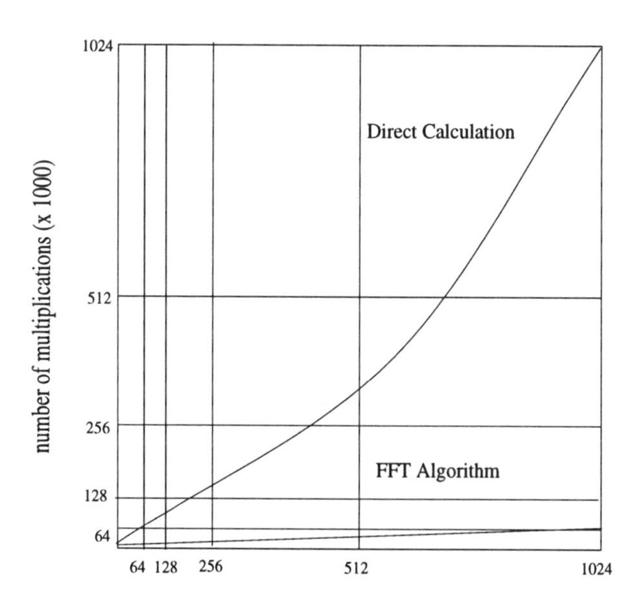

Figure 10.6 *Comparison of multiplier operations in DFTs and FFTs up to 1024-point transforms.*

Self assessment question 10.3: Calculate the number of DFT and radix-2 FFT multiplications required for an $N = 512$ point transform.

The major drawback of the FFT over the DFT is the control overhead required for its implementation. This is the price which must be paid for the increased computational efficiency. In some applications where hardware multipliers are readily available and time permits, then a simple DFT implementation may be preferable. For the $N = 1024$ FFT, the number of complex multiplications is 5120 while, in the DFT, there are 1024^2 multiplications, giving a 200-fold saving in complexity. There is a reduced 100-fold saving in the number of addition operations. Generally short transforms, $N = 16$, thus use the DFT, while large transforms, $N = 1024$, always use the FFT.

Self assessment question 10.4: What is the percentage saving in computation load when replacing 32, 256, 1024 and 8192-point DFTs with FFT processing?

10.3.2 Decimation-in-frequency algorithm

If the data are divided again into two sub-sequences then the DFT operations can be reconfigured for a decimation-in-frequency (DIF) algorithm. If the input data are accessed in bit reversed sequence order and split again into the even $N/2$ data points and the odd $N/2$ data points, etc., the resulting flowgraph for the DIF algorithm is very similar to Figure 10.5.

The fundamental difference between the DIT and DIF algorithms is that in the DIT algorithm, the input twiddle operation is incorporated on one of the butterfly inputs and hence the weighted values appears on *both outputs* from the butterfly operation. With the DIF algorithm, twiddles are incorporated on *one* of the butterfly outputs and hence the weighted values only appear on one of the two butterfly outputs. This is the way to distinguish between DIT and DIF, and *not* on the ordering of the input or output data samples. Figure 10.5 can thus be mapped directly to a DIF algorithm with bit reversed inputs by replacing the DIT input twiddle factors with appropriate DIF output twiddle factors. Now the W_8^0 to W_8^3 twiddles are incorporated into the first pass while only the $W_8^0 = 1$ twiddles are used on the final (third pass) calculation. The inputs in the DIF algorithm can again occur in natural order with the outputs in bit reversed order, i.e. $X(0)$, $X(4)$, $X(2)$, $\cdots$, etc., or both inputs and outputs can be in natural order [Lynn and Fuerst].

10.4 Implementation considerations

With the DIT algorithm, the multiplies are redundant on the first pass as $W^0 = 1$, and they grow to include all the Nth roots of unity on the last pass. This redundancy on the first pass allows the DIT data to be weighted with the window functions of section 9.3 when performing the spectrum analysis operation, which is a major practical attraction for the DIT over the DIF implementation!

10.4.1 Complex multiplier hardware

Consider the multiplication of the two complex numbers:

$$(a + j\,b)(c + j\,d) = (a\,c - b\,d) + j\,(a\,d + b\,c) \tag{10.15}$$

For fastest implementation this requires four real number multipliers. Experience shows that 16- and 24-bit fixed-point arithmetic provides adequate dynamic range (section 7.6.2) for many DFT based signal processing applications. Applications such as Kalman filtering, RLS adaptive algorithms and other sophisticated estimation operations normally require floating-point operation. The full dynamic range potential of a 16-bit fixed-point FFT processor (some 96 dB) will only be exploitable if a suitable data window is used on the time-domain input samples.

10.4.2 Alternative radix arithmetic approaches

Figure 10.4 showed the operations performed in the basic radix-2 arithmetic unit. The basic radix-4 (the 4-point DFT) unit takes four samples at a time and generates four outputs as shown in Figure 10.7. The operations were given previously in equation (10.13) and they are repeated here for naturally ordered data:

$$\begin{bmatrix} X(0) \\ X(1) \\ X(2) \\ X(3) \end{bmatrix} = \begin{bmatrix} 1 & 1 & 1 & 1 \\ 1 & -j & -1 & j \\ 1 & -1 & 1 & -1 \\ 1 & j & -1 & -j \end{bmatrix} \begin{bmatrix} x(0) \\ x(1) \\ x(2) \\ x(3) \end{bmatrix} \tag{10.16}$$

The advantage shared by radix-2 and radix-4 arithmetic units is that neither of them requires a complex multiplier. In the case of the radix-2 'butterfly', multiplication is by $+1$ or -1, as illustrated by the 2-point DFT matrix of equation (10.14) and Figure 10.4. In the radix-4 case, multiplication is by $+1, -1$ or $+j, -j$.

Self assessment question 10.5: Show how a radix-4 processor element can be designed without requiring any hardware multipliers by interchanging real and imaginary sample values to implement multiplication by $\pm j$. Represent the 4-sample input vector as $a + jb$, $c + jd$ etc.

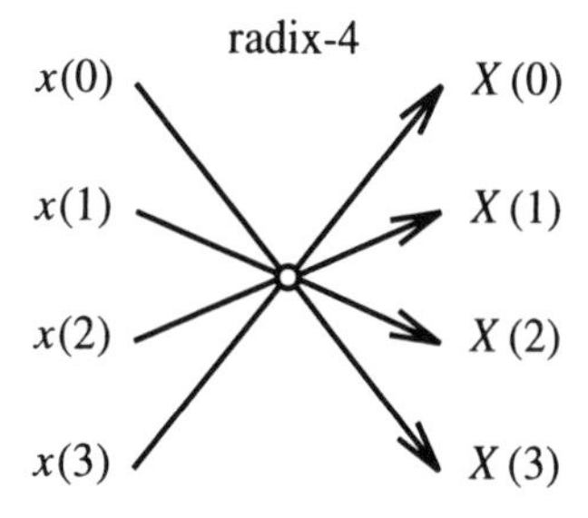

Figure 10.7 *Four-point transform operations.*

A radix-4 implementation is thus normally twice as fast as a radix-2 implementation, given the same hardware. This is because, in a radix-4 implementation, there are only half the number of memory accesses that there are in radix-2, since the radix-4 transform is achieved in half the number of passes ($\log_4(N)$ as opposed to $\log_2(N)$). Unfortunately, now only transform sizes which are powers of 4 can be accommodated. If, for instance, an 8-point or 32-point transform size is required, then a radix-2 or a mixed-radix solution must be employed.

Self assessment question 10.6: Repeat SAQ 10.1 for a radix-4 FFT to find the new number of passes.

This DFT partitioning of the 8-point transform into the product of smaller transforms applies to the factorisation of any N-point DFT transform where N is a non-prime number. Thus, for example, a 15-point transform can be factored as the product of 5-point and 3-point transforms in either order of occurrence. This factorisation operation has resulted in a rich field of research on DFT variants when N is not a power of 2 following work by Winograd, studies on Mersenne primes etc. [see Brigham for details]. Many of these algorithms have been investigated as alternatives to the radix-2 FFT to reduce particularly the number of computationally intensive multiplication operations. These are now less significant as, with the ready availability of integrated DSP chips with >50 MHz MAC processor rates, the *total* number of operations is now the more important parameter. Also code production time and subsequent code maintenance for an alternative DFT algorithm are no longer attractive, compared with the straightforward use of radix-2, radix-4 and split-radix FFT algorithms.

10.4.3 Real valued data

When the DFT is calculated for a real data sequence then the imaginary input sample values are all set to zero. The DFT provides a vector of complex output samples whose values were previously shown in equations (9.14) and (9.15). Figure 10.8(a) shows how the power spectrum $X(k)$ is obtained from the real input data. Thus the first $N/2 - 1$ values are the complex conjugate of the values from $N/2$ to $N - 1$. If the DFT output is input to a modulus function (to obtain the power spectrum) then the first $N/2 - 1$ sample values repeat in the second half of the DFT outputs, as for the aliasing on real sampled signals, see Figures 4.8 to 4.11.

When the DFT is calculated for a complex input data sequence then both the real and imaginary input DFT sample values are filled with valid data (Figure 10.8(b)). The DFT again provides a vector of complex output samples but here the complex processing recognises the phase in the input signal, and positive and negative phasors at the same angular frequency occupy distinct positions in the output data vector. Thus there is no longer a repeat in DFT outputs, as in the real valued case. As we have doubled the input data content by moving to complex input data, the output data content also doubles.

This capability can be used to compute simultaneously two real DFTs in a single complex DFT processor. For two real valued data sequences or vectors, $a(n)$ and $b(n)$, each of length N-samples, then one complex N-point DFT/FFT can calculate the two real valued transforms.

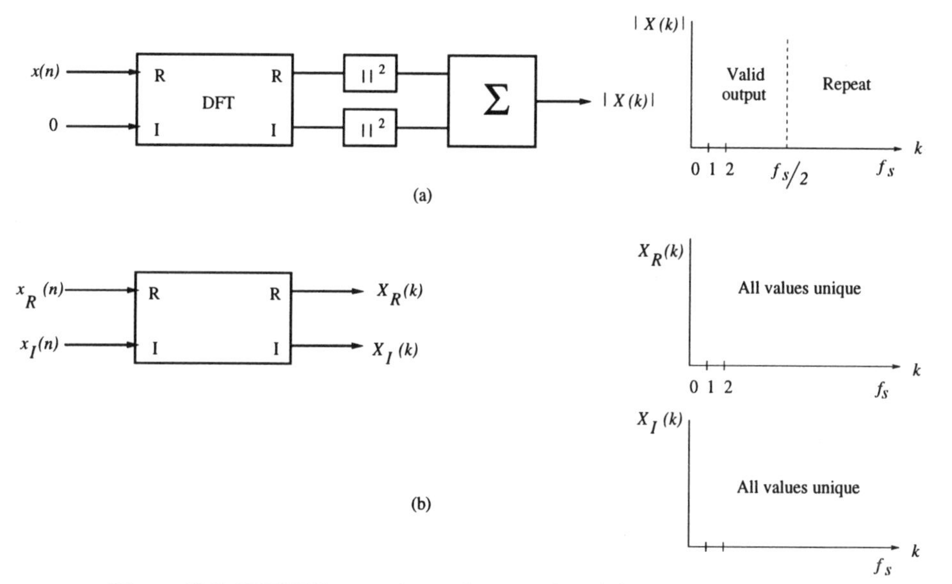

Figure 10.8 *DFT/FFT processing with: (a) real; and (b) complex input data.*

Given the signal $x(n) = a(n) + jb(n)$, where $a(n)$ and $b(n)$ are the individual real signals, the DFT of $x(n)$ is given by equation (9.12):

$$X(k) = \sum_{n=0}^{N-1} [a(n) + jb(n)] \exp\left(-\frac{jnk2\pi}{N}\right) \tag{10.17}$$

Recall that from equation (9.14(b)):

$$\exp\left(j\frac{2\pi n(N-k)}{N}\right) = \exp(j2\pi n)\exp\left(-\frac{jnk2\pi}{N}\right)$$
$$= \exp\left(-\frac{j2\pi nk}{N}\right)$$

since $\exp(j2\pi n) = 1$ for any value of n. Therefore:

$$X^*(N-k) = \sum_{n=0}^{N-1} [a(n) - jb(n)] \exp\left(\frac{j2\pi n(N-k)}{N}\right)$$
$$= \sum_{n=0}^{N-1} [a(n) - jb(n)] \exp\left(\frac{-jnk2\pi}{N}\right) \tag{10.18}$$

where X^* represents the complex conjugate of X. Adding $X(k)$ to $X^*(N-k)$, i.e. equation (10.17) to (10.18), gives:

$$X(k) + X^*(N-k) = 2\sum_{n=0}^{N-1} a(n) \exp\left(\frac{-j2\pi nk}{N}\right)$$

$$= 2A(k)$$

Therefore the Fourier transform of the real sequence $a(n)$ is given by:

$$A(k) = \frac{1}{2}[X(k) + X^*(N-k)] \qquad (10.19)$$

Similarly, subtracting $X^*(N-k)$ in equation (10.17) from $X(k)$ in equation (10.18) yields:

$$X(k) - X^*(N-k) = 2j \sum_{n=0}^{N-1} b(n) \exp\left(\frac{-j2\pi nk}{N}\right)$$

Therefore:

$$B(k) = \frac{1}{2j}[X(k) - X^*(N-k)] \qquad (10.20)$$

Therefore to compute the frequency spectra of each signal over the positive frequency range from DC to $f_s/2$ we use the complex DFT result as follows.

$X(k)$ is given directly as the first $N/2-1$ computed points. $X^*(N-k)$ is the complex conjugate of the computed points from $N/2$ to $N-1$. Note also that the computed output values are themselves complex, i.e.:

$$X(k) = X_R(k) + jX_I(k)$$

$$X(N-k) = X_R(N-k) + jX_I(N-k)$$

Hence from equation (10.19):

$$A(k) = \frac{1}{2}[X_R(k) + jX_I(k) + X_R(N-k) - jX_I(N-k)] \qquad (10.21(a))$$

and from equation (10.20):

$$B(k) = \frac{1}{2j}[X_R(k) + jX_I(k) - X_R(N-k) + jX_I(N-k)]$$

$$= \frac{1}{2}[-jX_R(k) + X_I(k) + jX_R(N-k) + X_I(N-k)] \qquad (10.21(b))$$

These equations provide the algorithm to compute the spectra of two N-point real time-series data records in a single N-point complex FFT. The only penalty is that the spectra of the individual signals must be unscrambled after the FFT of the combined signals has been computed.

Self assessment question 10.7: Draw the signal flow block diagram for this processing operation.

As a further development if one has a *single* real valued N-point data sequence to analyse, then this can be calculated as two $N/2$-point real DFTs using a single complex $N/2$-point FFT with a recombination operation based on the earlier DIT operations. Note that these clever approaches usually involve more sophisticated coding to implement the required algorithm operations.

10.4.4 Inverse transforms

There exists a duality between the DIT and DIF FFT algorithms. The DIT uses input twiddles while the DIF uses output twiddles, but both can have the same algorithm interconnection structure of Figure 10.5 for bit reversed inputs and naturally ordered outputs. The duality exists because the DIF realisation can be obtained from the DIT, and vice versa, simply by interchanging the inputs and outputs and reversing the process.

Now the IDFT operation involves scaling by $1/N$ and introducing a change in the sign of the sampled W phasor index from the forward DFT. This allows a given algorithm to easily process either time-domain or frequency-domain data (equation (9.13)) with only minor modifications. Thus the basic Figure 10.5 structure can readily perform either an 8-point DIT FFT or an 8-point DIF IFFT.

EXAMPLE 10.3

Examine the effect of zero-padding a 64-sample data record in a 256-point FFT.

Solution

Figure 10.9(a) is the basic 64-point transform operation for a sinusoid with three complete cycles in the 64-sample input data record. The output thus falls exactly on bin 3. In Figure 10.9(b) the 64 data points have been zero-padded by a factor of 4, to apparently increase the length of the data record to 256 points. The result is a decrease in quantisation in the frequency domain, showing the complete sinc x pattern underlying Figure 10.9(a).

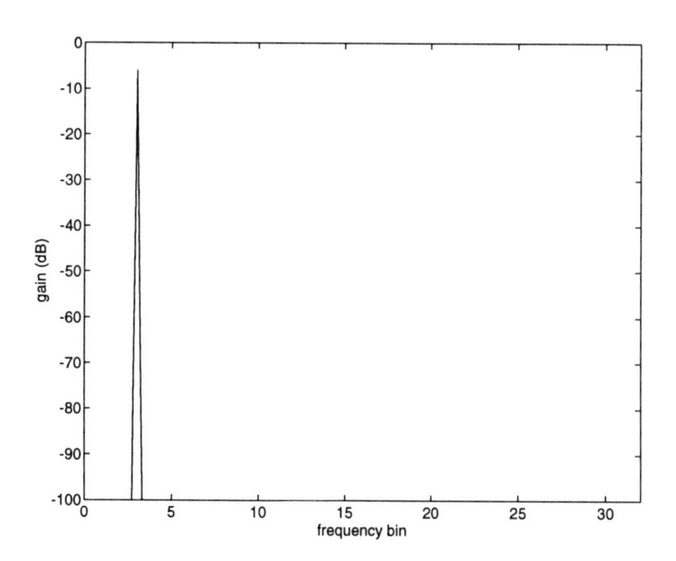

Figure 10.9(a) *FFT of 64-sample data record with an input sinusoid at* $3\omega_s/N = 3\omega_s/64$.

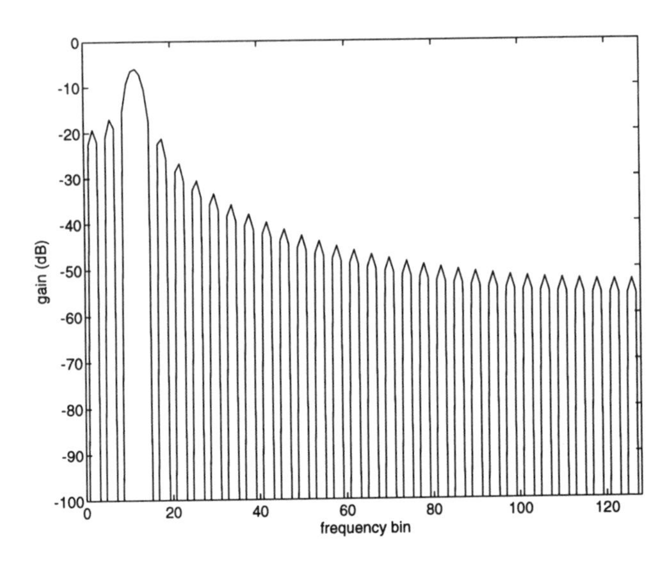

Figure 10.9(b) *FFT of 64-sample data record with an input sinusoid at* $3\omega_s/64$ *which has been zero-padded to 256 samples.*

Note that the resolution of the transform is unaltered. It is only possible to increase resolution by lengthening the duration of the signal record, i.e. by taking a longer input data sample set, i.e. by increasing N. This is shown in Figure 10.9(c) where the number of data samples N is increased to 256 and the resolution shows a decrease to the new $1/N\Delta t$ value. It is *not* possible to increase resolution by zero-padding. We can only achieve a decrease in the frequency-domain quantisation (equation (9.10)) by the use of an increased transform size.

10.5 Applications

There are two major applications for the FFT processors. These are in spectral analysis (Chapter 9) and in fast convolution.

Many modern radar and communication systems *demand* the linear phase response of the FIR matched filter (section 6.6) in the receiver to preserve phase information on the received signal (Figure 6.4(b)). However the complexity constraints of N multiplications and accumulations per input sample become a severe penalty in high bandwidth systems, particularly when N is greater than 256 points. The solution to this problem is to adopt a block processing approach where the convolution operation is replaced by multiplication in the frequency domain (Figure 10.10) [Brigham], as defined in section 2.3.

When convolution is implemented by taking the inverse DFT of the product of the DFTs of the individual time series, this only yields sensible results when the sampling period and length of both time-series are identical. Zero-padding can be used to

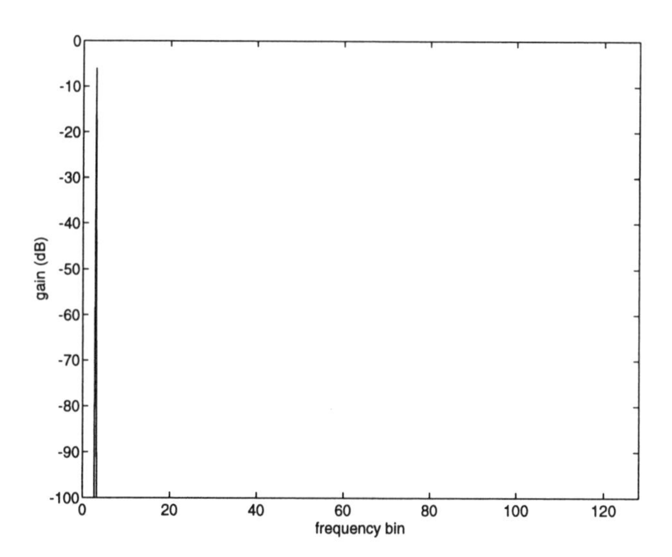

Figure 10.9(c) *FFT of full 256-sample data record with a sinusoid at* $3\omega_s/N = 3\omega_s/256$ *where output still occurs on bin 3, but now with the resolution of the full 256-point transform.*

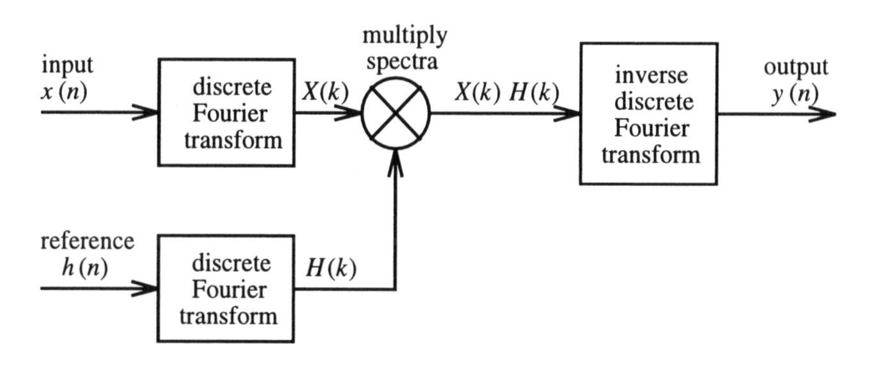

Figure 10.10 *FIR filter (convolution) realised by frequency-domain multiplication.*

equalise the time-series lengths if necessary. The DFT based convolution output is cyclical, repeating with the length of the constituent DFT processors. Thus the time-series data must be arranged in repeating closed loops, as shown in Figure 10.11.

Now the cyclic (or periodic) convolution of two N-element time series in an N-tap FIR filter yields an $2N-1$ element output (Figure 10.12(a)) of twice the time duration. Thus if two functions are convolved, enough zero-padding must be used in Figure 10.11 to ensure that one period of the result in Figure 10.12 is visible, and that it is isolated by sufficient zeros on either side.

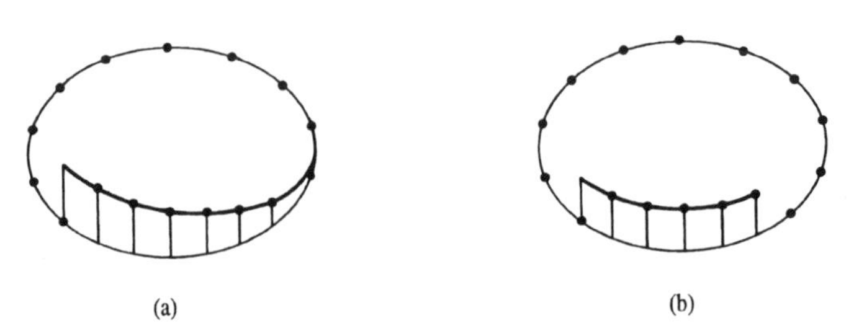

Figure 10.11 *Closed loop implementation of cyclic convolution.*

If insufficient leading and/or trailing zeros are present in the original time-series then the convolved sequences will have overlapping ends (Figure 10.12(b)), making normal interpretation difficult. To avoid this, the number of leading plus trailing zeros must be equal to the number of elements in the functions to be convolved from the first non-zero element to the last non-zero element. Thus in Figure 10.11, the waveforms to be convolved in (a) and (b) only occupy half the circumference of the circle, and in Figure 10.12(c) the convolved output is not distorted.

Thus for N-point data blocks this requires $2N$-point FFT processors but, with the arithmetic savings of Figure 10.6, it is still possible to accomplish an N-point convolution in much less than the N^2 multiplications required by the FIR filter approach of Figure 6.1. In this arrangement it is usual not to re-order the frequency data and hence a DIT forward FFT can most conveniently be followed by a DIF IFFT. Using this approach a 4096-point FIR filter or convolver, which operates with the frequency-domain fast convolution approach at the very impressive 40 Msample/s input data rate, was built in the late 1970s for a radar application! If the filter weights are

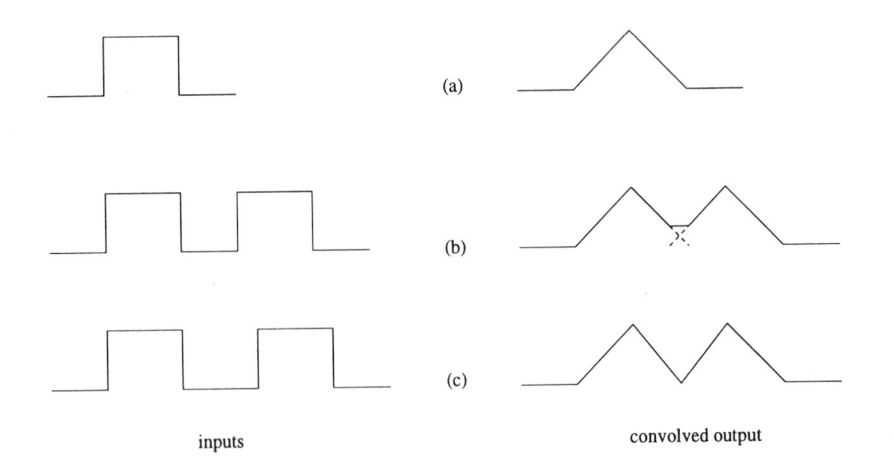

Figure 10.12 *Avoiding overlap in cyclic convolution: (a) linear convolution; (b) circular convolution with insufficient zero-padding; and (c) with just sufficient zero-padding.*

fixed then the $H(k)$ values in Figure 10.10 can be pre-computed and stored.

Self assessment question 10.8: How many complex multiplication operations are required to implement a 4096-point convolution operation via the FFT processing of Figure 10.10?

10.6 Summary

The DFT was derived in Chapter 9 and here the FFT has been developed from the DFT by exploiting many of the symmetries or repetitive operations within the DFT matrix calculation. Although the FFT involves an increase in the complexity of the control of the processor addressing, the saving in arithmetic operations is so impressive that it outweighs any of these detractions for large transform sizes. After deriving the FFT in its DIT form, the dual DIF form was introduced, the similarities between these were discussed, and their importance was highlighted for the practical implementation of FFT and IFFT operations.

Zero-padding was investigated as a means of taking a number of points which was not a power of two to the nearest power of 2 in order to implement a radix-2 FFT transform operation. Because zero-padding does not lengthen or extend the signal data record, there is no change in the resolution of the zero-padded DFT compared with the unpadded DFT. However the additional samples in the zero-padded data do provide more output samples in the display to give a better representation of the precise form of the overall DFT output. Finally, application of the FFT was explored for performing convolution of wide-band signals.

10.7 Problems

10.1 From the definition of a 4-point DFT, derive, from first principles, the radix-2 DIT algorithm and show the requisite flowgraph with the appropriate weights.

10.2 Derive the FFT decimation-in-time algorithm from the definition of the DFT for the specific case where $N = 8$. Indicate the twiddle values as complex number pairs expressed to 6 decimal digits. From the DIT flowchart for the 8-point transform (Figure 10.5), extend this into a 16-point DIT transform with W_{16} twiddle values.

10.3 For the DIT FFT of Figure 10.5, follow the input data samples $x(0), \cdots, x(7)$ through all the 2-point butterfly operations on each pass, and obtain the intermediate processed values for each pass in the FFT. Continue to obtain the final complete 8-point transformed output values $X(0)$ to $X(7)$, in terms of the products of twiddle values W_8^0 to W_8^3 with input sample values $x(0), \cdots, x(7)$, and confirm, by comparison with the DFT matrix, that your solution is correct.

10.4 For the radix-4 4-point DFT of Figure 10.7, with complex input sample values $x(0) = a + jb$, $x(1) = c + jd$, etc., derive the four complex transformed output values $X(0), \cdots, X(3)$.

10.5 The decimation-in-frequency FFT algorithm with bit reversed inputs and naturally ordered outputs flowgraph has the interconnection form shown in Figure 10.13. Now, by following the input data samples through all the 2-point butterfly operations on each pass, calculate the intermediate processed values on each pass to obtain the final complete 8-point transformed output values $X(0)$ to $X(7)$, in terms of the DIF twiddle values W_8^a to W_8^l.

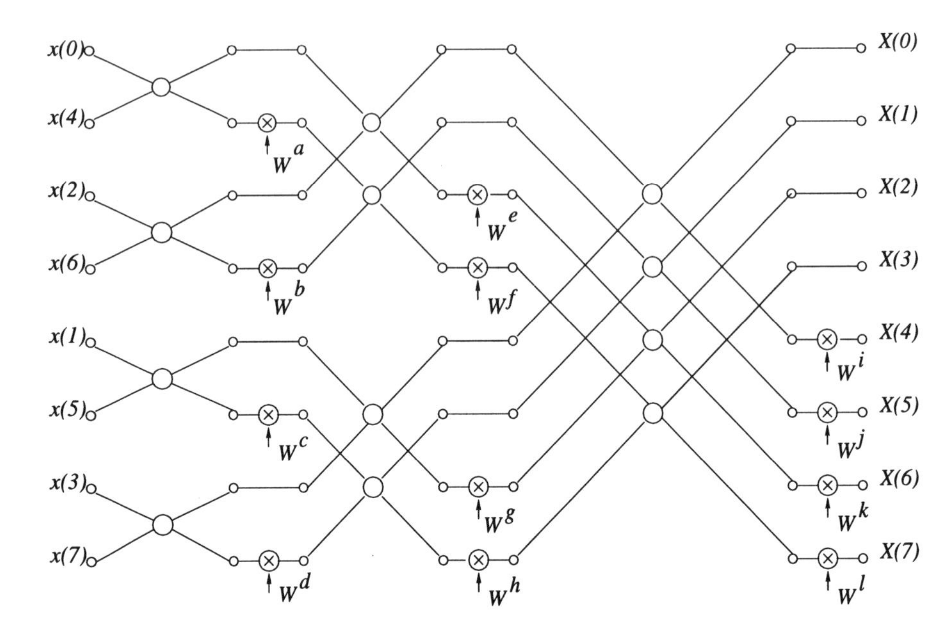

Figure 10.13 *Flowgraph for an 8-point radix-2 in-place decimation-in-frequency (DIF) FFT operation with the twiddle factors not fully identified.*

Use these equations to evaluate and obtain the actual values for the DIF twiddles W_8^a to W_8^l by substitution back into the DFT matrix of equation (10.2).

CHAPTER 11

Multirate signal processing

11.1. Introduction

Multirate systems differ from single rate processing systems in that the sample rate is altered at different places within the system. The motivation for this is to achieve the most appropriate value for the sampling rate, commensurate with the signal bandwidth, to minimise the total number of processor operations. As the bandwidth of a signal varies following filtering, changes can be made in the sample rate from point-to-point within the processor provided the Nyquist criterion is always satisfied. The aim is to operate, at each point in the processor, at the lowest sampling rate possible, without introducing aliasing effects.

Variable sample-rate processing is widely used in the discrete Fourier transform (DFT) [Brigham], Figure 9.7. As the bandwidth of each of the N parallel channels in a DFT filterbank is only $1/N$th of the total input signal bandwidth, the output sample rate in each channel can be reduced by the factor N – without incurring any loss of information. This process is somewhat erroneously called 'decimation', even though this strictly implies that N must equal 10.

Over the last two decades there have been major advances in the design of these multirate systems. Multirate systems find application in communications, speech processing, spectrum analysis, radar system and antenna systems. This final chapter initially reviews the basic concepts of multirate signal processing and the later sections are devoted to a discussion of the major multirate DSP applications.

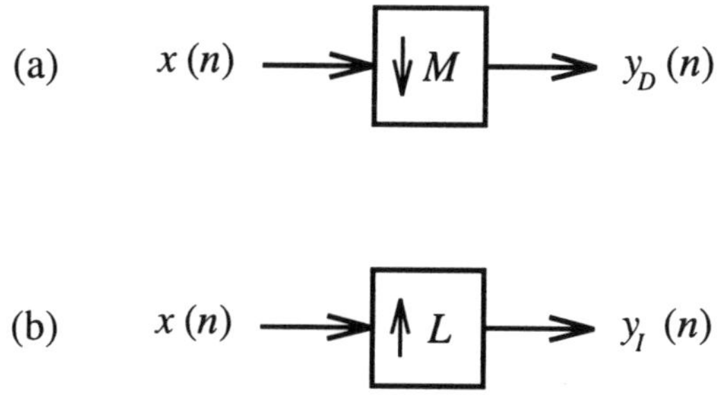

(a) $x(n) \rightarrow \boxed{\downarrow M} \rightarrow y_D(n)$

(b) $x(n) \rightarrow \boxed{\uparrow L} \rightarrow y_I(n)$

Figure 11.1 *Multirate processing: (a) M-fold downsampler; (b) L-fold upsampler.*

11.2 Decimation, interpolation, imaging and aliasing

Figure 11.1 shows building blocks for the basic decimation and interpolation components. Using the normal notation, the M-fold downsampler is characterised by the input–output relation:

$$y_D(n) = x(Mn) \tag{11.1}$$

which implies that the output at sample n is equal to the input at sample Mn. As a consequence, only the input samples with sample numbers equal to a multiple of M are retained at the output and the remainder are discarded.

EXAMPLE 11.1

Sampling-rate reduction by a factor of $M = 2$ is demonstrated in Figure 11.2 where every second sample is discarded. (This figure shows the input and output signals on the same absolute time scale as would be seen on an oscilloscope, rather than at a constant sample rate.)

Figure 11.3(a) shows a representation of the original input signal spectrum which is also conveniently drawn here with the same triangular shape as in Figure 4.8. The solid line component represents the original version of $X(\omega)$, comprising $x(n)$ in Figure 11.2. The dashed components are *some* of the aliasing terms. Downsampling by 2 (Figure 11.2(b)) causes the output sample rate to be halved and hence separation between the input signal spectrum and the alias component centred on the sampling frequency is also halved (Figure 11.3(b)). The reduced output sample rate of $f_{s(input)}/2$ thus moves the aliased frequency-domain components to half the frequency of their pre-downsampled position. The sample repeat at ω_s or 2π after downsampling is moved to half of its pre-downsampled value in Figure 11.3(b). In Figure 11.3(c) this separation

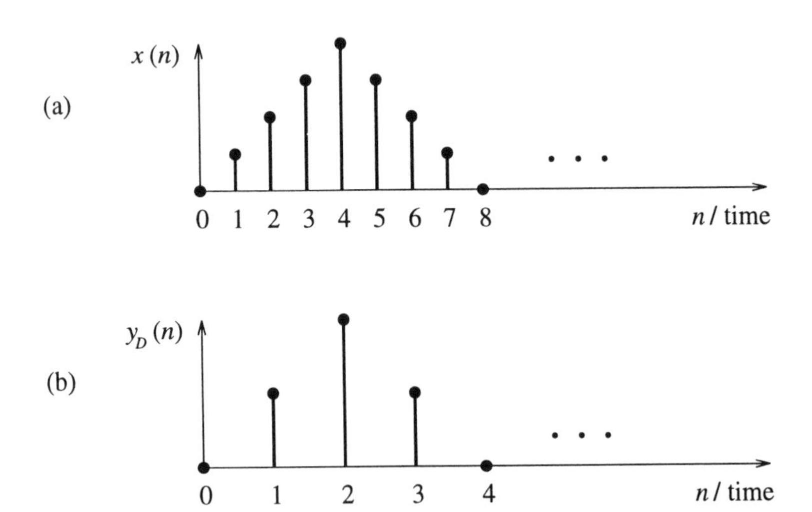

Figure 11.2 *Time-domain input and output signals after downsampling by 2.*

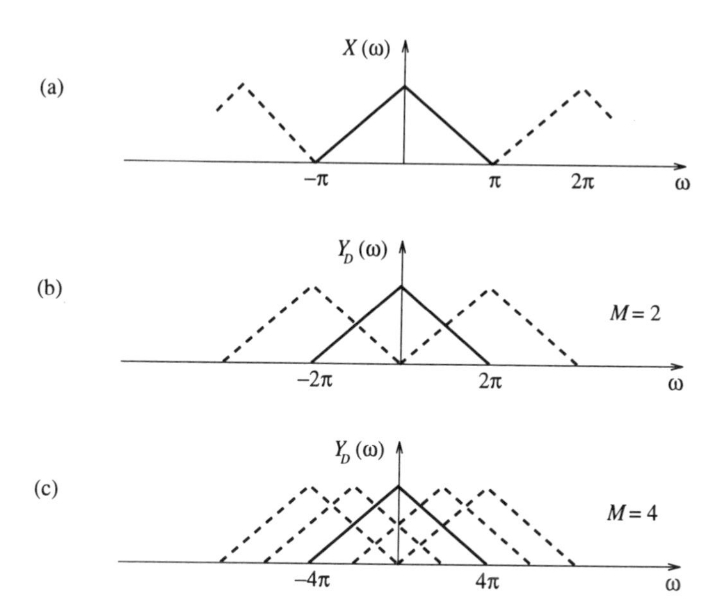

Figure 11.3 *Frequency-domain representation of a signal after downsampling by M=2 and M=4.*

has been quartered and the alias components centred on twice and three times the sampling frequency also overlap the basic information spectrum (solid curve). In the general downsampling by M operation, the $M-1$ uniformly shifted components will thus overlap the input signal spectrum. Figures 11.3(b) and (c) show this overlapping effect for decimation by the specific values of $M = 2$ and 4.

However if $X(\omega)$ is bandlimited to $|\omega| < \pi/M$, prior to the downsampling operation, there will be no actual overlap of the aliased signal components with the solid input signal spectrum shown in Figure 11.3. Overlaps are shown in Figure 11.3 because the signal has not been sufficiently bandlimited prior to performing the downsampling operation. The provision of these filtering operations to achieve the necessary bandlimiting of the input signal will be shown later in Figure 11.6.

Self assessment question 11.1: For a signal sampled at 10 kHz only comprising significant components from 0 to 1.5 kHz, what integer decimator rate can be used in this application to reduce the sample rate?

The z-transform of $y_D(n)$ changes to include the effect of the spectral aliases and is given as:

$$Y_D(z) = \frac{1}{M} \sum_{k=0}^{M-1} X(z^{1/M} W_M^k) \tag{11.2}$$

where again $W_M^k = \exp(-j2\pi k/M)$, as in Chapter 9.

In comparison, the L-fold upsampler (Figure 11.1(b)) generates an output $y_I(n)$ by inserting $L-1$ zero-valued samples between adjacent samples of $x(n)$.

EXAMPLE 11.2

Upsampling is demonstrated in Figure 11.4, for $L=2$, which increases the effective output sample rate. With the lower input sample rate shown in Figure 11.4, the input spectrum $X(\omega)$ repeats at multiples of the input sample rate, as shown in Figure 11.5(a). As the output is upsampled L times, the output sample rate ($Y_I(\omega)=2\pi$) now falls on to one of the repeated spectra at the frequency corresponding to L times the input sample rate, as shown in Figure 11.5(b). Thus, within the upsampled output Nyquist bandwidth, we have the original input spectrum plus the images arising directly from

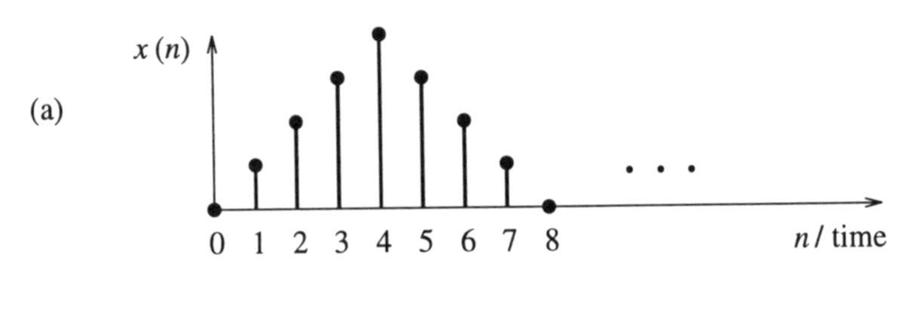

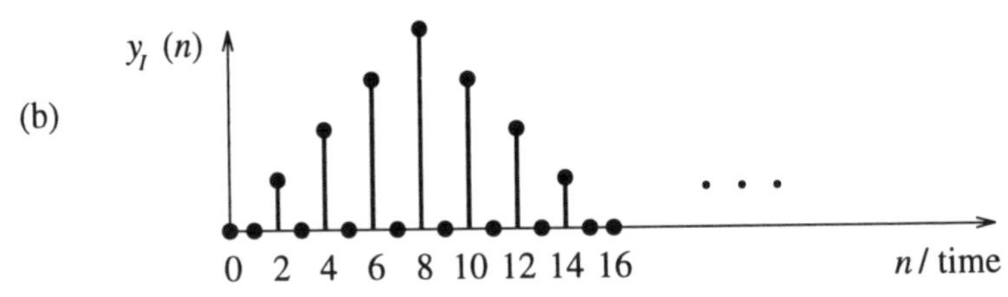

Figure 11.4 *Time-domain input and output signals after upsampling by L=2.*

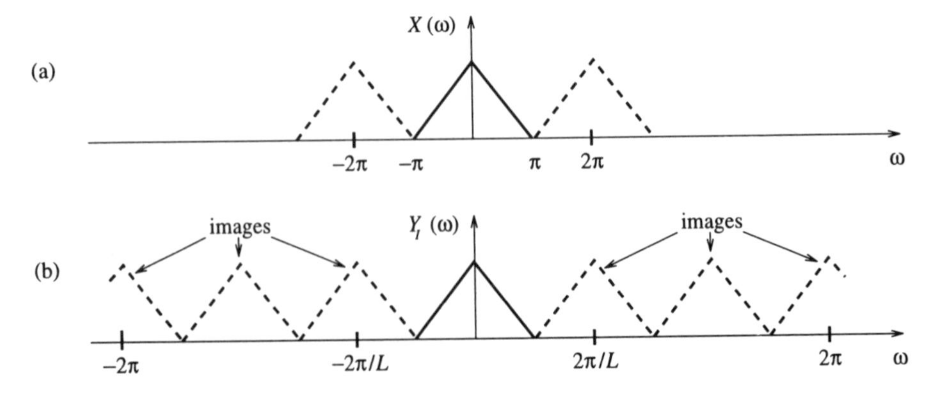

Figure 11.5 *Frequency-domain representation of a signal before and after interpolation by L=3.*

the upsampling operation.

The operation of the upsampler output $y_I(r)$, as shown in Figure 11.4, can thus be given as:

$$y_I(r) = \begin{cases} x(n), & \text{if } r = Ln \\ 0, & \text{otherwise} \end{cases} \tag{11.3(a)}$$

The z-transform of the upsampler output $y_I(r)$ is then given as:

$$Y_I(z) = X(z^L) \tag{11.3(b)}$$

This means that, within the sample rate of the processor, $Y_I(\omega)$ is an L-fold repeated version of $X(\omega)$, as demonstrated in Figure 11.5(b), with the faster output sample rate. Here we have simply increased the sample rate on the original signal spectrum, $X(\omega)$, to obtain $Y_I(\omega)$. The production of multiple copies of the basic spectrum, by the upsampler in Figures 11.4 and 11.5, is called the imaging effect.

The difference between aliasing and imaging is important to note. Aliasing can cause loss of information because of possible overlap of the (decimated) shifted versions of $X(\omega)$. On the other hand, as no time-domain samples are lost, imaging does *not* lead to any fundamental loss of information in the interpolation operation. The decimator and interpolator are both examples of linear systems, but they are not time invariant, hence they do not fall into the class of LTI systems.

11.2.1 Decimation

In a practical system, the actual signal bandwidth may be much smaller than half the sample rate and thus decimation can take place to optimise the efficiency. The downsampler must be preceded by a bandlimiting filter $H(z)$ whose purpose is to pass the required signal terms and, at the same time, remove any other signal components, to avoid the aliasing in Figure 11.3. For example, a low-pass filter stopband edge $\omega = \pi/M$ can serve as such a filter for an M-fold decimator. This cascade, shown in Figure 11.6(a), is commonly called a *decimation* filter. Figure 11.6 shows decimation by $M = 4$ where the input signal, $x(n)$, in Figure 11.6(b) has to be limited by $H(z)$ to $\pi/4$ (Figure 11.6(c)) to give the reduced bandwidth signal $x_1(n)$ (Figure 11.6(d)), before the decimation operation can take place. After decimation, the aliased components then fill in the parts of the spectrum (Figure 11.6(e)) which were previously suppressed by the decimation filter. Note that if the filter $H(z)$ had been omitted, we would experience the aliased overlap shown previously in Figures 11.3(b) and (c).

The bandlimiting filter in Figure 11.6(a) will normally be an FIR design, to achieve a linear phase characteristic. If this is so, the structure is very inefficient as the high sample rate input signal is processed and then $M-1$ of every M samples are discarded in the decimator! This inefficiency can be overcome by combining the decimator into the N-tap FIR filter as shown in Figure 11.6(f), where the FIR tap values, $h_0, \cdots, h_{N-1}$, are arranged for operation on separated input signal samples. Figure 11.6(f) shows the general decimation by M arrangement. This arrangement is more widely known as polyphase decomposition [Crochiere and Rabiner] and it is a well used technique to simplify the computational complexity of decimation and interpolation filterbanks.

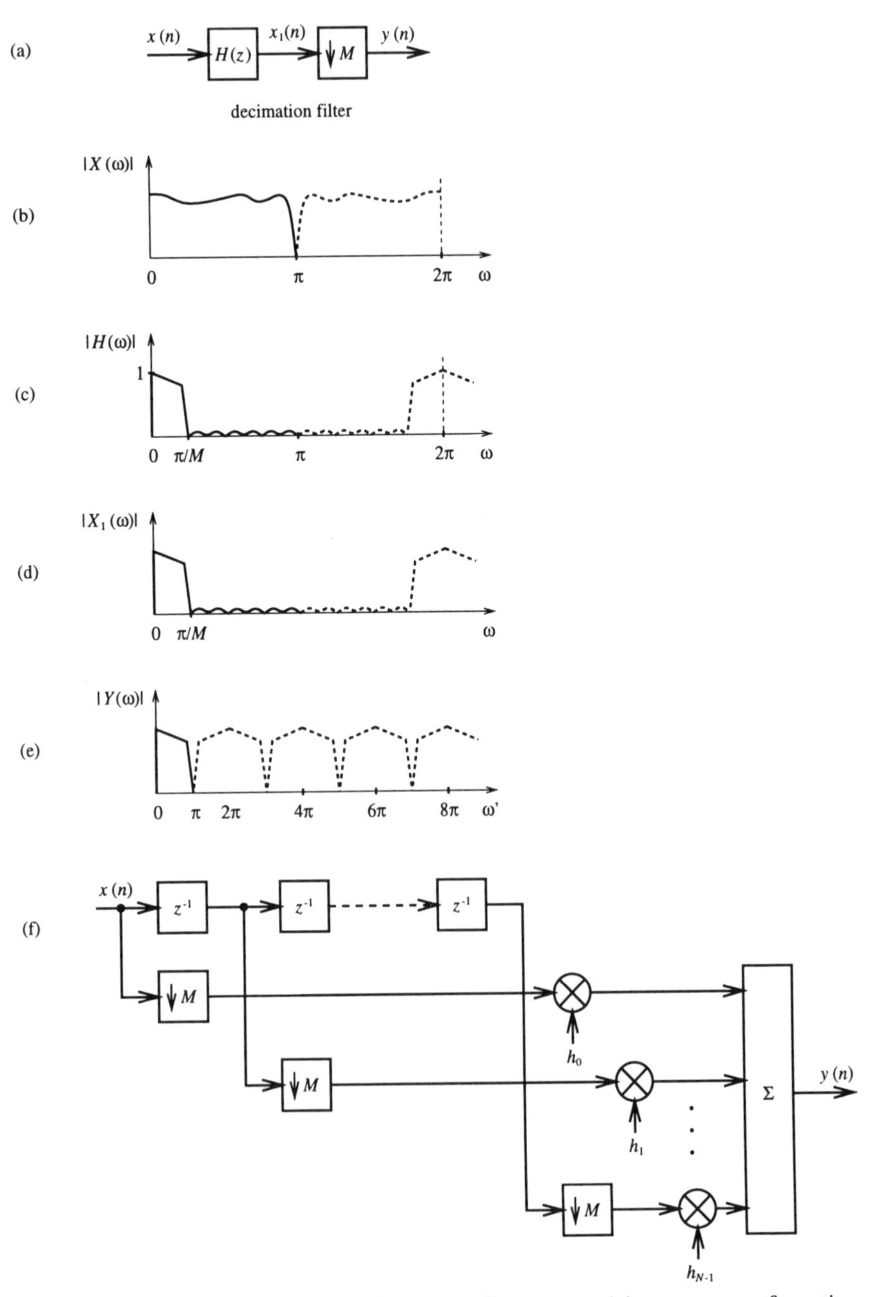

Figure 11.6 *Low-pass decimator and corresponding spectra: (a) processor configuration; (b) input spectrum; (c) decimation filter frequency response; (d) filtered signal; (e) decimated output signal; (f) decimator design which minimises the processing operations.*

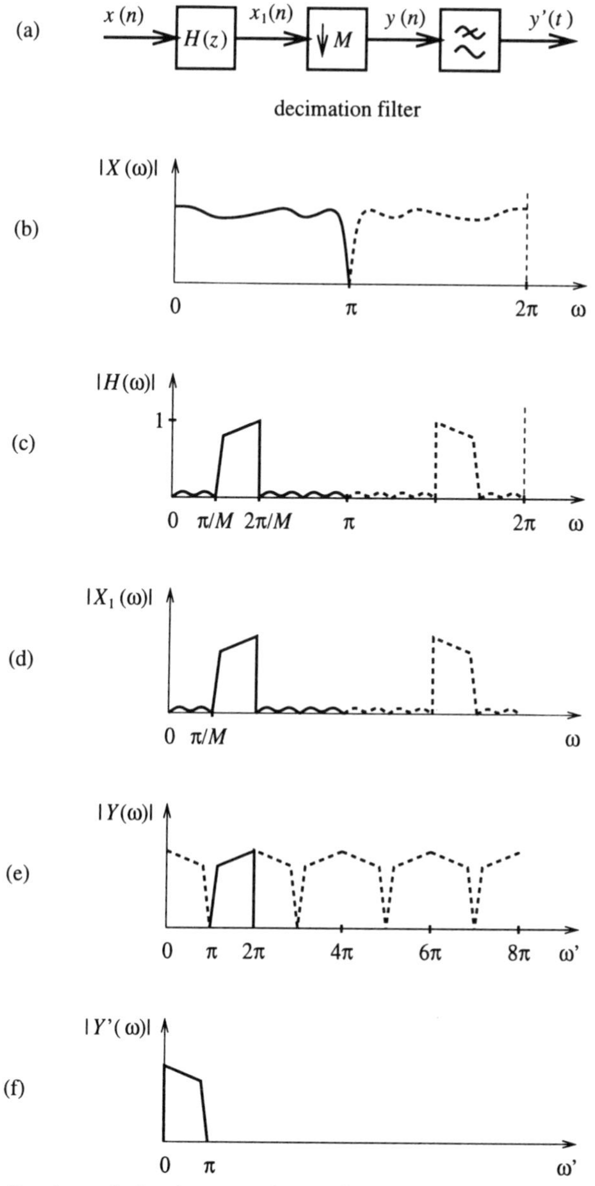

Figure 11.7 *Bandpass decimation operations and corresponding spectra: (a) processor configuration; (b) input spectrum; (c) decimation filter frequency response; (d) filtered signal; (e) decimated output signal; and (f) low-pass filtered output.*

The multiplications and summations are now occurring at the lower output sample rate so the total computation is reduced by the decimation factor M, greatly improving the overall efficiency, while still performing the same overall processing operation. The overall operation of this configuration is given by extending equation (6.1) as:

$$y(n) = \sum_{i=0}^{N-1} h(i)x(Mn - i) \tag{11.4}$$

In order for $x(n)$ to be recoverable it is not necessary for $X(\omega)$ to be restricted to $|\omega| < \pi/M$ by a *low-pass* operation. It is sufficient for the *total* bandwidth of $X(\omega)$ to be less than π/M. Thus a general bandpass signal with energy in the region $a \leq \omega \leq a + \pi/M$ can be decimated by M without creating overlap of the alias components, and the decimated signal in general is a full-band signal. There are further restrictions on the precise value of a which must equal $n\pi/M$ for the above equalities to be valid and the sampling to be accomplished by straightforward operations.

In the case of the bandpass decimator in Figure 11.7, $H(z)$ is shown as a bandpass filter with a passband response from π/M to $2\pi/M$ (Figure 11.7(c)), instead of the low-pass response from $-\pi/M$ to $+\pi/M$ in Figure 11.6(c), so that the *bandwidth* of the filtered signal $x_1(n)$ is identical to that shown in Figure 11.6(c). Thus we can again employ the same decimation by M to reduce the output sample rate to $1/M$ of the input rate, as shown in Figure 11.7 and achieve the full-band signal of Figure 11.7(e). This also folds, or aliases, the reduced information in the filtered spectrum to baseband (i.e. close to the zero frequency or DC term) to be consistent with the new lower output sample rate (Figure 11.7(f)). However, as the bandpass filter $H(z)$ has already removed the energy at this baseband frequency (Figure 11.7(c)), there is no problem with aliased responses.

Here the output can now be recovered with a lowpass filter following the decimation operation in Figure 11.7(a). Thus the frequency shifting to baseband has been accomplished by the decimation operation. Recall that this operation also occurred previously in the DFT chapter (Figure 9.7), to give baseband outputs on each of the bandpass DFT channels, $k = 1$ to $k = N - 1$.

11.2.2 Bandpass sampling

Nyquist's sampling theorem, if correctly interpreted, can be applied to any process be it a bandpass or baseband process. Normally a baseband signal has a defined –3 dB bandwidth which is only half the bandwidth, B, of a passband signal (Figure 11.8), because the passband signal is *doublesided.* Thus the direct application of Nyquist's theorem gives the minimum sampling rate as:

$$f_s \geq 2\,\frac{B}{2} = B \text{ Hz} \tag{11.5}$$

Equation (11.5) appears to be incorrect in that it suggests all the information present in a passband signal with bandwidth B Hz is preserved in only half the number of samples expected. However, for a complex baseband signal, there will be two real sample values for each sampling instant, i.e. an in-phase or real sample, and the quadrature or imaginary sample. The total number of real numbers characterising the passband signal is therefore the same.

In some applications we require to sample a bandpass signal in which the centre frequency is many times the signal bandwidth. Whilst we can sample this signal at twice the highest frequency component, $2f_c + B$ in Figure 11.8(a), and reconstruct the signal by low-pass filtering, it is normally possible to reconstruct the original signal by

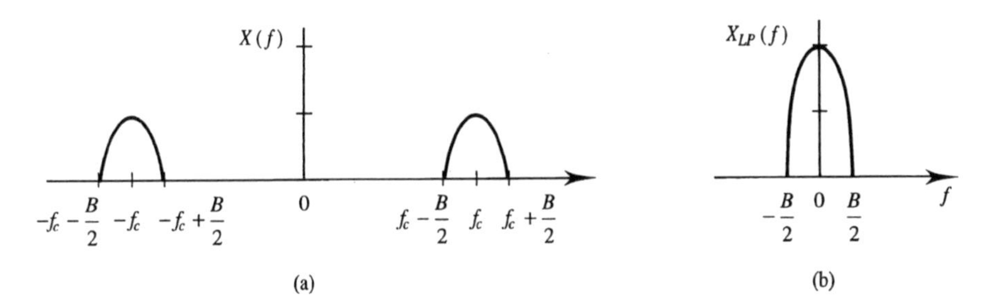

Figure 11.8 *(a) Passband and (b) equivalent baseband frequency spectra. (Source, Glover and Grant, 1998, see Bibliography, reproduced with the permission of Prentice-Hall.)*

sampling at a *much* lower rate and still retain all the information. Thus there exists one or more frequency *bands* in which the sampling frequency should lie. When sampling a bandpass signal there is generally an upper limit as well as a lower limit. The bandpass sampling criterion has been expressed as [Glover and Grant]:

A bandpass signal having no spectral components below f_L Hz or above f_H Hz is specified uniquely by its values at uniform intervals spaced $T_s = 1/f_s$ s apart provided that:

$$2B\left\{\frac{Q}{n}\right\} \le f_s \le 2B\left\{\frac{Q-1}{n-1}\right\} \tag{11.6}$$

where $B = f_H - f_L$, $Q = f_H/B$, n is a positive integer and $n \le Q$.

We next clarify the use of equation (11.6) and explain further the relationship between it and the normally used Nyquist (baseband) sampling criterion:

(i) If $Q = f_H/B$ is an integer then $n \le Q$ allows us to choose $n = Q$. In this case $f_s = 2B$ and the correct sampling frequency is exactly twice the signal bandwidth.

(ii) If $Q = f_H/B$ is not an integer then the lowest allowed sampling rate is given by choosing $n = int(Q)$ (i.e. the next lowest integer from Q). Lower values of n will still allow reconstruction of the original signal, the sampling rate will be unnecessarily high, but they provide for a wider allowed band of signal f_s.

(iii) If $Q < 2$ (i.e. $f_H < 2B$ or, equivalently, $f_L < B$) then $n \le Q$ means that $n = 1$. In this case:

$$2BQ \le f_s \le \infty \quad \text{(Hz)}$$

and since $BQ = f_H$ we have:

$$2f_H \le f_s \le \infty \quad \text{(Hz)}$$

This is a direct repeat of the Nyquist (baseband) sampling criterion.

The validity of the bandpass sampling criterion is most easily demonstrated using convolution (equation (6.1)) for the following special cases:

(i) When the spectrum of the bandpass signal $x(n)$ straddles nf_s, i.e. $X(f)$ straddles any of the lines in the spectrum of the sampling signal (Figure 11.9(a)), then

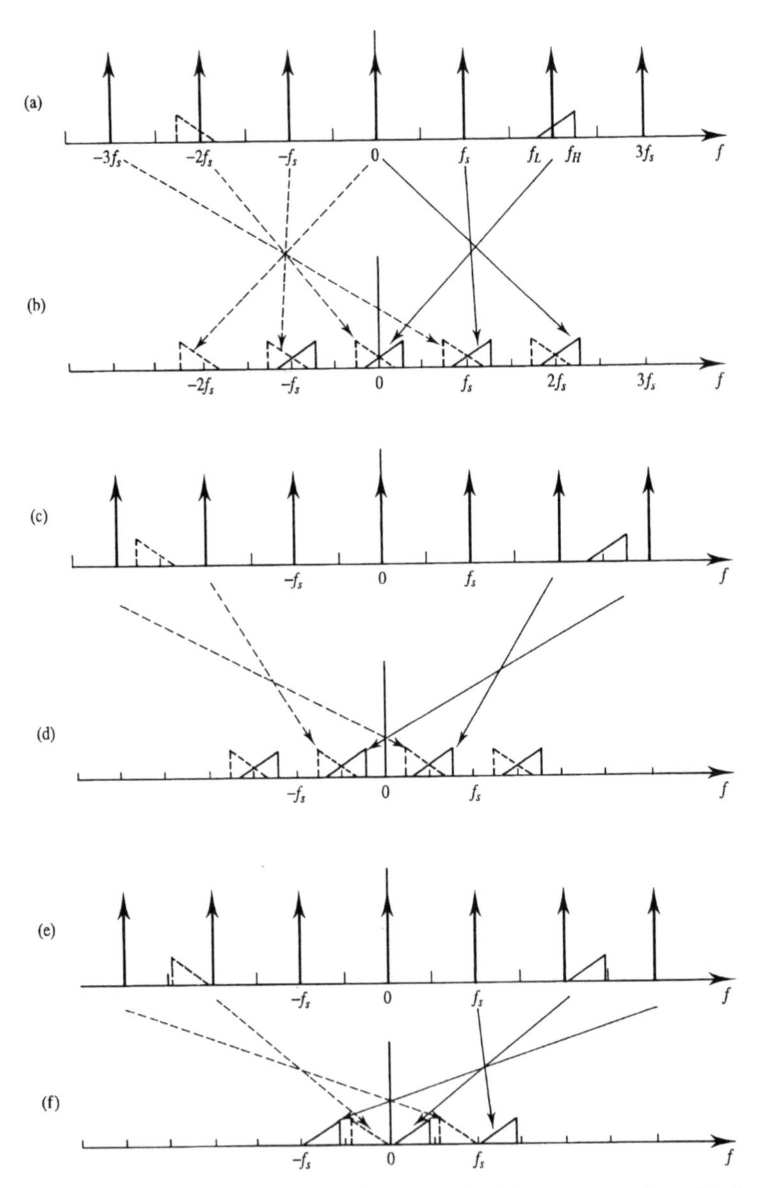

Figure 11.9 *Criteria for sampling of bandpass signals: (a) spectrum where $X(f)$ straddles $2 \times f_s$; (c) and straddles $2.5 \times f_s$; (b) & (d) overlapped spectra after bandpass sampling; (e) spectrum avoiding above straddling; and (f) recoverable baseband spectra. (Source: Glover and Grant, 1998, see Bibliography, reproduced with the permission of Prentice-Hall.)*

convolution results in interference between the positive and negative frequency spectral replicas (Figure 11.9(b)). (Figures 11.9 (a), (c) and (e) show the spectrum of a passband signal from f_L to f_H, with a much lower sample frequency f_s.)

(ii) When the spectrum of $x(n)$ straddles $(n + ½)f_s$, i.e. $X(f)$ straddles any odd integer multiple of $f_s/2$ (Figure 11.9(c)), then similar interference occurs (Figure 11.9(d)).

(iii) When the spectrum of $x(n)$ straddles neither nf_s nor $(n + ½)f_s$ (Figure 11.9(e)), then no interference between positive and negative frequency spectral replicas occurs (Figure 11.9(f)) and the spectrum can be recovered without aliasing.

If this is summarised then we have the correct conditions for sampling of bandpass signals when:

$$f_H \leq n \frac{f_s}{2} \quad \text{(Hz)} \tag{11.7}$$

$$f_L \geq (n-1) \frac{f_s}{2} \quad \text{(Hz)} \tag{11.8}$$

Using $f_L = f_H - B$ gives:

$$\frac{2}{n} f_H \leq f_s \leq \frac{2}{n-1} (f_H - B) \quad \text{(Hz)} \tag{11.9}$$

Defining $Q = f_H/B$ gives the bandpass sampling criterion of equation (11.6).

EXAMPLE 11.3

The following example, taken from Glover and Grant, illustrates the use and significance of equation (11.6). Select an appropriate minimum practical sample rate for the signal with centre frequency 9.5 kHz and bandwidth 1.0 kHz.

Solution

The highest and lowest frequency components in this signal are:

$$f_L = 9.0 \text{ kHz} \qquad f_H = 10.0 \text{ kHz}$$

Quotient Q is thus:

$$Q = \frac{f_H}{B} = \frac{10.0}{1.0} = 10.0$$

Applying the bandpass sampling criterion of equation (11.6):

$$2 \times 10^3 \left\{ \frac{10}{n} \right\} \leq f_s \leq 2 \times 10^3 \left\{ \frac{10-1}{n-1} \right\} \quad \text{(Hz)}$$

Since Q is an integer, the lowest allowed sampling rate is given by choosing $n = Q = 10$, i.e.:

$$2.0 \leq f_s \leq 2.0 \text{ (kHz)}$$

The significant point here is that there is zero tolerance in the sampling rate if distortion is to be completely avoided. If n is chosen to be less than its maximum value, e.g. $n = 9$, then:

$$2.222 \leq f_s \leq 2.250 \text{ (kHz)}$$

The sampling rate in this case would increase from 2 kHz to 2.236 ± 0.014 kHz, which is less attractive as the processor rate also increases. However there is now a permitted accuracy for the sampling clock of ± 0.63%.

Self assessment question 11.2: Repeat Example 11.3 with $n = 8$ and calculate the resulting percentage accuracy for the sampling clock frequency.

Self assessment question 11.3: Select an appropriate minimum sample rate for the signal with centre frequency 10.0 kHz and bandwidth 1.0 kHz.

Figure 11.10(a) shows how a bandpass signal is shifted to 0 Hz provided the sample rate ω_s is an integer sub-multiple of the signal centre frequency ω_0, i.e. $\omega_s = \omega_0/M$, as in Example 11.3. This operates provided the signal bandwidth $\Delta\omega$ is less than the decimator sample rate, ω_s. As the Nyquist criterion is not fully satisfied here with the single A/D converter stage, spectral folding occurs and the operation only functions for a symmetrical signal spectrum, such as that arising from AM or PSK modulation. One can also sample at a frequency above the highest signal component to move the bandpass signal close to 0 Hz.

Figure 11.10(b) shows how a complex asymmetrical signal, operating at the same centre frequency, ω_0, can be moved into a real (R) and imaginary (I) baseband representation by a complex (in-phase and quadrature) demodulation, to avoid the spectral inversion of the Figure 11.10(a) process. Thus Figure 11.10(b) follows the downconversion with the oscillator ω_c, by twin A/D converters operating at a sample rate, $\omega_s \geq \Delta\omega$, to digitise the signal of bandwidth $\Delta\omega$ while still satisfying the Nyquist criterion. The combined sample rate from the two A/D converters is $\omega_s + \omega_s = 2\omega_s$, so the Nyquist criterion is still satisfied. The baseband spectrum can be asymmetric (Figure 11.10(c)), as we are now holding a complex (i.e. with separate real and imaginary samples), rather than the purely real, signal representation of Figure 11.10(a). Although ω_s is ideally set at ω_0, any small variation or difference between these frequencies causes the output signal to be modulated by the difference frequency, and hence appear on both the R and I outputs.

11.2.3 Interpolation

Figure 11.11 shows the filtering operations associated with the upsampling operation. The filter, $H(z)$, which follows the upsampler (Figure 11.11(a)), is used to eliminate the previous images in Figure 11.5(b). Figure 11.11(c) shows the interpolated output $x_I(n)$ in the time and frequency domain. The cascade of upsampler and filter is, by convention, called the *interpolation* operation. Figure 11.11(d) shows the effect of a low-pass interpolation filter, which causes the frequency-domain images to be removed. When the output bandwidth is reduced, the time-domain interpolated zero samples now follow the narrow-band signal envelope. Figure 11.11(e) further shows the reduction in the aliased terms at the new output sample frequency when a zero-order-hold (ZOH) operation (section 4.4) follows the interpolation filter (Figure 11.11(a)), and the corresponding stepped time-domain output, as shown previously in Figures 4.2 and 4.4.

Just as in Figure 11.6(e), where we showed how the decimator computational efficiency could be improved, so also the interpolation filter can be simplified in a similar manner. Figure 11.11(f) shows a redesigned general interpolator and output filter where the filtering is performed first at the slow input sample rate, with the h_n coefficients, then the individual components are interpolated before the final high speed summation in the output delay line. In Figure 11.11(f), the input splitter is conveniently shown with the sum symbol. This is another example of the polyphase structure [Crochiere and Rabiner] which is widely used in multirate systems.

11.3 Applications of multirate systems

We now review a number of important applications of multirate filters and filter banks.

11.3.1 Transmultiplexers

One application is in the design of transmultiplexers (which are devices for conversion between frequency-division multiplexing (FDM) and time-division multiplexing (TDM) systems). These components translate between digital TDM and multiplexed analogue single-sideband modulated FDM communication systems. If we consider the TDM to be a FDM translation process then typically the input comprises 12 pulse code modulated (PCM) speech channels, each sampled at 8 kHz. These are then interpolated and each one is single-sideband (SSB) modulated on to separate carrier oscillators, which have 4 kHz spacings, to achieve the required 56 kHz to 112 kHz FDM spectrum. As the wide-band output requires a much higher sample rate than the input, the translator is fundamentally a multirate processor.

There is also a trend towards coded orthogonal FDM (COFDM) transmission for broadcast audio and video to mobile subscribers. Here a single high speed data stream is split into many (e.g. 1,000) parallel low data rate channels. This is done to minimise multipath propagation effects and it requires an IDFT (equation (9.13)) to generate the signal, which is realised with a DFT where the sampled complex phasors employ the opposite phase rotation from that of the forward DFT.

11.3.2 Analysis and synthesis filterbanks

A major application for multirate techniques is in filterbanks for spectrum analysis of a signal which might be used, for example, to achieve low bit-rate coding of the signal components. There are two basic types of filterbanks. The analysis bank is a set of contiguous filters $H_m(z)$ which splits a signal into M sub-band signals $x_m(n)$, as shown in Figure 11.12(a). Next, a synthesis bank (Figure 11.12(b)) consists of M synthesis filters $F_m(z)$, which combine the M signals $y_m(n)$ (possibly from an analysis bank) to form a reconstructed signal estimate $\hat{x}(n)$.

As the bandwidth in each filter is small compared with the input signal bandwidth, the overall processing in the analysis filterbank is similar to the processing in the DFT (Chapter 9). Remember that the DFT operated on a block of N data samples and is decimated by N on each channel output (Figure 9.7). The decimation operation which

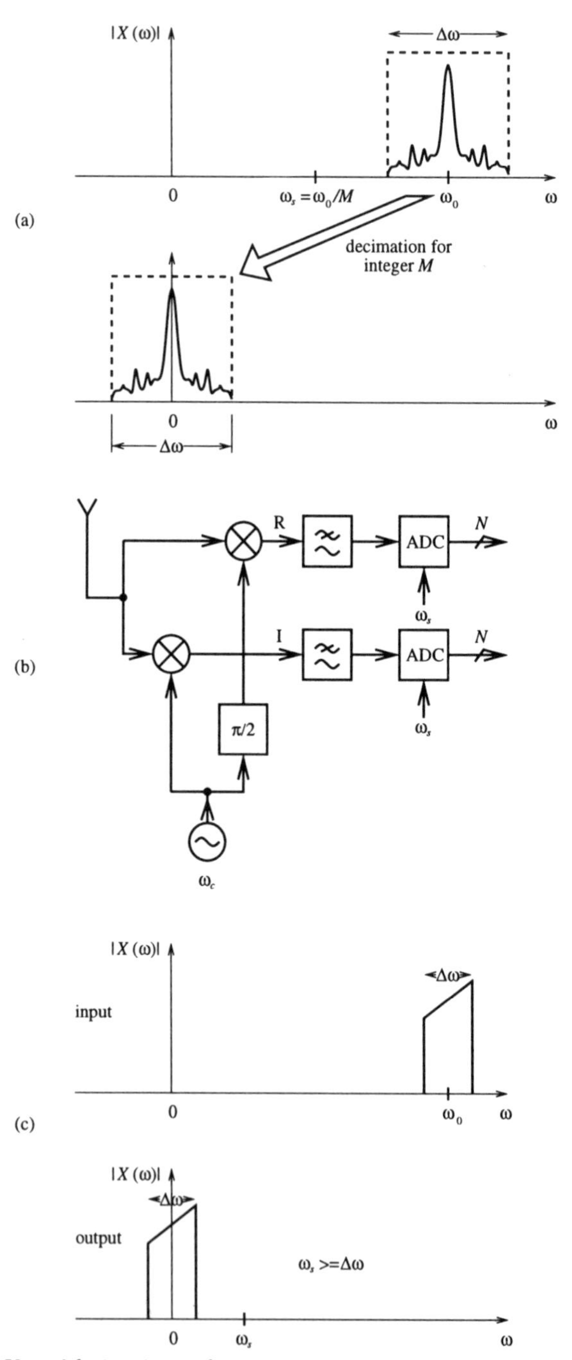

Figure 11.10 *Use of decimation to downconvert IF signals to baseband: (a) real sampling followed by decimation; (b) complex demodulation with in-phase and quadrature samplers; and (c) heterodyning to baseband with complex demodulation.*

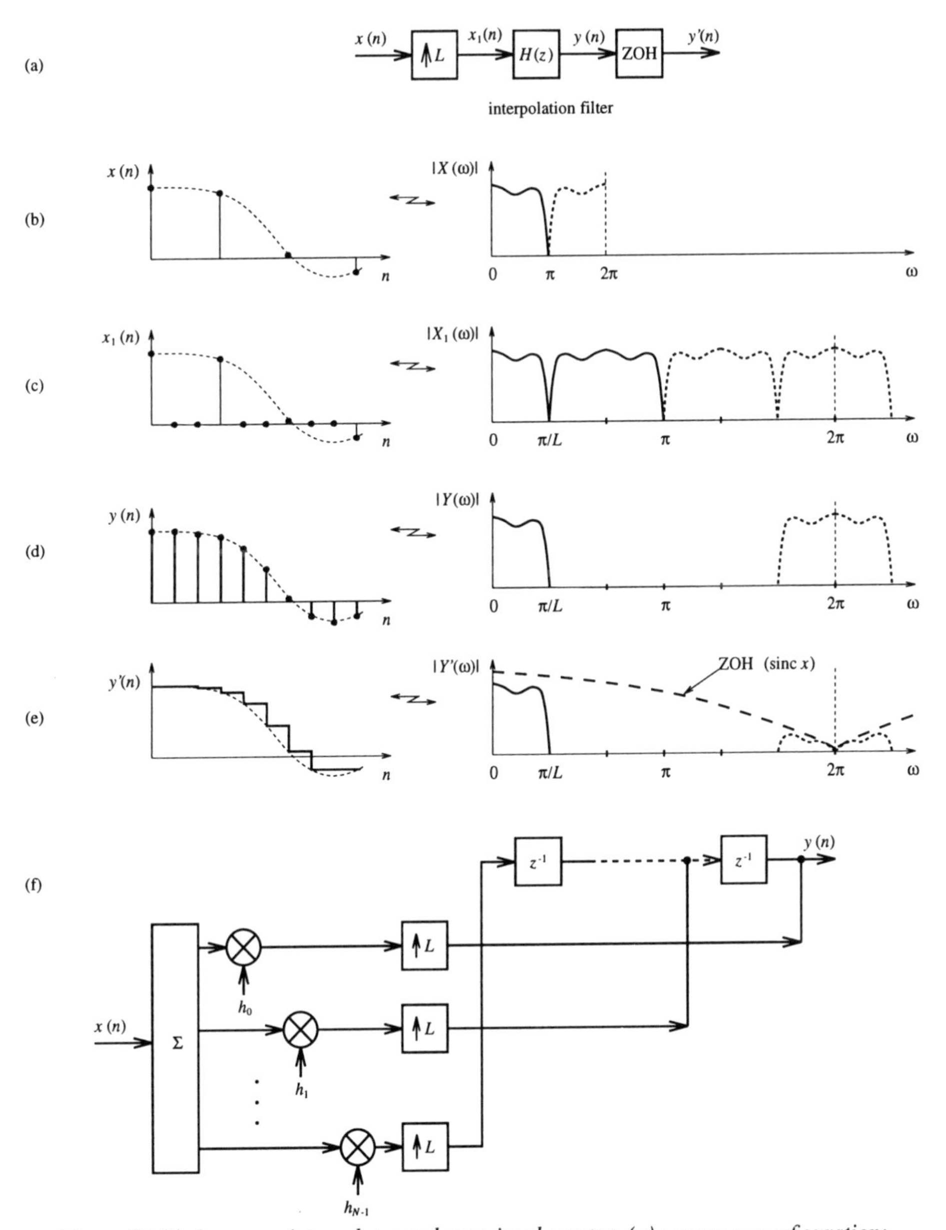

Figure 11.11 *Low-pass interpolator and associated spectra: (a) processor configuration; (b) input signal and spectrum; (c) upsampler output; (d) interpolated output; (e) output after zero-order-hold circuit; and (f) interpolator which minimises processing operations.*

follows the filters in Figure 9.7 and the analysis filters in Figure 11.12 not only reduces the sample rate in each of the channels but also 'folds' these individual narrow-band signals to baseband, as in the bandpass decimation of Figure 11.7. Thus a time-series of input signal samples, $x(0), x(1), \cdots, x(M-1)$, is 'transformed' or represented by the M parallel filterbank outputs $x_0(n)$ to $x_{M-1}(n)$, but the *total* number of samples is unchanged.

These analysers differ from the DFT operation in that there is more control over the response of the prototype filter, $H_0(z)$. In the filterbank approach, the frequency responses of $H_m(z)$ in Figure 11.12(a) are simply uniformly shifted versions of $H_0(\omega)$. The other filters $H_1(z)$ to $H_{M-1}(z)$ thus use the same prototype design as $H_0(z)$ but the FIR coefficients are subsequently modified by the complex exponential appropriate to their specific centre frequencies, see the following section.

The decimation and interpolation operations between the analysis and synthesis filterbanks are shown later (in Figure 11.20). The fact that the decimation operations in each channel move all the signals to baseband is not a problem, as the interpolators in the synthesis bank regenerate wide-band bandpass signals, as shown previously in Figure 11.5. Each synthesis filter then selects the components in the correct frequency band so that the final summation reconstitutes the appropriate signal estimate.

11.3.3 Filterbank design approaches

The concepts of decimation can be used to construct very efficient filterbanks when the component filters are all of equal bandwidth and the number of filters is a power of two. Figure 11.13 shows a tree approach to the design of an 8-channel filterbank. If the filters are FIR designs then the first or root stage filter pair splits the band into the upper or lower four channels. As this is a relatively wide-band filter it has a small number of FIR stages. As we progress down the tree, the sample rate reduces and the filters become narrower band but, usually, the percentage bandwidth is the same, as in the first or root stage, and the *same* prototype filter design can be *re-used* in each tree stage.

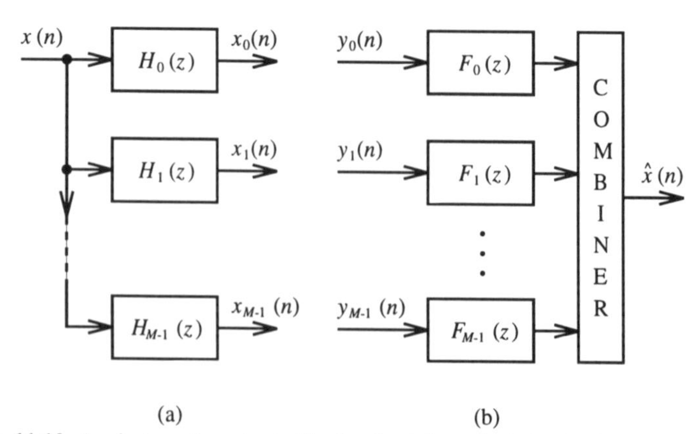

Figure 11.12 *Analysis and synthesis filterbanks: (a) analysis bank; (b) synthesis bank.*

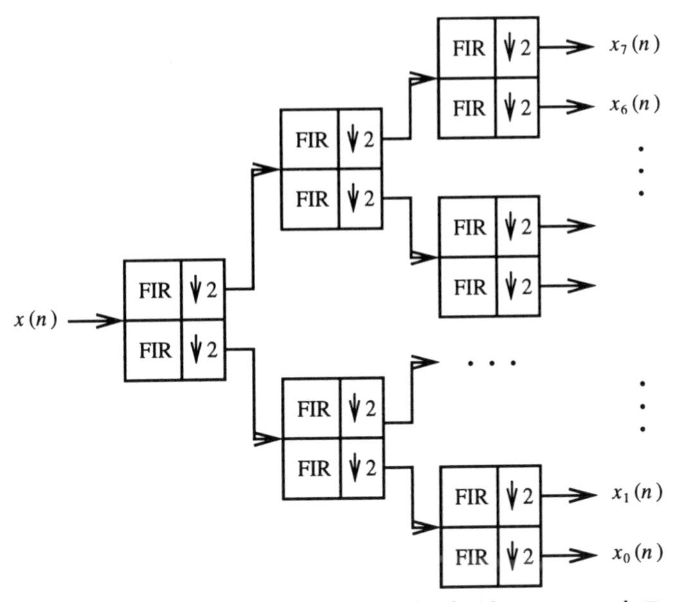

Figure 11.13 *Eight-channel filterbank realised with tree approach.* □

There is an interesting point to note about the two filters in the first stage. If these split the spectrum into signals 0 to $\omega_s/4$ from the lower filter and $\omega_s/4$ to $\omega_s/2$ from the upper filter, then the lower filter comprises a LPF centred at 0 rad/s and the upper filter is effectively a BPF centred at $\omega_s/2$. If both the filters have the same bandshape, they both use similar weights or FIR coefficient values. The only difference is that the bandpass filter weights must be derived from the low-pass weights by multiplication by the offset frequency $\omega_s/2$. But $\omega_s/2$ in sampled data form is simply the series +1, −1, +1, −1, etc. Thus, if the lower filter uses weight values h_0, h_1, h_2 etc., then the upper filter has weight values $h_0, -h_1, h_2, -h_3$ and they can both use the same set of multipliers with different summing operations. This makes for an extremely efficient filterbank design.

If the tree design of Figure 11.13 is implemented with 8-tap constituent FIR filters the successive decimation makes this equivalent, on each output $x_0(n), \cdots, x_7(n)$, to having an overall 32-tap filter capability at the input sample rate. Note in Figure 11.13 that, after each stage, the two (upper and lower) outputs both give low-pass signals in the band 0 to $\omega_s/4$, where ω_s is the input sample rate. The upper band of signals experiences a spectral inversion, as indicated by the aliased dashed component in Figure 11.3(b). This occurs when the input signal component from $\omega_s/4$ to $\omega_s/2$ is folded into the band $\omega_s/4$ to 0 by the decimation operation.

Self assessment question 11.4: Is there a fundamental difference in the processing operation performed when using the DFT and when using the multirate filterbank for signal analysis?

Self assessment question 11.5: Why is the multirate filterbank a more general approach to signal analysis than a DFT?

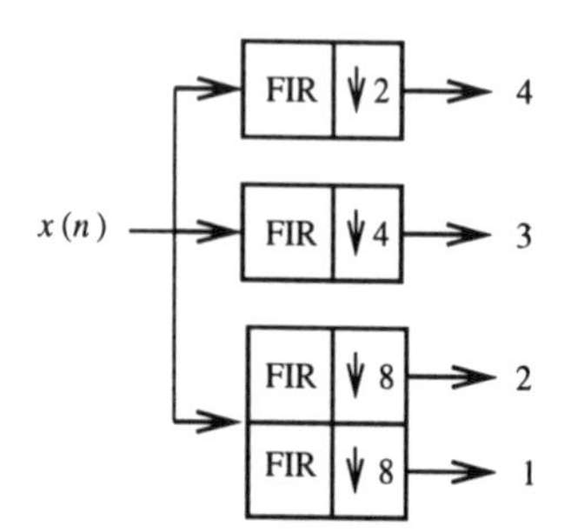

Figure 11.14 *Eight-channel filterbank realised by parallel filter (DFT) approach.*

The equivalent parallel filterbank (Figure 11.14) is not unlike a DFT as the outputs are also decimated, commensurate with the signal bandwidth in each individual channel. Note that the overall complexity of a parallel bank of eight 32-tap high sample rate filters (Figure 11.14) is reduced to only fourteen 8-tap filters in Figure 11.13. However considerable savings in overall complexity and processing operations are achieved when the multirate techniques are analysed in more detail.

When ascertaining the precise sample rate in each filter it is clear that there is in fact an identical processing computational load, at each stage in Figure 11.13. Although the 8-filter bank design of Figure 11.13 contains 14 distinct 8-tap filters, the overall processor rate is thus only $3 \times 8 = 24$ *multiplications per input data sample* to the processor when the above multiplier reductions are applied. (If the discarded samples in the decimators are not computed then this reduces to only 12 multiplications per input sample.) For large filterbanks, where the number of outputs M is a power of two, the number of multiplications equals the filter length (8 in this case) times the number of stages [Fliege], which is $\log_2 M$ or 3, in the Figure 11.13 example.

In comparison, the parallel 32-tap FIR bank of Figure 11.14 requires $8 \times 32 = 256$ multiplications per input sample. However the decimation operation throws away many of the filtered samples so only 32 of these multiplications are actually required to be computed, per input sample, to give the correct decimated output. Thus the filter trees of Figure 11.13 introduce computational savings, not unlike the DFT of Chapter 9.

However, the tree design with *similar filter lengths* at each stage gives a constant relative bandwidth while the absolute transition bandwidth varies between the channels [Fleige]! Constant absolute bandwidth to achieve identical filter *shapes* at each of the individual channels requires a halving of the number of filter coefficients, following the decimation operations at each stage in the tree, and, for this design, the tree and parallel filterbank structures would then possess a broadly comparable processor complexity. □
However, if polyphase decomposition [Crocheire and Rabiner] is incorporated into the parallel bank, its computation is simplified further, making it the preferred design approach for the realisation of large M filterbanks.

A key point about the tree approach is that there is enormous flexibility to shape or alter the responses within the filterbank. If we chose to analyse the signal with an unweighted DFT/FFT then the component filters would only possess the sinc x spectral response of Figure 11.15(a) with high crosstalk or spectral leakage. DFT weighting

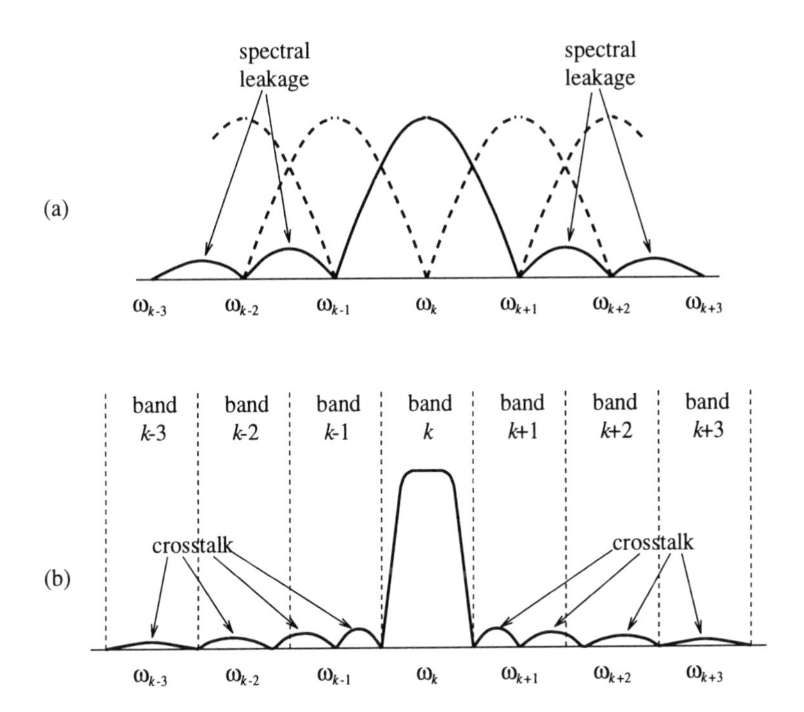

Figure 11.15 *Examples of filterbank designs with differing levels of overlap and crosstalk: (a) DFT; and (b) multirate.*

(section 9.3) widens and overlaps the individual filter passbands and reduces the interchannel crosstalk. The multirate approaches described here gives much more control over the frequency responses of the individual component filters, as the filter length can now be set independently of the DFT transform size! If long impulse response filters are used then we can obtain the more rectangular filterbank responses shown in Figure 11.15(b) with reduced interband leakage or crosstalk. However the longer impulse response duration may not be appropriate for a true spectrum analyser operation, see section 9.4, as it degrades the *quality* of the spectral estimates.

EXAMPLE 11.4

The decimated tree design of Figure 11.13 is also applicable to non-uniform filterbank realisation, as shown in Figure 11.16. Here the wide-band high frequency channels are output directly, while the lower part of the band is further split into sub-bands. This filterbank design is used in audio graphic equalisers to obtain linear phase, constant Q-filter designs where the ratio of bandwidth to centre frequency is approximately constant across the filterbank (equation (11.6) defined the Q value).

If the filter design of Figure 11.16 was attempted with a single large FFT, to obtain the resolution of the narrowest channel, then this would be inappropriate to the wider filters, as the FFT forces us to process a long duration data record to achieve the resolution of the narrowest bandwidth channel. This is particularly unattractive for

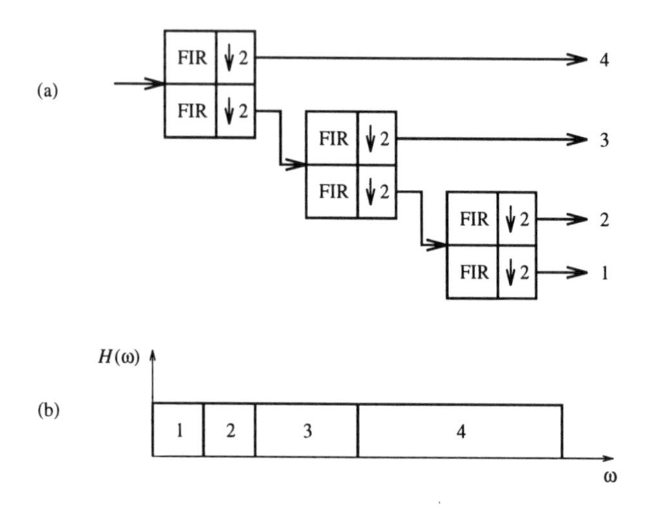

Figure 11.16 *Four-channel non-uniform filterbank.*

analysing non-stationary time varying signals as the signals may alter over the long duration data block window. This deficiency or drawback is overcome in Figure 11.16 as the length of each filter employed, or more accurately the duration of the impulse response, is carefully matched to the required channel bandwidth. The structure of Figure 11.16 can also be extended to include two or more filters within each FIR block to further subdivide the individual channels.

The nonlinear scaling of Figure 11.16(b) is also used in other spectrum analysers based on wavelets [Cohen, Chui, Massopust]. The wavelet comprises a windowed Fourier transform where the sets of sampled complex phasor values or basis vectors are all derived by time scaling or dilating a single prototype filter or mother wavelet. Thus wide-band analysis uses a short data record, while narrow-band analysis uses the same filter impulse response time scaled to a longer duration. In this multiscale transform the wavelet basis vectors control the design of the filters shown in Figure 11.16(a).

EXAMPLE 11.5

Another application of decimation is in the high resolution 'zoom' FFT. When analysing a signal with, say, a 102.4 kHz bandwidth in a 1024-point FFT, the resolution is 200 Hz. If higher resolution is required then the usual approach is to increase the length of the FFT. This is expensive in computational power and, in some situations, another option is available.

If, after an initial FFT, the signal band of interest is seen to be much less than the full 100 kHz FFT bandwidth, one can 'zoom' in to examine a restricted band in much greater detail without increasing the FFT length (Figure 11.17(b)). Here we assume that the signal of interest lies somewhere between bins 51 and 64 of the high sample rate FFT, i.e. between 10.2 and 12.8 kHz. This is achieved by employing an input FIR filter with a centre frequency set at the desired signal centre frequency (11.5 kHz), with a restricted bandwidth, covering only the spectrum we wish to examine in greater detail.

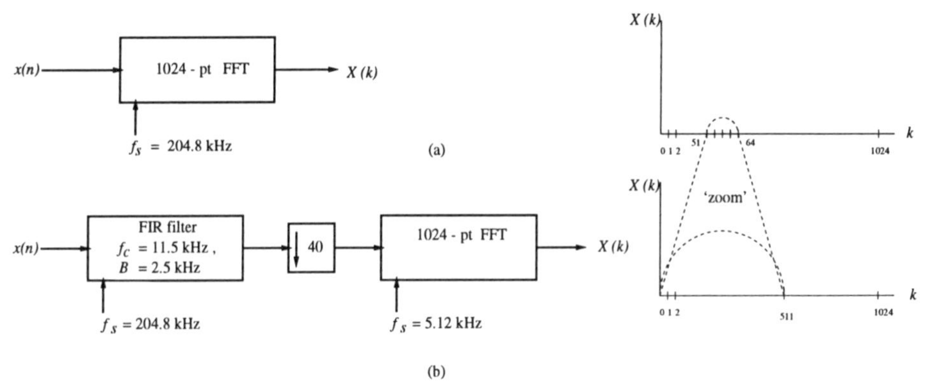

Figure 11.17 *(a) Conventional and (b) zoom FFT processor operation, which increases the spectral resolution across a limited band of frequencies.*

Following the FIR filter one can decimate by 100/2.5, i.e. 40, as we have now restricted the signal bandwidth from 102.4 to 2.5 kHz. The length of this filter and associated processing overhead can be estimated from equation (6.17) and Figure 6.12. Recall that the bandpass decimation filter of Figure 11.7 also moves the spectrum of interest to baseband. The new lower sample rate of 5.12 kHz is input sample rate to the original 1024-point FFT. The final resolution will then be increased 40-fold from 200 Hz to 5 Hz.

This saves the need for increased FFT transform size to achieve the lower required resolution and, further, the lower clock speed FFT in Figure 11.17(b) operates at a reduced arithmetic rate. Note that, if the input spectrum had been previously restricted to only signals close to 11.5 kHz, then the FIR filter in front of the decimator could be dispensed with and the bandpass decimator of Figure 11.9 employed. The precise values of the input signal bandwidth and sample rates must satisfy the equation (11.6) and Figure 11.9(e) conditions.

Self assessment question 11.6: Why in Example 11.5 is the Figure 11.17(a) resolution 200 Hz?

Self assessment question 11.7: If the input signal was known to lie between 9.9 and 12.4 kHz in Example 11.5, compare the output sample rates in the Figure 11.17(b) solution with that of the bandpass decimator of Figure 11.9.

11.4 Audio and speech processing

Speech signals have a characteristically shaped spectrum with the spectral components having different levels of significance when attempting to achieve accurate signal reconstruction. This permits filterbank techniques to be used to more effectively reduce the overall sample rate which is especially important for speech signal transmission.

EXAMPLE 11.6

In digital audio studios, it is a common requirement to change the sampling rates of bandlimited sequences. This arises, for example, when an analogue music waveform $x_a(t)$ is to be digitised. Assuming that the significant information is in the band $0 \leq f \leq 22$ kHz, a minimum sampling rate of 44 kHz is required (Figure 11.18(a)). It is, however, necessary to perform analogue filtering before sampling to eliminate aliasing and of out-of-band noise. Specifications for the analogue filter $H_a(f)$ (Figure 11.18(b)) are very stringent: it should have a fairly flat passband (so that $X_a(f)$ is not distorted) and a narrow transition band (so that only a small amount of unwanted energy passes on to the output). Optimal filters for this purpose (such as the Chebyshev filters in Chapter 5) have a very nonlinear phase response around the band edge (i.e. around 22 kHz), see Figure 5.4(b).

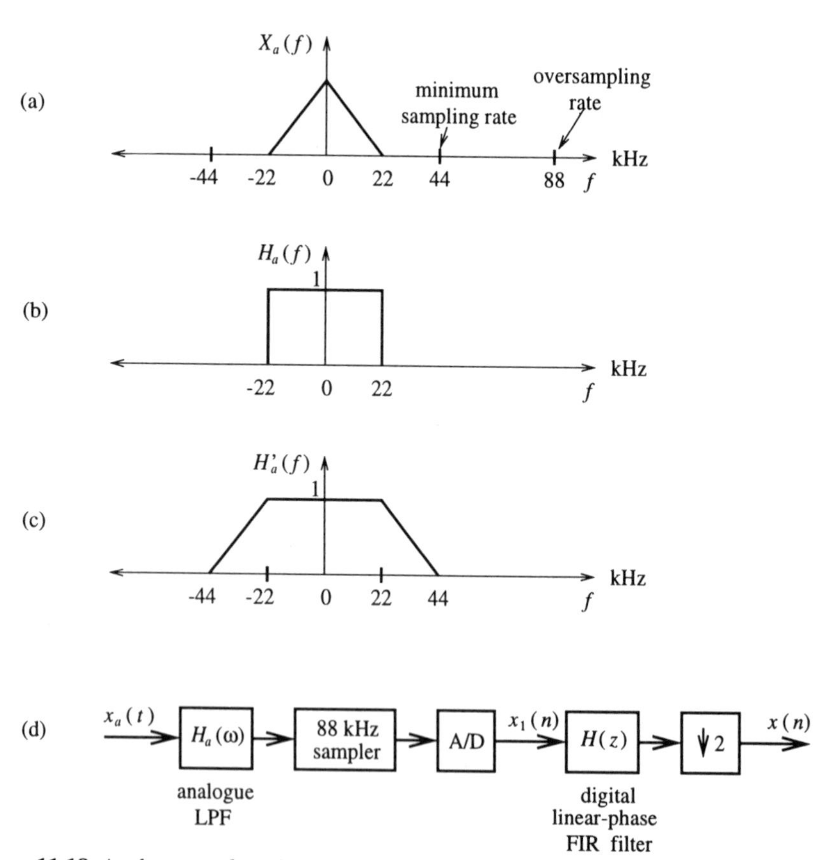

Figure 11.18 *Analogue-to-digital conversion: (a) input spectrum and two possible sample rates; (b) required anti-aliasing filter response for sampling at 44 kHz; (c) required anti-aliasing response for sampling at 88 kHz; and (d) improved scheme for data conversion.*

Solution

A common strategy to solve this problem is to oversample $x_a(t)$ by a factor of two (and often four). The filter $H'_a(f)$ now has a much wider transition band, as shown in Figure 11.18(c), so that the phase-response nonlinearity will be much lower. A simple analogue Bessel filter (Chapter 5), which has linear phase in the passband can be used in practice. The filtered and digitised sequence, $x_1(n)$, is then low-pass FIR filtered (Figure 11.18(d)) by a digital filter $H(z)$ and decimated by the same factor of two to obtain the final digital signal, $x(n)$. Since the required $H(z)$ is digital, it can be designed as a linear phase filter (Chapter 6), to provide the desired degree of sharpness, but it does necessitate a large number of filter coefficients (e.g. 256) if a steep roll-off is required (Figure 6.12).

In digital audio, high quality copying of material from one medium to another is difficult as the sampling rates used for various media are often quite different from each other. It is therefore necessary, in studios, to design efficient non-integral sample rate converters, providing further applications of multirate filterbanks.

Normal digital (mono) audio is sampled at 22.05 kHz with 8-bit quantisation accuracy. CD quality stereo sound doubles the sample rate to 44.1 kHz and requires 16-bit quantisation. The highest normal audio sample rate, in digital audio tape recorders, is 48 kHz.

High quality audio (hi-fi stereo) coders combine nonlinear filterbanks with variable rate quantisation to achieve an efficient low-rate coder design. In the moving pictures experts group (MPEG) audio coder, the quantisation level is set by a masking threshold which measures the strongest signals which are presented and examines, by FFT analysis, the threshold at which other nearby audio tones or signals will be detected, by the human ear, as separate signals.

The quantisation accuracy is then set at this threshold level so that the high level of quantisation noise is masked or hidden by the stronger signals. This auditory perceptual analysis [Furui and Sondhi] forms the basis of the MPEG stereo music coder. Stereo music normally requires 2×16 bit samples at 44.1 ksample/s, i.e. 1400 kbit/s, but this is reduced to 384 kbit/s in MPEG I or only 128 kbit/s (i.e. 1.5 bit/sample) for the more sophisticated MPEG III coders. This shows the effectiveness of the nonlinear filterbank and sophisticated quantisation accuracy control to achieve significant bit-rate compression in the encoder with no loss in audio quality.

11.4.1 Speech and image signal coding

One often encounters signals with significant energy located in a particular region of the frequency spectrum, and in other parts of the spectrum there is much less energy. This provides another application for sub-band coding.

EXAMPLE 11.7

An example of this is shown in Figure 11.19(a). The information in $|\omega| > \pi/2$ is not small enough to be discarded and we cannot decimate $x(n)$ without causing significant aliasing. It does seem unfortunate that the small fraction of energy in the high frequency region should prevent us from obtaining any signal compression capability.

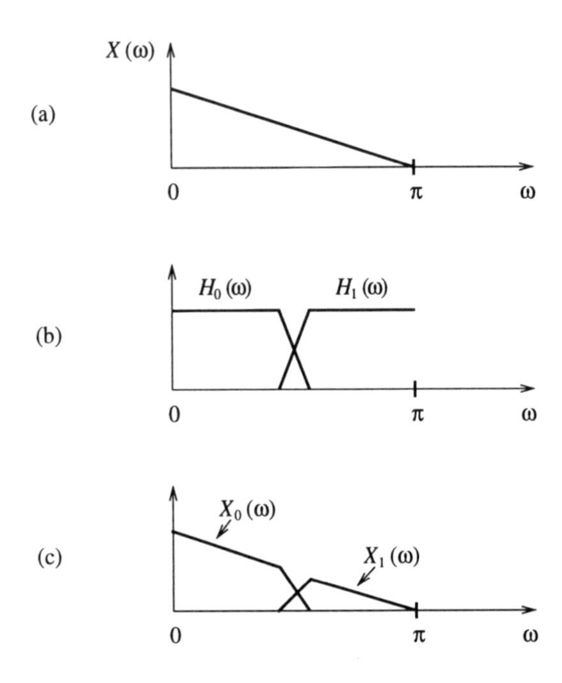

Figure 11.19 *Splitting of a signal into two sub-bands: (a) input spectrum; (b) analysis filterbank with two bands; and (c) resulting energy in each band.*

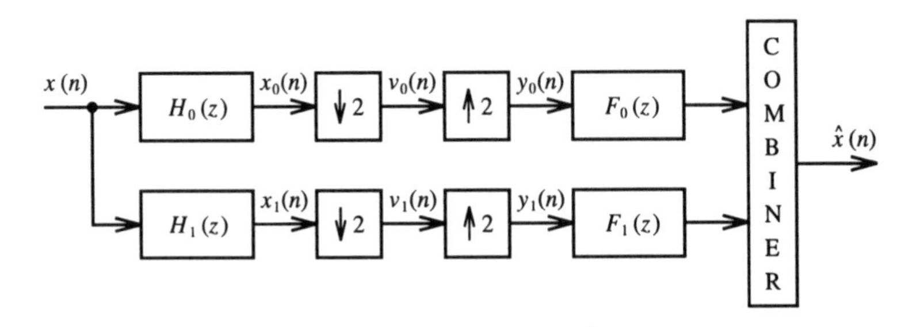

Figure 11.20 *Analysis and synthesis scheme for sub-band coding.*

However, the signal can be conveniently split into two frequency bands by using an analysis bank (Figure 11.19(b)), as shown in Figure 11.20. For this case the low-pass FIR filter design can be readily transformed to a high-pass design by moving the DC response to $\omega_s/2$. This is accomplished by multiplying the low-pass FIR coefficients by the appropriate sampled exponential at $\omega_s/2$ which, in this case, is again the magnitude sequence +1, −1, +1, −1, etc. Thus both filters $H_0(z)$ and $H_1(z)$ use the same FIR weight values and the product terms are summed appropriately to realise the two separate filters.

The sub-band signal $x_1(n)$ now has much less energy than $x_0(n)$, and so can be encoded with many fewer bits than $x_0(n)$. As an example, let $x(n)$ be a 10 kHz bandwidth signal normally requiring 8-bit quantisation so that the resulting data rate is 160 kbit/s. Let us assume that the sub-band signals $v_0(n)$ and $v_1(n)$ in Figure 11.20 can be represented with 8 bits and 4 bits per sample, respectively. Because these signals are also decimated by two (Figure 11.20), the data rate now works out to be 10 kbit/s $\times 8 + 10$ kbit/s $\times 4 = 120$ kbit/s, which is a compression by 4/3. This is the basic principle of sub-band coding (SBC) [Crochiere and Rabiner]: split the signal into two or more sub-bands; decimate each sub-band signal; and allocate the quantiser bits to samples in each sub-band, depending on the energy content of each sub-band.

Self assessment question 11.8: Repeat Example 11.7 with four equal width sub-bands using quantiser accuracy of 8, 6, 4, and 2 bits per band to find the overall bit rate required for this filterbank design.

In order to reconstruct the signal, a pair of reconstruction filters $F_0(z)$ and $F_1(z)$ are used in Figure 11.20, and the filters are designed such that the output signal $\hat{x}(n)$ is identical to the input signal $x(n)$. This is known as the condition for *perfect reconstruction* (PR). Finding good filters requires careful design of the individual filters, $H_0(z)$, $H_1(z)$, $F_0(z)$, $F_1(z)$, to achieve trade-off between spatial and frequency-domain characteristics while satisfying the PR condition.

In Figure 11.20, the process of decimation and interpolation by 2:1 at the outputs of $H_0(z)$ and $H_1(z)$ sets all odd samples of these signals to zero. For the low-pass branch, this is equivalent to multiplying $x_0(n)$ by ½ $\{1 + (-1)^n\}$. Hence $X_0(z)$ is converted to ½$\{X_0(z) + X_0(-z)\}$. Similarly $X_1(z)$ is converted to ½$\{X_1(z) + X_1(-z)\}$. Thus:

$$\begin{aligned} Y(z) &= \tfrac{1}{2}\{X_0(z) + X_0(-z)\}\, F_0(z) + \tfrac{1}{2}\{X_1(z) + X_1(-z)\}\, F_1(z) \\ &= \tfrac{1}{2} X(z)\{H_0(z)\, F_0(z) + H_1(z)\, F_1(z)\} \\ &\quad + \tfrac{1}{2} X(-z)\{H_0(-z)\, F_0(z) + H_1(-z)\, F_1(z)\} \end{aligned} \tag{11.10}$$

The first PR condition requires aliasing cancellation and forces the above term in $X(-z)$ to be zero. Hence $H_0(-z)F_0(z) + H_1(-z)\, F_1(z) = 0$, which can be achieved if:

$$H_1(z) = z^{-k}\, F_0(-z) \tag{11.11(a)}$$

and:

$$F_1(z) = z^{k}\, H_0(-z) \tag{11.11(b)}$$

where k must be odd (usually $k = \pm 1$).

The second PR condition is that the transfer function from $X(z)$ to $Y(z)$ should be unity, i.e. $H_0(z)\, F_0(z) + H_1(z)\, F_1(z) = 2$. If we define a product filter $P(z) = H_0(z)\, F_0(z)$ and substitute the results from equation (11.11), then this condition becomes:

$$H_0(z)\, F_0(z) + H_1(z)\, F_1(z) = P(z) + P(-z) = 2 \tag{11.12}$$

This needs to be true for all z and, since the odd powers of z in $P(z)$ cancel out, this implies that $p_0 = 1$ and that $p_n = 0$ for all n even and nonzero.

$P(z)$ is the transfer function of the low-pass (0) branch in Figure 11.20 (excluding the effects of the decimator and interpolator) and $P(-z)$ is that of the high-pass (1) branch. For compression applications in Example 11.7, $P(z)$ should be zero-phase to minimise distortions [Crochiere and Rabiner, Jayant and Noll]. This design technique ensures perfect reconstruction and is applicable both to filter designs and to wavelet filterbands, which also use the Figure 11.16 structure [Vetterli and Kovacevic].

The clever point about the above analysis is that it realises a combined all-pass characteristic by using a quadrature mirror perfect reconstruction filter where $H_0(z)$ comprises the image of the $H_1(z)$ response about $\pi/4$ – hence the name 'quadrature'. The aliased distortions due to interband leakage and images are then exactly compensated when the individual channels are summed in the all-pass design [Jayant and Noll].

There is a subtle distinction between PR and quadrature mirror filterbank designs. PR filters ensure minimal time-domain deviations in the resonstructed filter outputs while the quadrature mirror design provides for maximum selectivity between the individual filters. The designer must choose which criterion he wishes to adopt when undertaking this filterbank design.

EXAMPLE 11.8

In sub-band coding, the signal can be split into M sub-bands (Figure 11.21), with each sub-band signal decimated by M and independently quantised. The coding in each sub-band is typically more sophisticated than just quantisation. Figure 11.21 provides typical numbers of bit/sample for 16 to 32 kbit/s transmission rate speech coding

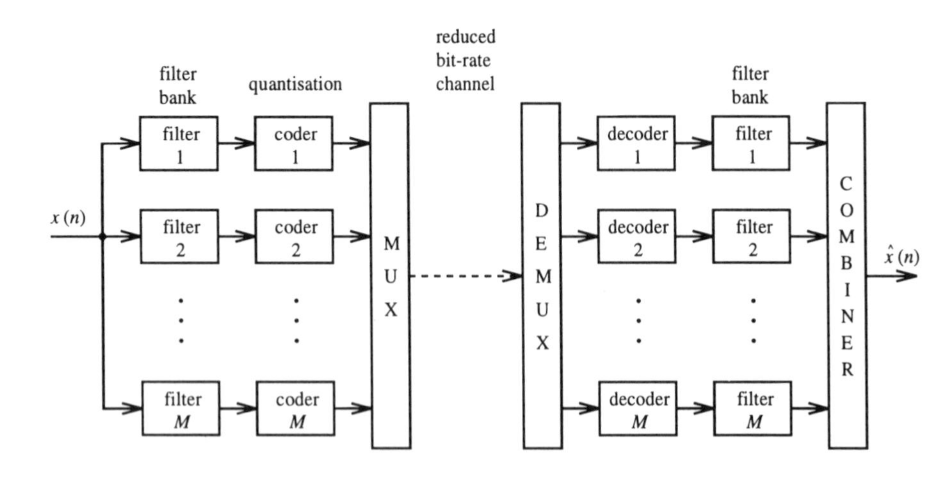

Figure 11.21 *Practical M-channel sub-band speech coder details.*

systems. Sub-band coding techniques are also being applied to obtain higher fidelity wideband, >4 kHz, telephony transmission while still using only 64 kbit/s or lower transmission rates. This is achieved by coding over wider analogue bandwidths, such as 8 kHz, and splitting this into 0 to 4 kHz and 4 to 8 kHz sub-bands for subsequent quantisation. This technique is used in the ITU G.722 high quality audio coder which employs the two sub-bands to code a wide-band 8 kHz signal at 64/56/48 kbit/s. This is a low complexity coder, implemented in one DSP microprocessor, which can operate at an bit error rate of 10^{-4} for teleconferencing and loudspeaker telephone applications.

Two dimensional sub-band coding is also used for data compression of video and other image transmission systems. This is important for the digital storage of broadcast quality signals, for example high definition TV. These filterbanks can also be used for other applications requiring a *channelised* signal band. One other use is to incorporate an array of slow speed A/D converters, after the analysis filterbank, to simplify the normal high speed, high accuracy, analogue-to-digital conversion requirements.

11.5 Summary

This final chapter has provided an introduction to the concept of downsampling, decimation, upsampling and interpolation in multirate DSP systems. Here the sample rate is adjusted commensurate with the signal bandwidth in an attempt to minimise the number of processing operations and hence processor complexity. The spectral images and aliases created by these operations have been described, and methods of minimising them have been reported. The precise sample rate requirement for bandpass signals has also been defined. The use of these techniques, to encode efficiently speech and other signal types, has been described. It has been shown that considerable simplification in computational complexity can be achieved with multirate techniques, when implementing high performance, high rate, signal samplers and filterbanks.

11.6 Problems

11.1 A data telemetry system is oversampled at 12 kHz and the sample rate has to be reduced to 2 kHz for subsequent processing. The pre-decimation filter must not introduce any phase distortion. Discuss the design options for this filter and select the preferred design approach.

11.2 If the system in problem 11.1 contained bandpass data from 6.5 kHz to 7 kHz, show the effect of reducing the sample rate from 12 to 6 to 2 kHz and also from 12 to 4 to 2 kHz by drawing corresponding spectra at the intermediate and final sample rates. Are the two solutions identical?

11.3 You are required to design a non-uniform filterbank for encoding a speech signal which has four bands (1 – 4) as defined below. Discuss with the aid of diagrams how the filterbank can be realised with multirate processing techniques, using: (a) a bank of parallel filters; and separately (b) a tree approach.

Clearly state how many distinct filter designs are required to achieve the two solutions. Estimate the filter complexity, in operations per input sample value, for (a) and (b) such

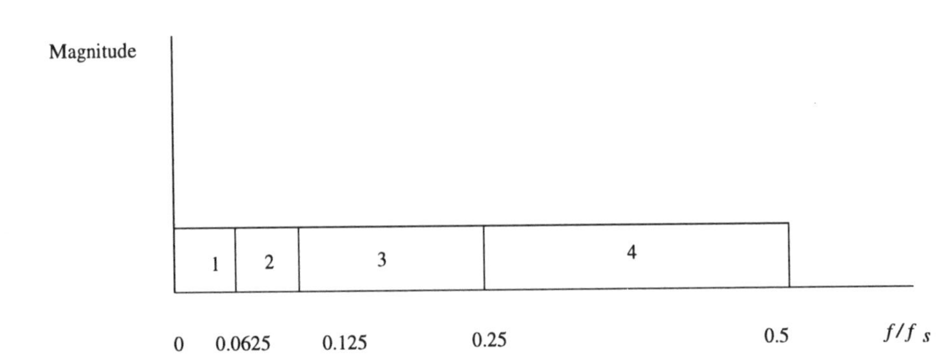

that you can compare the two potential solutions.
If the input signal is sampled at 8 kHz and the bands are quantised as:

Band 1	–	8 bit/sample
Band 2	–	6 bit/sample
Band 3	–	4 bit/sample
Band 4	–	2 bit/sample

what will be the output bit-rate of the filterbank encoder and what is the saving over directly quantising the input at 8 bit/sample?

11.4 You are required to implement a multi-channel FIR based filterbank, with 32 linearly spaced channels, for a signal sampled at 1 MHz. The preferred approach is to use multirate techniques with a filter tree design, and incorporate filters on each channel, which provide an effective 128-tap capability at the input sample rate. Draw the block diagram for the filter tree processor indicating sample rates and filter length at each stage. What is the major advantage of this approach over the DFT/FFT filterbank operation?
Calculate the overall complexity of the filter tree approach in operations per second. If you have available a DSP microprocessor with a multiply and accumulate cycle time of 30 ns, can the processor achieve a real-time implementation of the filterbank? Can you suggest any improvements that could simplify the processing? Comment also on the possibility of this DSP microprocessor being applied to implement real-time 32-point DFT or FFT operations on the same data set.

11.5 An analogue bandpass signal has a bandwidth of 40 kHz and a centre frequency of 10 MHz. What is the minimum theoretical sampling rate which will avoid aliasing? What would be the best practical choice for the nominal sampling rate? Would an oscillator with a frequency stability of 1 part in 10^6 be adequate for use as the sampling clock?
[80.16 kHz; 80.1603 kHz; yes]

Appendix A - Matrix theory revision

The purpose of this appendix is to present the basic matrix conventions and operations which are used in this text. Dealing first with column matrices or vectors, we define a vector of order N as a column vector composed of N elements:

$$\mathbf{x} = \begin{bmatrix} x_o \\ x_1 \\ . \\ . \\ x_{N-1} \end{bmatrix} \tag{A.1}$$

The transpose of this vector is a row vector of order N:

$$\mathbf{x}^T = [x_o x_1 \cdots x_{N-1}] \tag{A.2}$$

The scalar, inner or dot product of two such vectors is given by:

$$\mathbf{x}^T \mathbf{y} = \sum_{i=0}^{N-1} x_i \, y_i = \mathbf{y}^T \mathbf{x} \tag{A.3}$$

The vector or outer product of these two vectors is given by:

$$\mathbf{x} \, \mathbf{y}^T = \mathbf{y} \, \mathbf{x}^T = \mathbf{A} \tag{A.4}$$

where the matrix $\mathbf{A}$ is an $(N \times N)$ square matrix. In general such an $(N \times N)$ square matrix is defined as follows:

$$\mathbf{A} = \begin{bmatrix} a_{00} & a_{01} & .. & a_{0,N-1} \\ a_{10} & a_{11} & .. & . \\ . & . & & . \\ . & . & & . \\ a_{0,N-1} & a_{N-1,1} & .. & a_{N-1,N-1} \end{bmatrix} \tag{A.5}$$

Transposition of the matrix $\mathbf{A}$ is achieved by reflection about the leading diagonal.

EXAMPLE A.1

Find the transpose of matrix $\mathbf{A}$ below:

$$\mathbf{A} = \begin{bmatrix} a_{00} & a_{01} & a_{02} \\ a_{10} & a_{11} & a_{12} \\ a_{20} & a_{21} & a_{22} \end{bmatrix}$$

$$\mathbf{B} = \mathbf{A}^T = \begin{bmatrix} a_{00} & a_{10} & a_{20} \\ a_{01} & a_{11} & a_{21} \\ a_{02} & a_{12} & a_{22} \end{bmatrix}$$

The matrix **A** is termed symmetric when $\mathbf{A}^T = \mathbf{A}$. The operation of transposition has the following properties:

$$(\mathbf{A}^T)^T = \mathbf{A}$$

$$(\mathbf{A} + \mathbf{B})^T = \mathbf{A}^T + \mathbf{B}^T$$

$$(\mathbf{A}\mathbf{B})^T = \mathbf{B}^T \mathbf{A}^T \tag{A.6}$$

Multiplication of two square $(N \times N)$ matrices leads to a product matrix which is also square and $(N \times N)$:

$$\mathbf{C} = \mathbf{A}\,\mathbf{B} = \sum_{k=0}^{N-1} \begin{bmatrix} (a_{0k} & b_{k0}) & (a_{0k}\,b_{k1}) & (a_{0k} & b_{k,N-1}) \\ \cdot & \cdot & \cdot & \cdot & \cdot & \cdot \\ \cdot & \cdot & \cdot & \cdot & \cdot & \cdot \\ (a_{N-1,k}\,b_{k,N-1}) & \cdot & \cdot & (a_{N-1,k}\,b_{k,N-1}) \end{bmatrix} \tag{A.7}$$

The properties of multiplication are:

$$\mathbf{A}\,\mathbf{B} \neq \mathbf{B}\,\mathbf{A}$$

$$\mathbf{A}\,\mathbf{B} = \mathbf{A}\,\mathbf{C} \textit{ does not imply } \mathbf{B} = \mathbf{C}$$

$$\mathbf{A}\,(\mathbf{B}\,\mathbf{C}) = (\mathbf{A}\,\mathbf{B})\,\mathbf{C}$$

$$\mathbf{A}\,(\mathbf{B} + \mathbf{C}) = \mathbf{A}\,\mathbf{B} + \mathbf{A}\,\mathbf{C} \tag{A.8}$$

Every square matrix **A** has associated with it a determinant which is a scalar defined as $|\mathbf{A}|$. It is defined as the sum of the products of the elements in any row or column of **A** with the respective cofactors c_{ij} of that element a_{ij}. The cofactor, c_{ij}, is defined as $(-1)^{i+j} m_{ij}$ where m_{ij} is the determinant of the $(N-1) \times (N-1)$ matrix formed by removing the ith row and jth column of **A**. The properties of the determinant of **A** are

$$|\mathbf{A}| = 0 \text{ if } \textit{one row is zero or two rows are identical}$$

$$|\mathbf{A}\,\mathbf{B}| = |\mathbf{A}|\,|\mathbf{B}|$$

$$|\mathbf{A}^{-1}| = |\mathbf{A}|^{-1} \tag{A.9}$$

EXAMPLE A.2

Find the determinant of matrix **B** below.

$$\mathbf{B} = \begin{bmatrix} 3 & 1 & 0 \\ -1 & 2 & 4 \\ 5 & 6 & -2 \end{bmatrix}$$

$$|\mathbf{B}| = 3\,(-28) - 1\,(-18) + 0.0$$

$$= -66$$

The inverse of a matrix **A** is defined by:

$$\mathbf{A}^{-1}\,\mathbf{A} = \mathbf{A}\,\mathbf{A}^{-1} = \mathbf{I} \tag{A.10}$$

where:

$$\mathbf{I} = \begin{bmatrix} 1\,0\,0\,0 \\ 0\,1\,0\,0 \\ 0\,0\,1\,0 \\ 0\,0\,0\,1 \end{bmatrix}$$

is the identity matrix.

$$\mathbf{A}^{-1} = \frac{1}{|\mathbf{A}|}\, adj\,(\mathbf{A}) \tag{A.11}$$

where $adj(\mathbf{A})$ is the adjoint matrix of **A** and is the transpose of the matrix of the cofactors of **A**. The properties of the inverse are:

$$(\mathbf{A}^{-1})^{-1} = \mathbf{A}$$

$$(\mathbf{A}\,\mathbf{B})^{-1} = \mathbf{B}^{-1}\,\mathbf{A}^{-1}$$

$$(\mathbf{A}^{-1})^{T} = (\mathbf{A}^{T})^{-1} \tag{A.12}$$

For a square matrix **A**, a non-zero vector **a** is an eigenvector if a scalar λ exists such that:

$$\mathbf{A}\,\mathbf{a} = \lambda\,\mathbf{a}$$

$$(\mathbf{A} - \lambda\,\mathbf{I})\,\mathbf{a} = 0 \tag{A.13}$$

where λ is an eigenvalue of **A**. Equation (A.13) yields a non-trivial eigenvector only if:

$$|\mathbf{A} - \lambda\,\mathbf{I}| = 0 \tag{A.14}$$

This is the so-called characteristic equation for **A** and its roots are the eigenvalues of **A**. The properties of eigenvalues are:

$$|\mathbf{A}| = \prod_{i=0}^{N-1} \lambda_i$$

$$trace\,(\mathbf{A}) = \sum_{i=0}^{N-1} \lambda_i \tag{A.15}$$

More details on matrix theory can be found in texts such as [Broyden].

Appendix B - Signal transforms

Selected Laplace transforms

$x(t)$ $(t \geq 0)$	$X(s)$
$\delta(t)$	1
$\delta(t-\alpha)$	$\exp(-\alpha s)$
1 (unit step)	$\frac{1}{s}$
t (unit ramp)	$\frac{1}{s^2}$
$\exp(-\alpha t)$	$\frac{1}{s+\alpha}$
$t\exp(-\alpha t)$	$\frac{1}{(s+\alpha)^2}$
$\sin(\alpha t)$	$\frac{\alpha}{s^2+\alpha^2}$
$\cos(\alpha t)$	$\frac{s}{s^2+\alpha^2}$
$e^{-\alpha t}\sin(\omega t)$	$\frac{\omega}{(s+\alpha)^2+\omega^2}$
$e^{-\alpha t}\cos(\omega t)$	$\frac{s+\alpha}{(s+\alpha)^2+\omega^2}$

Selected z-transforms

$x(n) \quad (n \geq 0)$	$X(z)$
$\delta(n)$ (unit pulse)	1
$\delta(n-m)$	z^{-m}
1 (unit step)	$\dfrac{z}{z-1}$
n (unit ramp)	$\dfrac{z}{(z-1)^2}$
$\exp(-\alpha n)$	$\dfrac{z}{(z-e^{-\alpha})}$
$n\exp(-\alpha n)$	$\dfrac{e^{-\alpha}z}{(z-e^{-\alpha})^2}$
$\sin(\beta n)$	$\dfrac{z\sin(\beta)}{z^2-2z\cos(\beta)+1}$
$\cos(\beta n)$	$\dfrac{z^2-z\cos(\beta)}{z^2-2z\cos(\beta)+1}$
$e^{-\alpha n}\sin(\beta n)$	$\dfrac{z\,e^{-\alpha}\sin(\beta)}{z^2-2ze^{-\alpha}\cos(\beta)+e^{-2\alpha}}$
$e^{-\alpha n}\cos(\beta n)$	$\dfrac{z^2-ze^{-\alpha}\cos(\beta)}{z^2-2ze^{-\alpha}\cos(\beta)+e^{-2\alpha}}$

Four classes of Fourier transform

Fourier series – periodic and continuous in time, discrete in frequency	
$x(t) = \sum_{n=-\infty}^{+\infty} X_n \exp(jn\omega_0 t)$	$X_n = \dfrac{1}{T}\int_{-T/2}^{T/2} x(t)\exp(-jn\omega_0 t)\,dt$
Fourier transform – continuous in time, continuous in frequency	
$x(t) = \dfrac{1}{2\pi}\int_{-\infty}^{\infty} X(\omega)\exp(\,j\omega t\,)\,d\omega$	$X(\omega) = \int_{-\infty}^{\infty} x(t)\exp(-j\omega t)\,dt$
Discrete-time Fourier transform – discrete in time, continuous and periodic in frequency	
$x(n\Delta t) = \dfrac{\Delta t}{2\pi}\int_{-\pi/\Delta t}^{\pi/\Delta t} X(\omega)\exp(-jn\Delta t\omega)\,d\omega$	$X(\omega) = \sum_{n=-\infty}^{\infty} x(n\Delta t)\exp(-j\omega n\Delta t)$
Discrete Fourier transform – discrete and periodic in time and in frequency	
$x(n) = \dfrac{1}{N}\sum_{k=0}^{N-1} X(k)\exp\left(\dfrac{jnk2\pi}{N}\right)$	$X(k) = \sum_{n=0}^{N-1} x(n)\exp\left(\dfrac{-jnk2\pi}{N}\right)$

Solutions to self assessment questions

Chapter 1

SAQ 1.1 The energy in the signal is defined by equation (1.2):

$$E = \int_{-\infty}^{\infty} x^2(t)\, dt = \int_{0}^{1} (5t)^2 dt = 25/3$$

SAQ 1.2 Using equation (1.13), $T = 1$, $\omega_0 = 2\pi$, thus:

$$X_n = \int_{-1/2}^{1/2} 4t \exp(-jn2\pi t)\, dt$$

Integration by parts yields:

$$X_n = \left[\frac{2j\pi nt \exp(-2j\pi nt) + \exp(-2j\pi nt)}{\pi^2 n^2} \right]_{-1/2}^{1/2} = \frac{2}{\pi^2 n^2} (\sin(\pi n) - n\pi \cos(\pi n)$$

SAQ 1.3 Using equation (1.16):

$$X(\omega) = \int_{1}^{4} 4 \exp(-j\omega t)\, dt = \left[-\frac{4}{j\omega} \exp(-j\omega t) \right]_{1}^{4} = -\frac{4}{j\omega} (\exp(-j\omega) - \exp(-4j$$

This comprises two impulses at the above frequencies.

SAQ 1.4 Using equation (1.20):

$$X(s) = \int_{0^-}^{\infty} \exp(-t/5) \exp(-st) dt = \left[-5 \frac{\exp(-st) \exp(-t/5)}{1 + 5s} \right]_{0^-}^{\infty} = \frac{5}{5s + 1}$$

SAQ 1.5 Using KCL:

$$\text{current} = C \frac{d}{dt} (\exp(jn\omega_0 t) - y) = \frac{y}{R}$$

Hence:

$$C\exp(jn\omega_0 t)\,jn\omega_0 - C\frac{dy}{dt} = \frac{y}{R}$$

thus:

$$y + RC\frac{dy}{dt} = RCjn\omega_0\exp(jn\omega_0 t)$$

Adopt an assumed solution, i.e.: $y_n(t) = K\exp(jn\omega_0 t)$. Substitute back into equation:

$$K\exp(jn\omega_0 t) + RCKjn\omega_0\exp(jn\omega_0 t) = RCjn\omega_0\exp(jn\omega_0 t)$$

Solving for K:

$$K + RCKjn\omega_0 = RCjn\omega_0 \Rightarrow K = \frac{RCjn\omega_0}{1 + RCjn\omega_0}$$

SAQ 1.6 Using method (i):

$$C\frac{d}{dt}(x(t) - y(t)) = \frac{y}{R} \Rightarrow RC\frac{dx}{dt} = RC\frac{dy}{dt} + y$$

Take Laplace transforms and assume initial conditions are zero:

$$RCsX(s) = RCsY(s) + Y(s) \Rightarrow H(s) = \frac{RCs}{RCs + 1}$$

Method (ii), transform circuit:

$$\frac{X(s) - Y(s)}{1/(sC)} = \frac{Y(s)}{R} \Rightarrow H(s) = \frac{RCs}{RCs + 1}$$

Chapter 2

SAQ 2.1

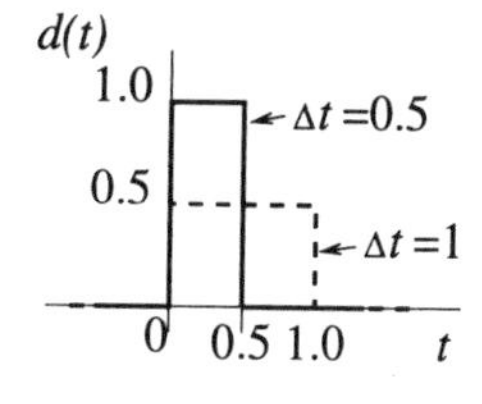

SAQ 2.2

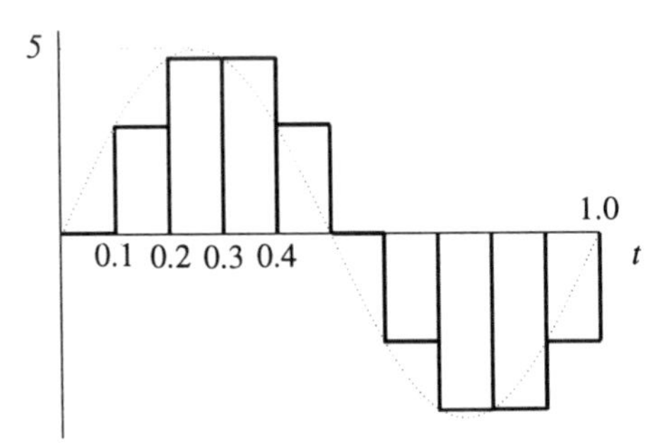

First pulse: $5\sin(2\pi 0) \times d(t-0) \times 0.1 = 0$
Second pulse: $5\sin(2\pi \times 0.1) \times d(t-0.1) \times 0.1 = 0.294d(t-0.1)$
Third pulse: $5\sin(2\pi \times 0.2) \times d(t-0.2) \times 0.1 = 0.476d(t-0.2)$

SAQ 2.3 Transfer function:

$$H(s) = \frac{sL}{sL+R} = \frac{s}{s+R/L}$$

This is not a proper fraction so divide the numerator by the denominator to give:

$$H(s) = 1 - \frac{R/L}{s+R/L}$$

Taking inverse Laplace transforms yields:

$$h(t) = \delta(t) - \frac{R}{L}\exp\left(\frac{-t}{R/L}\right)$$

SAQ 2.4 Solution:

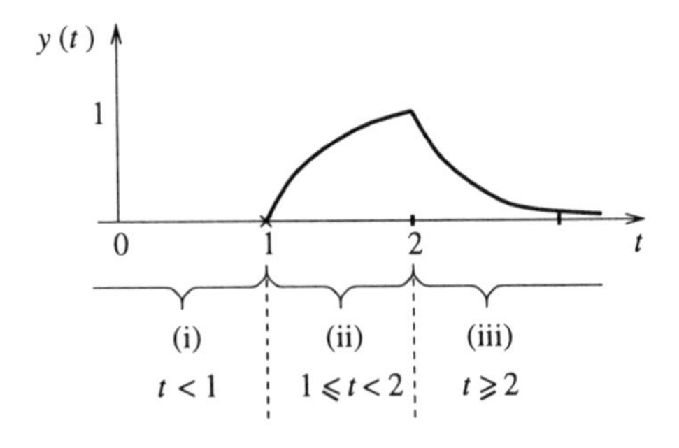

Region (i): $y(t) = 0$
Region (ii):

$$y(t) = 1 - \exp\left(-\frac{t-1}{RC}\right)$$

Region (iii):

$$y(t) = \exp\left(-\frac{t-1}{RC}\right)\left[\exp\left(\frac{1}{RC}\right) - 1\right]$$

Chapter 3

SAQ 3.1 Two zeros at $s = -2$ and poles at $s = -7 \pm j7.14$:

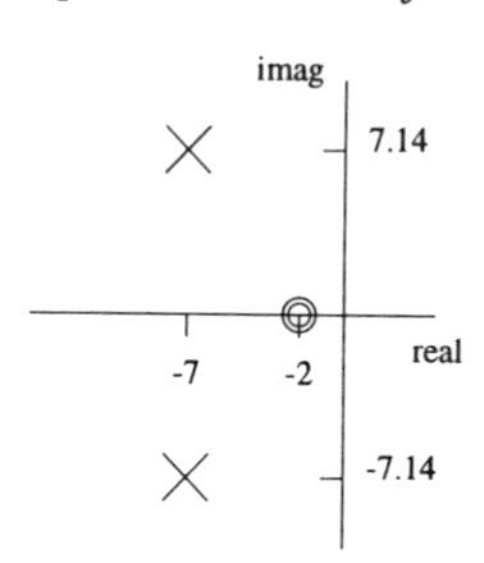

SAQ 3.2

$$H(s) = 1/(s^2 + 4s + 4) \Rightarrow H(j\omega) = 1/((j\omega)^2 + 4(j\omega) + 4)$$

SAQ 3.3 Zeros at $s = 0.1 \pm j$, poles at $s = 0.2 \pm 3j$ and 0.

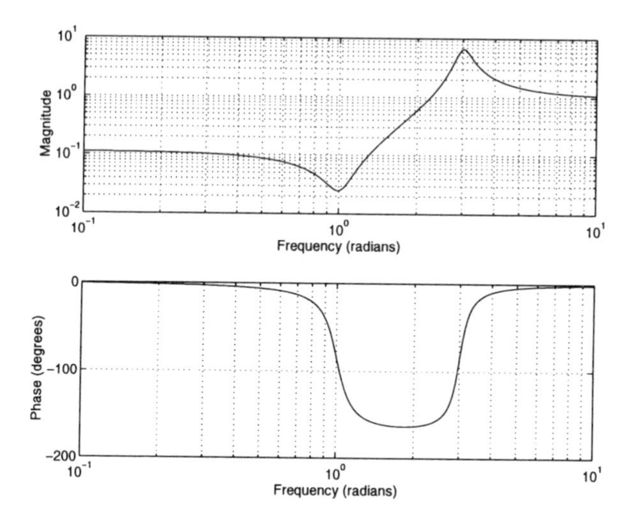

SAQ 3.4 Exact Bode plot:

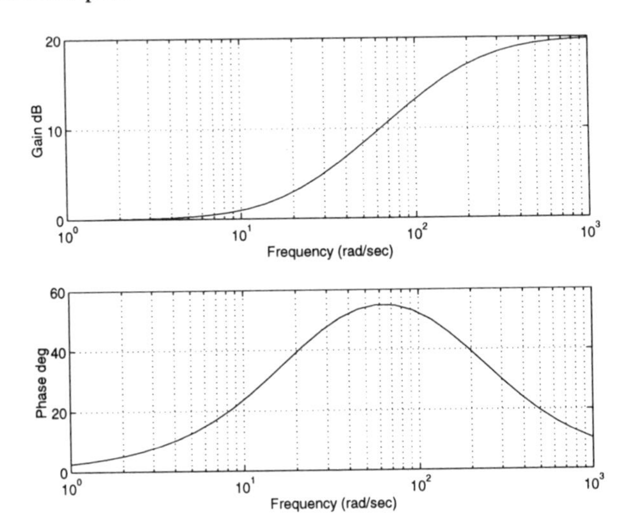

SAQ 3.5 Transfer function:

$$H(s) = 1/(s^2 + 0.2s + 1.01)$$

Poles at $s = -0.1 \pm j$ and hence system is stable. Partial fraction expansion:

$$H(s) = \frac{0.5j}{s + 0.1 + j} - \frac{0.5j}{s + 0.1 - j}$$

Impulse response: $h(t) = \exp(-0.1\, t)\sin(t)$; $t \geq 0$; period of oscillation is 2π seconds and time constant of decay is 10 seconds. Impulse response with envelope of delay shown below:

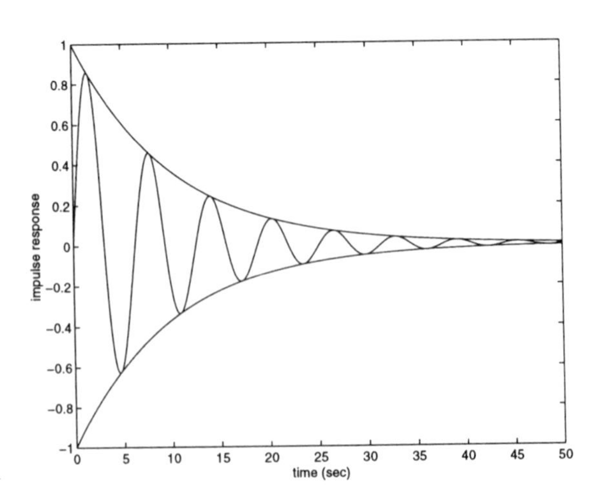

SAQ 3.6 Undamped natural frequency $\omega_0 = 2$ rad/s and damping factor $\zeta = 1/2$.

Chapter 4

SAQ 4.1 The frequencies which are present at the output of the D/A are: $3 + 10n$ kHz where n 1, 2, $\cdots$, etc. and $7 + 10n$ kHz where n 1, 2, $\cdots$, etc.

SAQ 4.2 Using (4.6):

$$X(z) = \sum_{n=0}^{\infty} \alpha^n z^{-n} = \sum_{n=0}^{\infty} (\alpha z^{-1})^n = \frac{1}{1 - \alpha z^{-1}} = \frac{z}{z - \alpha}$$

SAQ 4.3 Not a proper fraction, thus:

$$\frac{X(z)}{z} = \frac{(z-1)}{(z^2 - 5z + 6)} = \frac{2}{z-3} - \frac{1}{z-2}$$

and:

$$X(z) = \frac{2z}{z-3} - \frac{z}{z-2}$$

From tables in Appendix B:

$$x(n) = 2(3^n) - 2^n \quad n \geq 0$$

SAQ 4.4 At $n = 2$ the convolution equation becomes:

$$y(2) = \sum_{m=0}^{2} h(m)\, x(2-m) = 3(1) + 2(0.5) + 1(0.25) = 4.25$$

At $n = 3$ the convolution equation becomes:

$$y(3) = \sum_{m=0}^{2} h(m)\, x(3-m) = 0(1) + 3(0.5) + 2(0.25) = 2$$

SAQ 4.5 At a quarter of the sampling frequency, $\omega = (2\pi/\Delta t)/4$ and hence $\omega\Delta t = \pi/2$. Frequency response: $H(\omega) = 2 + \exp(j\omega\Delta t)$. Thus the gain is: $|2 + \exp(j\pi/2)| = 2.236$.

SAQ 4.6 Zero at the origin and pole at -1. Frequency response:

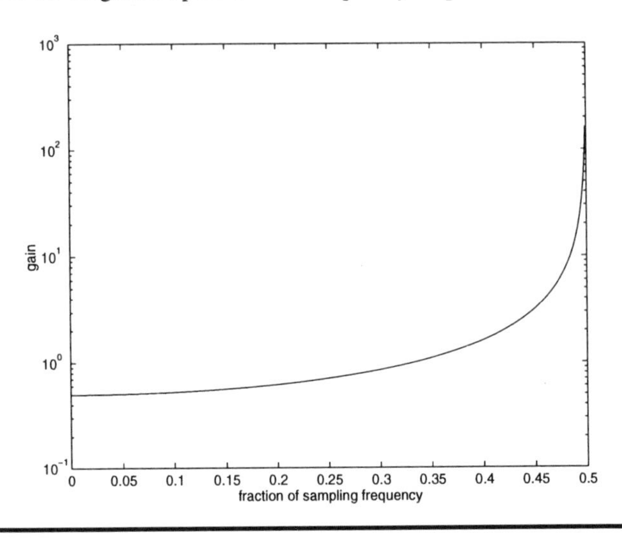

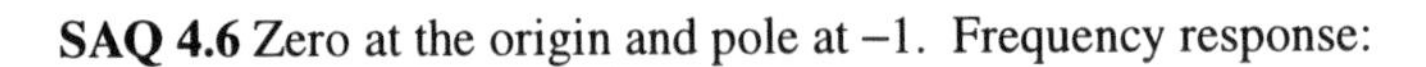

Chapter 5

SAQ 5.1 From the 'brick wall' frequency response by Fourier transform theory, the impulse response must have a sinc (x) shape.

SAQ 5.2 The number of poles is independent of the cut-off frequency and depends only on the filter order. It is 2 poles per order, thus for this question the answer is 8 poles.

SAQ 5.3 For a 4th order filter, the roll-off rate is $4 \times 20 = 80$ db/decade or $4 \times 6 = 24$ dB/octave. Thus at 2 kHz there is 24 dB attenuation, at 10 kHz there is 80 dB and at 100 kHz there is 160 dB.

SAQ 5.4 Stopband is specified at 12/3 or 4 times the cut-off frequency. Thus for a single stage filter the attenuation is $4 \times 6 = 24$ dB. For two stages the attenuation is 48 dB and for 3 stages it is 72 dB. Thus 3 stages will be required to meet the design specification of –60 dB.

SAQ 5.5 Chebyshev provides fastest roll-off or best rejection for a given number of stages of filter order and thus it is very efficient from a computation standpoint. The disadvantage is that there is ripple in the passband (Figure 5.4(a)), and also the phase response is highly non-linear.

SAQ 5.6 Solution is as in Figure 5.8 but a_1 and b_1 are zero so these paths do not exist. $a_1 = 1$, $a_2 = 10$ and $b_2 = 5$.

SAQ 5.7 The theoretical duration is infinity.

SAQ 5.8 All bilinear z-transform filters have infinite attenuation at half the sampling frequency, i.e. 25 kHz in this case.

SAQ 5.9

$$H(z) = \frac{z^2 + 2z + 1}{10.2z^2 - 9.62z + 3.37} = \frac{(z+1)^2}{z + \dfrac{9.6 \pm \sqrt{92.16 - 137.5}}{20.4}}$$

$$= \frac{(z+1)^2}{(z + 0.5 + 0.33)\,(z + 0.5 - j\,0.33)}$$

This has two zeros at $z = -1$ and poles at $-0.5 \pm j0.33$.

SAQ 5.10 Replace $\{z\}$ by $\{-z\}$ for LP to HP transform:

$$H(z) = \frac{z^2 - 2z + 1}{10.2z^2 + 9.62z + 3.37} = \frac{(z-1)^2}{(z - 0.5 + j0.33)\,(z - 0.5 - j0.33)}$$

Thus the poles and zeros reflect about the vertical axis:

$$H(z) = \frac{0.098 - 0.195\,z^{-1} + 0.098\,z^{-2}}{1 + 0.942\,z^{-1} + 0.333\,z^{-2}}$$

In comparison with the LPF of Figure 5.11, the signs of the a_1 and b_1 weights alter but the numerical values are all unchanged.

SAQ 5.11 Compared with SAQ 5.7, the finite precision causes the impulse response to be of finite duration. Because of the limit cycle effects in 5.5.2, there can also be an oscillating output for no input signal.

Chapter 6

SAQ 6.1 In this FIR filter $H(z) = 1 + z^{-1} = (z + 1)/z$. This has one zero at $z = -1$ and a pole at $z = 0$. The frequency response, which can be obtained by applying the Figure 4.24(a) constructions, peaks at DC and f_s and has a minimum value of zero at $f_s/2$.

SAQ 6.2 For (a): $H(z) = 1 + ½\,z^{-1} = (z + ½)/z$. This has a zero at $z = -½$ and a pole at $z = 0$. The frequency response peaks at DC and f_s with a magnitude of 1.5 and at $f_s/2$ the minimum value is 0.5. The phase response is double humped with zero values at $f = 0, f_s/2, f_s$. For (b): $H(z) = ½ + z^{-1} = (½z + 1)/z$. This has its zero at $z = -2$. The frequency response is as in (a) with maximum value of 3 and a minimum value (at $f_s/2$) of 1. The phase response is 0° at DC, −180° at $f_s/2$ and 0° again at f_s.

SAQ 6.3 Linear phase filtering is essential in radar and communications to enhance pulsed signal returns and measure accurately the timing of the received signals.

SAQ 6.4 This introduces a symmetrical impulse response.

SAQ 6.5 The coefficient values would each be scaled to $10\times$ their values.

SAQ 6.6 For this increased duration impulse response, the –3 dB bandwidth is reduced to one tenth of the sample frequency.

SAQ 6.7 For half the original bandwidth, by Fourier theory, the impulse response main-lobe width will be doubled in duration. Now if the gain of both filters is identical then the values will have to be scaled by ½ as there are twice as many of them.

SAQ 6.8 For the Hamming filter the transition bandwidth is:

$$\frac{3.3}{N\Delta t} = \frac{3.33}{21\times 10^{-3}} = 0.157 \text{ kHz}$$

SAQ 6.9 For the Hanning filter design, the transition bandwidth must be only 50 Hz. Thus for filter order N:

$$50 = \frac{3.1}{N\Delta t} = \frac{3.1}{N\times 10^{-3}} \quad \text{i.e.} \quad N = \frac{3.1}{0.05} = 62$$

The filter order is thus 62. Note that as the transition bandwidth is one third that of SAQ 6.7 then the filter order is three times larger!

SAQ 6.10 From Figure 6.10 find δ_1 for a passband ripple of ½ dB. Substitution into the equation yields $\delta_1 = 0.06$. Also for –40 dB stopband attenuation, $\delta_2 = 0.01$. Now using equation (6.17):

$$\hat{N} = \frac{2}{3}\frac{1}{\Delta t\Delta f}\log_{10}\left(\frac{1}{10\times 0.06\times 0.01}\right) = \frac{2}{3}\frac{10^3}{50}\log_{10}\frac{100}{0.6}$$

This yields $\hat{N} = 30$.

SAQ 6.11 For an input variance of 1/12, the output variance will be 1/36 in this 3-stage FIR averager, see example 7.3.

SAQ 6.12 The frequency response of the matched filter is the complex conjugate of the expected signal.

SAQ 6.13 The optimum filter in white Gaussian noise is a matched filter.

SAQ 6.14 Clearly this has peak values of ± 3 every 3 samples as the signal propagates through the FIR receiver.

Chapter 7

SAQ 7.1 By definition:

$$\sigma_x^2(n) = E[(x(n) - m_x(n))^2]$$

Start by multiplying out all the term inside the expectation operator:

$$\sigma_x^2(n) = E[x^2(n) - 2x(n)m_x(n) + m_x^2(n)]$$

As in equations (7.1) and (7.3), expectation is an integration operation and hence we can re-order the expectation and addition operations:

$$\sigma_x^2(n) = E[x^2(n)] - 2E[x(n)m_x(n)] + E[m_x^2(n)]$$

The mean is a constant and hence can be taken outside the expectation operator:

$$\sigma_x^2(n) = E[x^2(n)] - 2E[x(n)]m_x(n) + m_x^2(n)$$

From the definition of the mean in equation (7.1):

$$\sigma_x^2(n) = E[x^2(n)] - m_x^2(n)$$

SAQ 7.2 Let $\{x(n)\} = \{1.3946, 1.6394, 1.8742, 2.7524, 0.6799\}$. Using equation (7.2) an estimate of the mean is:

$$\hat{m}_x = \frac{1}{5} \sum_n x(n) = 1.6681$$

An estimate of the variance would then be:

$$\hat{\sigma}_x^2 = \frac{1}{5} \sum_n (x(n) - \hat{m}_x)^2 = 0.4541$$

SAQ 7.3 The power at the input is given by the product of the PSD and the bandwidth of the signal – PB_s. The power at the output is given by the product of the PSD, the filter gain and the bandwidth of the filter PGB_f. Since the gain is unity, the power at the output of the filter will be less than at the input by a factor of 10.

SAQ 7.4 The PSD describes how the average power in a signal is distributed in the frequency domain. The pdf describes how the instantaneous value of a signal is distributed.

SAQ 7.5

$$S_{yy}(z) = H(z)\, H(z^{-1})\, S_{xx}(z) = (0.5 + 0.75\, z^{-1})\, (0.5 + 0.75\, z)\, \sigma_x^2$$
$$= (0.375\, z + 0.8125 + 0.375\, z^{-1})\, \sigma_x^2$$

The autocorrelation at the output is the inverse z-transform of this: $\phi_{yy}(-1) = 0.375\sigma_x^2$ $\phi_{yy}(0) = 0.8125\sigma_x^2$ $\phi_{yy}(1) = 0.375\sigma_x^2$. The variance of the output is $0.8125\sigma_x^2$ with a corresponding RMS value of $\sqrt{0.8125}\sigma_x$. The PSD:

$$S_{yy}(\omega) = (0.375\, e^{-j\omega\Delta t} + 0.8125 + 0.375\, e^{j\omega\Delta t})\, \sigma_x^2$$
$$= (0.8125 + 0.75 \cos(\omega\Delta t))\, \sigma_x^2$$

SAQ 7.6 At output of filter:

$$S_{yy}(z) = (0.1 - 0.8z^{-1})(0.1 - 0.8z)$$

with zeros at $z = 8$ and 1/8. The term $(0.1 - 0.8z)$ is minimum since it has a root at 1/8 but it is non-causal because of the positive power of z. Rewriting we have: $(0.1 - 0.8z) = z(0.1z^{-1} - 0.8)$ and $(0.1 - 0.8z^{-1}) = (0.1z - 0.8)z^{-1}$. Thus:

$$S_{yy}(z) = (0.1z - 0.8)(0.1z^{-1} - 0.8)$$

The minimum phase causal filter is $(0.1z^{-1} - 0.8)$. The whitening filter is the inverse of this, i.e.: $1/(0.1z^{-1} - 0.8)$. This is NOT the inverse of $H(z)$.

SAQ 7.7 In section 7.6.2 the dynamic range was calculated on the basis of a sine wave signal occuppying the whole dynamic range. In this example the signal amplitude is a factor of 2 smaller than the max. voltage – hence signal power is reduced by a factor of 4. Thus signal-to-quantisation noise is:

$$1.76 + 6M - 10\log_{10}(4) = -88.3 \text{ dB}$$

Chapter 8

SAQ 8.1 Using equation (8.4):

$$\xi = E[x^2(n)] - 2\,\mathbf{h}^T\,\boldsymbol{\Phi}_{yx} + \mathbf{h}^T\,\boldsymbol{\Phi}_{yy}\,\mathbf{h}$$

Then using equation (8.8):

$$\xi = E[x^2(n)] - 2\,\mathbf{h}^T\,\boldsymbol{\Phi}_{yx} + \mathbf{h}^T\,(\boldsymbol{\Phi}_{yy}\,\mathbf{h}) = E[x^2(n)] - 2\,\mathbf{h}^T\,\boldsymbol{\Phi}_{yx} + \mathbf{h}^T\,\boldsymbol{\Phi}_{yx}$$
$$= E[x^2(n)] - \mathbf{h}^T\,\boldsymbol{\Phi}_{yx}$$

SAQ 8.2 Thus:

$$\boldsymbol{\Phi}_{yy} = \begin{bmatrix} 1.35 & 0.5 & 0 \\ 0.5 & 1.35 & 0.5 \\ 0 & 0.5 & 1.35 \end{bmatrix}$$

Since the lag d is 2, the z-transform of the cross-correlation sequence is:

$$S_{yx'}(z) = z^{-2}\,(z + 0.5) = z^{-1} + 0.5z^{-2}$$

Hence the cross-correlation terms are: $\phi_{yx'}(0) = 0$, $\phi_{yx'}(1) = 1$ and $\phi_{yx'}(2) = 0.5$. The cross-correlation vector is: $\boldsymbol{\Phi}_{yx'} = [0.0\ \ 1.0\ \ 0.5]^T$. Solving the three simultaneous equations yields the impulse response vector of the equaliser: $\mathbf{h} = [-0.3081\ \ 0.8318\ \ 0.0623]^T$. The minimum MSE can then be obtained using equation (8.9):

$$E[e^2] = \sigma_x^2 - \mathbf{h}^T\,\boldsymbol{\Phi}_{yx'} = 0.14$$

SAQ 8.3 Using equation (8.14):

$$\mathbf{R}_{yy}(n) = \sum_{k=0}^{n} \mathbf{y}(k)\, \mathbf{y}^T(k) = \sum_{k=0}^{n-1} \mathbf{y}(k)\, \mathbf{y}^T(k) + \mathbf{y}(n)\mathbf{y}^T(n)$$
$$= \mathbf{R}_{yy}(n-1) + \mathbf{y}(n)\mathbf{y}^T(n)$$

SAQ 8.4 Using equation (8.15):

$$\mathbf{r}_{yx}(n) = \sum_{k=0}^{n} \mathbf{y}(k)\, x(k) = \sum_{k=0}^{n-1} \mathbf{y}(k)\, x(k) + \mathbf{y}(n)x(n)$$
$$= \mathbf{r}_{yx}(n-1) + \mathbf{y}(n)x(n)$$

SAQ 8.5 A 2-tap filter with a white noise input, hence: $\Phi_{yy} = 2I_2$ where I_2 is a (2×2) identity matrix. Using equation (8.23):

$$\nabla = 4\begin{bmatrix} 2.9 \\ 4.0 \end{bmatrix} - 2\begin{bmatrix} 3 \\ 1 \end{bmatrix} = \begin{bmatrix} 5.6 \\ 14.0 \end{bmatrix}$$

SAQ 8.6 i) LMS: $N = 4$, 8 multiples and 8 additions; $N = 128$, 256 multiplies and 256 additions. ii) BLMS: $N = 4$, 28 multiples and 60 additions; $N = 128$, 78 multiplies and 135 additions. The BLMS is more expensive than the LMS for short filters but more computationally efficient for long filters.

SAQ 8.7 The RLS has faster initial convergence than the LMS and its convergence rate is much less dependent on the statistics of the input signal. The RLS is computationally more expensive than the LMS.

SAQ 8.8 As the bandwidth of the filter is decreased, the signal at its output becomes more correlated from sample to sample and hence more predictable. At the end of the day the signal is more predictable than it was at the beginning of the day.

Chapter 9

SAQ 9.1 The eight roots of unity are:

$W_8^0 = 1$
$W_8^1 = 0.7 - j\,0.7$
$W_8^2 = -j$
$W_8^3 = -0.7 - j\,0.7$
$W_8^4 = -1$
$W_8^5 = -0.7 + j\,0.7$
$W_8^6 = +j$
$W_8^7 = 0.7 + j\,0.7$

SAQ 9.2 This involves the following 8 multiply and sum operations:
$(1\times 1) + (0.7+j0.7)\times -j + (j\times -1) + (-0.7+j0.7)\times j - (1\times 1) +$
$(-0.7-j0.7)\times -j\,(-j\times -1) + (0.7-j0.7)\times j = 0$

SAQ 9.3 In the first channel we have an N-tap FIR where all the weights are unity. This is a low-pass filter, centred on 0 Hz, whose impulse response is N samples long. Fourier theory tells us that such a signal has sinc x response with a mainlobe width of $1/N\Delta t$ Hz to the first null. sinc x responses can then be overlapped by arranging the next filter in the bank to have a sinc x response, centred on $1/N\Delta t$ Hz, and the filterbank then overlaps at the –3 dB points, as shown in Figure 9.8.
Thus the first low-pass channel has a narrow bandwidth and, as in Figure 4.10(c) the output sample rate can be set at $1/N\Delta t$ Hz. As the input sample rate is $1/\Delta t$ Hz, the downsampling is by the factor N.

SAQ 9.4 As in SAQ 9.3, the sinc x responses with $\Delta t = 10^{-3}$ and an $N = 8$ transform have zeros at 1/8, 1/4, 3/8, 1/2, 5/8, 3/4, 7/8, 1 kHz. These are the bin frequencies in the 8-point DFT.

SAQ 9.5 Resolution $\Delta f = 1/N\Delta t$ Hz. For a 1024-point DFT sampled at 512 kHz:

$$\Delta f = \frac{1}{1024\left(\dfrac{1}{512\times 10^3}\right)} = \tfrac{1}{2}\times 10^3 = 500\ \text{Hz}.$$

SAQ 9.6 For 16-bit arithmetic, dynamic range is 6 dB per bit. Some allowance must be given for the peak-to-mean ratio of the signal as the quantiser must not clip the large signal transients. Thus typically the SNR is 6 dB per bit minus 6–10 dB to allow for this. Thus for 16-bit arithmetic, the dynamic range is 85–90 dB.

SAQ 9.7 The level of leakage can be calculated by noting that the signal at the bin 16 position is given by the sinc$x$ value where x is equal to $11\pi/2$, i.e. it falls on to the top of the fifth sidelobe in the sinc$x$ response in Figure 10.11(a). This magnitude is $1/(5.5\pi)$ which equals 0.0579 or -24.8 dB, as shown in Figure 10.11(b).

SAQ 9.8 This can only be achieved by using *a priori* knowledge to change either the sample rate or the transform length to accommodate the signal. For the 128-point DFT sampled at 128 kHz, the resolution is 100 Hz and the signal is at 1.05 kHz. Changing the sampling frequency to $12.8\times 10/10.5$, i.e. to 121,904.76 Hz will put the frequency on to bin 10. Note that this is signal specific and it is not a GENERAL solution to the problem.

SAQ 9.9

$$\Delta f = \frac{1}{256}\left(\frac{1}{51.2\times 10^3}\right) = \frac{2\times 10^3}{10} = 200\ \text{Hz}.$$

Signal is at 63 Hz, the zeros on the sinc x are every 200 Hz and bin 7 occurs at 1400 Hz.

Thus on the sinc x response, where zeros occur every π rad, bin 7 is $\pi(1400-63)/200$ rad from the peak, i.e. 6.685π. Evaluating $\text{sinc}(x) = (\sin(\pi x))/(\pi x)$ where $x = 6.685$ gives the value of $0.835/(6.685\pi) = 0.0398$, or –28 dB.

SAQ 9.10 For the 64-sample Hanning window $w_0 = 0.5 - 0.5 = 0$, $w_{32} = 0.5 - 0.5\cos(2\pi\ 32/64) = 0.5 + 0.5\cos\pi = 1$, $w_{64} = 0.5 - 0.5 = 0$.

SAQ 9.11 For the 64-sample Hamming window $w_0 = 0.54 - 0.46 = 0.08$, $w_{32} = 0.54 + 0.46 = 1$, $w_{64} = 0.54 - 0.46 = 0.08$.

SAQ 9.12 For –40 dB sidelobe level, the Hamming window is the most appropriate function. For –55 dB there are several choices but the Dolph–Chebychev has the narrowest width for the main peak and so gives the best spectrum analyser capabilities.

SAQ 9.13 Power spectral density is given by equation (9.24) as the Fourier transform of the autocorrelation function. The power spectrum measure is simply the squared modulus of the amplitude spectrum or DFT of the signal and it does not use correlation measures – see later equation (9.26) and Figure 9.16.

SAQ 9.14 Classical spectrum analysers are general processors that are not signal dependent. They offer poor resolution. Modern analysers offer superior resolution but they assume a signal type and this is the basis of the model employed.

Chapter 10

SAQ 10.1 The number of passes is $\log_2 N$. For $N = 256$ then $\log_2 512 = 8$ and so there are eight passes.

SAQ 10.2 If the binary addresses of the input locations 0 to 7 are reflected 'mirror' fashion with the MSB being regarded as the LSB etc., then the resulting number indicates which data sample should be stored in any given location. For example location 1 (address 001_2) would hold data sample 4 (sample 100_2). Thus data re-ordering for a DIT FFT processor is easily achieved by simply adjusting the address decoder.

SAQ 10.3 The number of DFT complex multipliers is N^2 and for $N = 512$ this is 512^2 or 262,144. The number of FFT complex multipliers is $(N/2)\log_2 N$ which is $256\log_2 512$ or 2304.

SAQ 10.4 Extending SAQ 10.3, we can make the following table for the number of complex multiply operations:

N	DFT	FFT	% saving
32	1024	80	92.19
256	65,536	1024	98.44
512	262,144	2304	99.12
1024	1,048,576	5120	99.51
8192	67,174,416	49,152	99.93

SAQ 10.5 In the radix-4 case, multiplication is by $+1, -1$ or $+j, -j$. Thus from equation (10.16):

$$\begin{bmatrix} X(0) \\ X(1) \\ X(2) \\ X(3) \end{bmatrix} = \begin{bmatrix} 1 & 1 & 1 & 1 \\ 1 & -j & -1 & j \\ 1 & -1 & 1 & -1 \\ 1 & j & -1 & -j \end{bmatrix} \begin{bmatrix} a+jb \\ c+jd \\ e+jf \\ h+ji \end{bmatrix}$$

Multiplication by the complex number $\pm j$ may be simply achieved by interchanging real and imaginary parts and changing the sign of the real or imaginary parts of the resultant respectively. Thus:

$$\begin{aligned} X(0) &= a+c+e+h \ + j(b+d+f+i) \\ X(1) &= a+d-e-i \ + j(b-c-f+h) \\ X(2) &= a-c+e-h \ + j(b-d+f-i) \\ X(3) &= a-d-e+i \ + j(b+c-f-h) \end{aligned}$$

and no actual multiply operations are required!

SAQ 10.6 The number of passes is $\log_4 N$ and, for $N = 256$, there are four passes.

SAQ 10.7

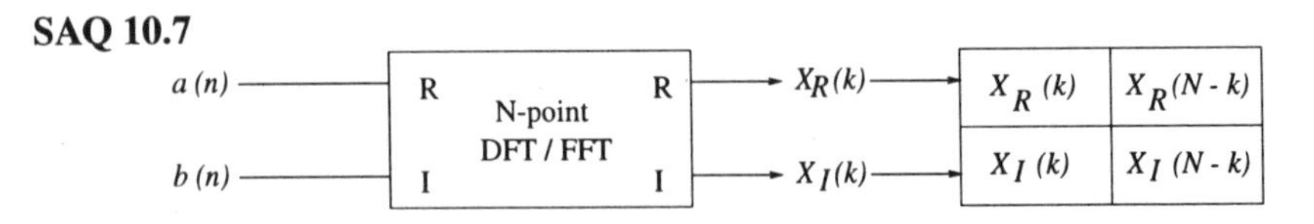

Output vectors

$A(k) = 1/2 \times$

$X_R(k)$	$+X_R(N-k)$
$+j\,X_I(k)$	$-j\,X_I(N-k)$

$B(k) = 1/2 \times$

$X_I(k)$	$X_I(N-k)$
$-j\,X_R(k)$	$+jX_R(N-k)$

SAQ 10.8 For a 4096-point convolution, Figure 11.9 thus involves $8192 \times 13 \times 2$ complex FFT MAC operations plus a further 8192 complex multiplications for the $X(k)H(k)$ operation. The total number of MAC operations is thus 8193×27 which is a considerable saving over the $4096^2 \approx$ 16M operations of Figure 6.1.

Chapter 11

SAQ 11.1 For this signal the normal minimum (Nyquist) sample rate would be 3 kHz. At input sample rate 10 kHz, decimation by 3 would then give 3.33 kHz output.

SAQ 11.2 For $n = Q = 8$ we get: $2500 < f_s < 2571$. The sample rate is thus 2535.5 Hz with an accuracy of 1.4%.

SAQ 11.3 The quotient $Q = 10.5$ is now no longer an integer. The lowest allowed sampling rate is therefore given by $n = int(Q) = 10.0$. The sampling rate is now bound by:

$$2.100 \leq f_s \leq 2.111 \text{ (kHz)}$$

This gives a required sampling rate of 2.106 ± 0.006 kHz or 2.106 kHz $\pm 0.26\%$.

SAQ 11.4 Now there is no difference between filterbank and DFT.

SAQ 11.5 Because in the N-channel multirate filterbank the filter length or impulse duration is not constrained to be N samples long, as in the DFT. Thus making the length longer than N gives the improved filterbank performance shown in Figure 11.15.

SAQ 11.6 If the signal bandwidth is 102.4 kHz then the sample rate is 204.8 kHz. Resolution is $f_s/N = 200$ Hz.

SAQ 11.7 For Figure 11.17(b), f_c in the FIR filter is now 11.15 kHz but the signal bandwidth is still 2.5 kHz so the solution is as before. If the Figure 11.9 bandpass sampling criterion had been used, then:

$$Q = \frac{f_H}{B} = \frac{12.4}{2.5} = 4.96$$

Thus the n value must be 4 to avoid the spectral aliases. Now:

$$5 \times 10^3 \left(\frac{4.96}{4} \right) \leq f_s \leq 5 \times 10^3 \left(\frac{4.96 - 1}{4 - 1} \right)$$

This gives a range for the f_s value of 6.2 to 6.6 kHz. Note that as the input FIR filter of Figure 11.17(b) is not present, a higher sample rate is required to stop the straddling of Figure 11.9 occurring.

SAQ 11.8 For 4 equal width sub-bands covering the 0 kHz wide spectrum, i.e. 2.5 kHz wide bands, the output sample rate in each band is 5.0 kHz. The total bit rate for the output signal is thus $5(8 + 6 + 4 + 2) = 5 \times 20 = 100$ kbit/s. The compression is thus $160/100 = 1.6$ times.

Bibliography

1. Background theory

Assefi, T., *Stochastic Processes and Estimation Theory with Applications*, Wiley, New York, 1979.

Balmer, L., *Signals and Systems* (2nd edn), Prentice-Hall, Hemel Hempstead, 1997.

Bozic, S.M., *Digital and Kalman Filtering*, Edward Arnold, London, 1979.

Broyden, C.G., *Basic Matrices*, Macmillan Press, Basingstoke, 1975.

Spiegel, M.R., *Mathematical Handbook for Formulas and Tables*, Schaum's Outline Series Publication, McGraw-Hill, New York.

Wax, N., *Selected Papers on Noise and Stochastic Processes* [Reprints of S.O. Rice BSTJ Classic Papers], Dover, New York, 1954.

Wiener, N., *Extrapolation, Interpolation and Smoothing of Stationary Time Series*, Wiley, New York, 1949.

2. Stochastic processes

Kay, S.M., *Fundamentals of Statistical Signal Processing*, Prentice-Hall, Englewood Cliffs, NJ, 1988.

McGillem, C.D. and Cooper, G.R., *Continuous and Discrete Signal and System Analysis*, Holt, Rinehart and Winston, New York, 1984.

Mendel, J.M., *Lessons in Digital Estimation Theory*, Prentice-Hall, Englewood Cliffs, NJ, 1987.

Papoulis, A., *Signal Analysis*, McGraw-Hill, New York, 1977.

Scharf, L.L., *Statistical Signal Processing*, Addison-Wesley, Reading, MA, 1991.

3. General DSP techniques

Bellanger, M., *Digital Processing of Signals*, Wiley, New York, 1984.

Candy, J.V., *Signal Processing – The Model Based Approach*, McGraw-Hill, New York, 1986.

DeFatta, D.J., Lucas, J.G. and Hodgkiss, W.S., *Digital Signal Processing*, Wiley, New York, 1988.

Huang, T.S., *Two-Dimensional Digital Signal Processing*, Springer-Verlag, Berlin, 1981.

Ifeachor, E. and Jervis, B.W., *DSP – A Practical Approach*, Addison-Wesley, Maidenhead, 1993.

Jackson, L.B., *Signals, Systems and Transforms*, Addison-Wesley, Reading, MA, 1991.

Kuc, R., *Introduction to DSP*, McGraw-Hill, New York, 1988.

McClellan, J.H. and Rader, C.M., *Number Theory in Digital Signal Processing*, Prentice-Hall, Englewood Cliffs, NJ, 1979.

Oppenheim, A.V., Willsky, A.S. and Young, I.T., *Signals and Systems*, Prentice-Hall, Englewood Cliffs, NJ, 1983.

Oppenheim, A.V. and Schafer, R.W., *Discrete-time Signal Processing*, Prentice-Hall, Englewood Cliffs, NJ, 1989.

Oppenheim, A.V., Willsky, A.S. with Nawab, S.H., *Signals and Systems* (2nd edn), Prentice-Hall, Englewood Cliffs, NJ, 1997.

Papoulis, A., *Circuits and Systems*, Holt Rinehart and Winston, London, 1980.

Peled, A. and Liu, B., *Digital Signal Processing: Theory Design and Implementation*, Wiley, New York, 1976.

Porat, B., *A Course in Digital Signal Processing*, Wiley, New York, 1997.

Proakis, J.G. and Manolakis, D., *Digital Signal Processing*, Macmillan Inc., New York, 1989.

Proakis, J.G., Rader, C.M., Ling, F. and Nikias, C.L., *Advanced Digital Signal Processing*, Macmillan Inc., New York, 1992.

Rabiner, L.R., and Gold, B., *Theory and Application of Digital Signal Processing*,

Prentice-Hall, Englewood Cliffs, NJ, 1975.

Strum, R.D. and Kirk, D.E., *Discrete Systems and DSP*, Addison-Wesley, Reading, MA, 1988.

Van Den Enden, A.W.M., *Discrete Signal Processing*, Prentice-Hall, Englewood Cliffs, NJ, 1988.

4. Digital filters

Hamming, R.M., *Digital Filters*, Prentice-Hall, Englewood Cliffs, NJ, 1983.

Jackson, L.B., *Digital Filters and Signal Processing*, Kluwer Academic, Norwood, MA, 1986.

McClellan, J.H. *et al., FIR Linear Phase Filter Design Program*, in *Programs for Digital Signal Processing*, IEEE Press, New York, 1979.

Parks, T.W. and Burrus, C.S., *Digital Filter Design*, Wiley, New York, 1987.

Williams, C.S., *Design of Digital Filters*, Prentice-Hall, Englewood Cliffs, NJ, 1986.

Zverev, A.I., *Handbook of Filter Synthesis*, Wiley, New York, 1967.

5. Discrete and fast Fourier transforms

Bracewell, R., *The Fourier Transform and its Applications*, McGraw-Hill, New York, 1965.

Brigham, E.O., *The Fast Fourier Transform*, Prentice-Hall, Englewood Cliffs, NJ, 1974.

Cooley, J.W. and Tukey, J.W., The Fast Fourier Transform Algorithm ..., in Rabiner, L.R. and Rader, C.M. (eds.), *Digital Signal Processing*, IEEE Press, New York, 1972, pp. 271–293.

Harris, F.J., On the Use of Windows for Harmonic Analysis with the Discrete Fourier Transform, *Proc. IEEE*, Vol. 66, No. 1, pp. 51–83, January 1978.

Lynn, P.A. and Fuerst, W., *Digital Signal Processing with Computer Applications*, Wiley, New York, 1994.

Nussbaumer, H.J., *Fast Fourier Transform and Convolution Algorithms*, Springer-Verlag, Berlin, 1981.

Rao, K.R., (ed.), *Discrete Transforms and their Applications*, Van Nostrand Reinhold, New York, 1985.

6. Spectral analysis

Kay, S.M., *Modern Spectral Estimation*, Prentice-Hall, Englewood Cliffs, NJ, 1988.

Marple, S.L., *Digital Spectral Analysis*, Prentice-Hall, Englewood Cliffs, NJ, 1987.

7. Adaptive filters

Bellanger, M.G., *Adaptive Digital Filters and Signal Processing*, Marcel Dekker, 1987.

Bozic, S.M., *Digital and Kalman Filtering*, Edward Arnold, London, 1979.

Cowan, C.F.N. and Grant, P.M., *Adaptive Filters*, Prentice-Hall, Englewood Cliffs, NJ, 1985.

Haykin, S.S., *Adaptive Filter Theory*, Prentice-Hall, Englewood Cliffs, NJ, 1991.

Honig, M.L. and Messerschnmitt, D.G., *Adaptive Filters: Structures, Algorithms and Applications*, Kluwer Academic, Norwood, MA, 1984.

Kalouptsidis, N. and Theodoridis, S., *Adaptive System Identification and Signal Processing Algorithms*, Prentice-Hall, Englewood Cliffs, NJ, 1993.

Mulgrew, B. and Cowan, C.F.N., *Adaptive Filters and Equalisers*, Kluwer Academic, Norwood, MA, 1988.

Regalia, P., *Adaptive IIR Filtering in Signal Processing and Control*, Marcel Dekker, 1994.

Widrow, B. and Stearns, S.D., *Adaptive Signal Processing*, Prentice-Hall, Englewood Cliffs, NJ, 1984.

8. Multirate DSP

Chui, C.K., Montefusco, L. and Puccio, L., *Wavelets*, Academic Press, Boston, MA, 1995.

Cohen, L., *Time-frequency Analysis*, Prentice-Hall, Englewood Cliffs, NJ, 1995.

Crochiere, R.E. and Rabiner, L.R., *Multirate Digital Signal Processing*, Prentice-Hall, Engelwood Cliffs, NJ, 1983.

Fliege, N.J., *Multirate Digital Signal Processing*, Wiley, New York, 1994.

Malvar, H.S., *Signal Processing with Lapped Transforms*, Artech House, Boston, MA, 1992.

Massopust, P.R., *Fractal Functions, Fractal Surfaces and Wavelets*, Academic Press, Boston, MA, 1995.

Vaidyanathan, P.P., *Multirate Systems and Filterbanks*, Prentice-Hall, Englewood Cliffs, NJ, 1993.

Vetterli, M. and Kovacevic. J., *Wavelets and Subband Coding*, Prentice-Hall, Englewood Cliffs, NJ, 1995.

9. DSP simulation

Buck, J.R., Daniel, M.M. and Singer, A.C., *Computer Explorations in Signals and Systems*, Prentice-Hall, Englewood Cliffs, NJ, 1997.

Burrus, C.S., McClellan, J.H., Oppenheim, A.V., Parks, T.W., Schafer, R.W. and Schuessler, H.W., *Computer Based Exercises for Signal Processing*, Prentice-Hall, Englewood Cliffs, NJ, 1994.

Embree, P.M. and Kimble, B., *C Language Algorithms for Digital Signal Processing*, Prentice-Hall, Englewood Cliffs, NJ, 1991.

Embree, P.M., *C Algorithms for Real-Time Digital Signal Processing*, Prentice-Hall, Englewood Cliffs, NJ, 1995.

Etter, D.M., *Introduction to MATLAB for Engineers and Scientists*, Prentice-Hall, Englewood Cliffs, NJ, 1996.

Stearns, S.D. and David, R.A., *Signal Processing Algorithms in MATLAB*, Prentice-Hall, Englewood Cliffs, NJ, 1996.

10. System applications of DSP techniques

Boyd, I., Speech Coding for Telecommunications, *Electronics and Communication Engineering Journal*, Vol. 4, No. 5, pp. 273-283, October 1992.

Cook, C.E. and Bernfeld, M., *Radar Signals*, Academic Press, New York, 1967.

Deller, J.R., Proakis, J.G. and Hansen, J.H.L., *Discrete-time Processing of Speech Signals,* Macmillan, Basingstoke, 1992.

Furui, S. and Sondhi, M.M. (eds.), *Advances in Speech Signal Processing,* Marcel Dekker, 1992.

Glover, I.A. and Grant, P.M., *Digital Communications*, Prentice-Hall, Hemel Hempstead, 1998.

Gray, A.H. and Markel, J.D., *Linear Prediction of Speech*, Springer-Verlag, Berlin, 1976.

Jayant, N.S. and Noll, P., *Digital Coding of Waveforms*, Prentice-Hall, Englewood Cliffs, NJ, 1984.

Lewis, B.L., Kretschmer, F.F. and Shelton, W., *Aspects of Radar Signal Processing*, Artech House, Boston, MA, 1986.

Oppenheim, A.V., *Applications of Digital Signal Processing*, Prentice-Hall, Englewood Cliffs, NJ, 1978

Westall, F.A. and Ip, S.F.A., *DSP in Telecommunications*, Chapman & Hall, London, 1993.

11. DSP Hardware

Lee, E.A., Programmable DSP Architectures, see *IEEE ASSP Magazine*, Vol. 5, No. 4 and Vol. 6, No. 1, 1988/89.

Motorola DSP 56000/560001, *Digital Signal Processor User Manual*, Phoenix, AZ.

Index